Extracellular Matrix

Unilever Research Central Library
Colworth House

Unilever

This book is due back on the date stamped below

B6·2

If you wish to extend the loan please ring extension 2230

The Practical Approach Series

SERIES EDITORS

D. RICKWOOD
Department of Biology, University of Essex
Wivenhoe Park, Colchester, Essex CO4 3SQ, UK

B. D. HAMES
Department of Biochemistry and Molecular Biology
University of Leeds, Leeds LS2 9JT, UK

Affinity Chromatography
Anaerobic Microbiology
Animal Cell Culture
  (2nd Edition)
Animal Virus Pathogenesis
Antibodies I and II
Basic Cell Culture
Behavioural Neuroscience
Biochemical Toxicology
Biological Data Analysis
Biological Membranes
Biomechanics—Materials
Biomechanics—Structures and
  Systems
Biosensors
Carbohydrate Analysis
  (2nd Edition)
Cell–Cell Interactions
The Cell Cycle
Cell Growth and Division
Cellular Calcium
Cellular Interactions in
  Development
Cellular Neurobiology

Centrifugation (2nd Edition)
Clinical Immunology
Computers in Microbiology
Crystallization of Nucleic Acids
  and Proteins
Cytokines
The Cytoskeleton
Diagnostic Molecular Pathology
  I and II
Directed Mutagenesis
DNA Cloning: Core Techniques
DNA Cloning: Expression
  Systems
Drosophila
Electron Microscopy in Biology
Electron Microscopy in
  Molecular Biology
Electrophysiology
Enzyme Assays
Essential Developmental
  Biology
Essential Molecular Biology I
  and II
Experimental Neuroanatomy

Extracellular Matrix
A Practical Approach

Edited by

M. A. HARALSON

Vanderbilt University

and

JOHN R. HASSELL

University of Pittsburgh

OXFORD UNIVERSITY PRESS
Oxford New York Tokyo

Oxford University Press, Walton Street, Oxford OX2 6DP

Oxford   New York

Athens   Auckland   Bangkok   Bombay
Calcutta   Cape Town   Dar es Salaam   Delhi
Florence   Hong Kong   Istanbul   Karachi
Kuala Lumpur   Madras   Madrid   Melbourne
Mexico City   Nairobi   Paris   Singapore
Taipei   Tokyo   Toronto
and associated companies in
Berlin   Ibadan

Oxford is a trade mark of Oxford University Press

Published in the United States
by Oxford University Press Inc., New York

A catalogue record for this book is available from the British Library

Library of Congress Cataloging-in-Publication Data
Extracellular matrix: a practical approach/edited by M.A. Haralson
and John R. Hassell.—Ed. 1.
p. cm.—(The Practical approach series; 151)
Includes bibliographical references.
1. Extracellular matrix—Research—Laboratory manuals.
I. Haralson, M. A.   II. Hassell, John R.   III. Series.
QH603.E93E94   1995   574.87—dc20   94–33106
ISBN 0 19 963221 9 (Hbk)
ISBN 0 19 963220 0 (Pbk)

Typeset by Footnote Graphics, Warminster, Wilts
Printed in Great Britain by Information Press Ltd, Eynsham, Oxon.

Preface

The past two decades have witnessed an explosion of information about the extracellular matrix. Landmark discoveries leading to the expanded interest in and importance of this field in contemporary biomedical research include: (1) the appreciation that each major type of extracellular matrix macromolecule represents a family of related proteins that exhibit a degree of tissue-specific expression; (2) the role played by the interaction of matrix components with the cell in maintaining and determining tissue architecture; (3) the importance of matrix molecules in normal tissue development and repair; and (4) the significant changes that occur in the extracellular matrix during many acquired and inherited diseases. A number of books and reviews have summarized the basic concepts about the structure and function of the extracellular matrix molecules. Many of the methods used in matrix research differ, however, from traditional biochemical, immunological, and molecular biological approaches. Moreover, a number of the more recent advances in this field reflect changes in existing methodologies or the development of entirely new methods. In this book we have tried to focus not only on the most current but also the best established experimental approaches used in extracellular matrix research.

We have included the major methodologies used for investigating most aspects of extracellular matrix biochemistry, molecular biology, and cell biology in vertebrates. Space obviously precludes inclusion of methodologies for investigating every genetic type of collagen, of each recognized structural glycoprotein and of each proteoglycan, as well as the matrix components of invertebrates. Nor have we covered the methodologies used for studying the receptors—both transmembrane proteins (the integrins) and proteoglycans (the syndecans)—for the matrix components because proper coverage of the methods in this area alone would require an entire volume. Despite these exclusions, it is our hope that the methodologies included in this book will provide the user with a basis for investigating those molecules to which specific chapters were not devoted and to be a starting point for investigations in those areas not covered.

In preparing this volume, we are indebted to many individuals. First, and foremost, we would like to express our appreciation to the authors of each chapter, without whose contributions this volume would not have been possible. Second, we would like to thank the editors of the Oxford University Press, who have guided us throughout the preparation of the volume. Third, we are indebted to Ms Jean McClure and Mr Danny Riley for their assistance in the preparation of this volume, to Dr Gregory Sephel for his suggestions and input and to Ms Sonya Bowling-Brown and Dr Samuel DiMari for their

Michael A. Haralson and John R. Hassell

careful reading and helpful comments in the preparation of these chapters. Finally, we would like to acknowledge all of the investigators, past and present, in this area whose scientific expertises have contributed to the many discoveries that form the basis of this field.

Nashville and Pittsburgh   Michael A. Haralson
September 1995   John R. Hassell

Contents

2. Types I, III, and V collagen and total collagen from cultured cells and tissues   31

Beverly Peterkofsky, Michael A. Haralson, Samuel J. DiMari, and Edward J. Miller

Contents

Contents

3. The collagens of cartilage (types II, IX, X, and XI) and the type IX-related collagens of other tissues (types XII and XIV)   73

Richard Mayne, Michel Van Der Rest, Peter Bruckner, and Thomas M. Schmid

4. Matrix components (types IV and VII collagen, entactin, and laminin) found in basement membranes   99

Billy G. Hudson, Sripad Gunwar, Albert E. Chung, and Robert E. Burgeson

Contents

Contents

10. Isolation and characterization of proteoglycan core protein

Vincent C. Hascall, Masaki Yanagishita, Anthony Calabro, Ronald Midura, Jody A. Rada, Shukti Chakravarti, and John R. Hassell

Contents

Contents

13. Use of extracellular matrix and its components in culture 289

Hynda K. Kleinman, Maura C. Kibbey, Frances B. Cannon, Benjamin S. Weeks, and Derrick S. Grant

14. Immunological identification of extracellular matrix components in tissues and cultured cells 303

Peter S. Amenta and Antonio Martinez-Hernandez

Contents

List of contributors

STEVEN K. AKIYAMA
Laboratory of Developmental Biology, National Institute of Dental Research, National Institutes of Health, Building 30, Room 405, Bethesda, MD 20892, USA.

PETER S. AMENTA
UMDNJ-Robert Wood Johnson Medical School, Department of Pathology, MEB 228, One Robert Wood Johnson Place, New Brunswick, NJ 08903-0019, USA.

RICHARD BERG
Collagen Corporation, 2500 Faber Place, Palo Alto, CA 94303, USA.

PAUL BORNSTEIN
Department of Biochemistry, University of Washington, SJ-70, Seattle, WA 98195, USA.

GINA BRISCOE
Department of Cell Biology, Duke University Medical Center, Durham, NC 27710, USA.

PETER BRUCKNER
Institute of Physiological Chemistry and Pathophysiology, Westfälische Wilhelms-Universität, Waldeyerstrasse 15, D-48149 Münster, Germany.

ROBERT E. BURGESON
MGH-EAST-CBRC, Building 149, 13th Street, Charleston, MA 02129, USA.

ANTHONY CALABRO
Proteoglycan Chemistry Section, Bone Research Branch, National Institute of Dental Research, National Institutes of Health, Building 30, Room 106, Bethesda, MA 20892, USA.

FRANCES B. CANNON
Laboratory of Developmental Biology and Anomalies, National Institute of Dental Research, National Institutes of Health, Building 30, Room 407, Bethesda, MA 20892, USA.

SHUKTI CHAKRAVARTI
Department of Genetics, Rm.609B; Bldg. BRB, Case Western Reserve University, 10900 Euclid Avenue, Cleveland, OH 44106-4944, USA.

ALBERT E. CHUNG
Department of Biological Science, University of Pittsburgh, A519 Langley Hall, Pittsburgh, PA 15260, USA.

List of contributors

JEFFREY M. DAVIDSON
Department of Pathology, Vanderbilt University, School of Medicine, Nashville, TN 37232-2561, USA, and VA Medical Center, Nashville, TN 37212-2561, USA.

BENOIT DE CROMBRUGGHE
Department of Molecular Genetics, University of Texas, M.D. Anderson Cancer Center, 1515 Holcombe Boulevard, Box 11, Houston, TX 77030, USA.

SAMUEL J. DIMARI
Department of Pathology, Vanderbilt University School of Medicine, Nashville, TN 37232-2561, USA.

HAROLD P. ERICKSON
Department of Cell Biology, Duke University Medical Center, Durham, NC 27710, USA.

SUSAN J. FISHER
Departments of Somatology, Anatomy and Pharmaceutical Chemistry, University of California, San Francisco, San Francisco, California 94143-0512, USA.

JOHN T. GALLAGHER
University of Manchester, Department of Medical Oncology, Christie CRC Research Centre, Wilmslow Road, Manchester M20 9BX, UK.

MARIA GABRIELLA GIRO
Istituto di Istologia, Universita di Padova, Padova, Italy.

DERRICK S. GRANT
Cardeza Foundation, Jefferson Medical College, Philadelphia, PA 19107, USA.

SRIPAD GUNWAR
Department of Biochemistry, School of Life Science, Central University of Hyderabad, Gachibowli, Hyderabad -500134, A.P., India.

MICHAEL A. HARALSON
Department of Pathology, Vanderbilt University School of Medicine, Nashville, TN 37232-2561, USA.

VINCENT C. HASCALL
Biomedical Engineering Department, The Cleveland Clinic Foundation, 9500 Euclid Avenue, Cleveland, Ohio 44195, USA.

JOHN R. HASSELL
The Eye and Ear Institute, The University of Pittsburgh School of Medicine, Department of Ophthalmology, 23 Lothrop Street, Room 1025, Pittsburgh, PA 15213, USA.

List of contributors

BILLY G. HUDSON
Department of Biochemistry and Molecular Biology, University of Kansas Medical Center, 39th and Rainbow Boulevard, Kansas City, Kansas 66160-7421, USA.

MAURA C. KIBBEY
Laboratory of Developmental Biology and Anomalies, National Institute of Dental Research, National Institutes of Health, Building 30, Room 407, Bethesda, MA 20892, USA.

HYNDA K. KLEINMAN
Laboratory of Developmental Biology and Anomalies, National Institute of Dental Research, National Institutes of Health, Building 30, Room 407, Bethesda, MA 20892, USA.

M. LYON
University of Manchester, Department of Medical Oncology, Christie CRC Research Centre, Wilmslow Road, Manchester M20 9BX, UK.

ANTONIO MARTINEZ-HERNANDEZ
VAMC and University of Tennessee, Department of Pathology, 1030 Jefferson Avenue, Memphis, TN 38104-2193, USA.

RICHARD MAYNE
Department of Cell Biology, Box 302 Volker Hall, The University of Alabama Medical Center in Birmingham, Birmingham, AL 35294-0019, USA.

ROBERT P. MECHAM
Departments of Cell Biology and Medicine, (Respiratory and Critical Care Division), Washington University School of Medicine, St Louis, MO, USA.

DR RONALD MIDURA
Department of Orthopaedic Surgery, University of Iowa, 180ML Iowa City, IA 52232, USA.

EDWARD J. MILLER
Department of Biochemistry, The University of Alabama Medical Center in Birmingham, University Station, Birmingham, AL 35294-0019, USA.

BEVERLY PETERKOFSKY
Chief Section of Biological Interactions, Laboratory of Biochemistry, National Cancer Institute, National Institutes of Health, Building 37, Room 4C18, Bethesda, MD 20892, USA.

JODY A. RADA
The Eye and Ear Institute, The University of Pittsburgh School of Medicine Department of Ophthalmology, 203 Lothrop Street, Room 1025, Pittsburgh, PA 15213, USA.

List of contributors

LINDA J. SANDELL
Department of Orthopedics and Biochemistry, RK-10, University of Washington, Seattle, WA 98195, USA.

E. HELENE SAGE
Department of Biological Structure, University of Washington, SM-20, Seattle, WA 98195, USA.

THOMAS M. SCHMID
Department of Biochemistry, Rush-Presbyterian—St. Luke's Medical Center, Chicago, IL 60612, USA.

DAVID TOMAN
Collagen Corporation, 2500 Faber Place, Palo Alto, California 94303, USA.

J. E. TURNBULL
University of Manchester, Department of Medical Oncology, Christie CRC Research Centre, Wilmslow Road, Manchester M20 9BX, UK.

MICHEL VAN DER REST
Institute de Biologie Structurale, 41 Avenue des Martyrs, F-38027, Grenoble Cx, France.

BENJAMIN S. WEEKS
Department of Biology, Hamilton College, Clinton, New York, 13323, USA.

ZENA WERB
Department of Anatomy, Laboratory of Radiobiology and Environmental Health, LR 102, University of California, San Francisco, CA 94143-0750, USA.

KENNETH M. YAMADA
Laboratory of Developmental Biology, National Institute of Dental Research, National Institutes of Health, Building 30, Room 421, Bethesda, MA 20892, USA.

MASAKI YANAGISHITA
Proteoglycan Chemistry Section, Bone Research Branch, National Institute of Dental Research, National Institutes of Health, Building 30, Room 106, Bethesda, MD 20892, USA.

Abbreviations

ABM	alveolar basement membrane
AEC	aminoethylcarbazole
AMP	ampicillin
APMA	4-aminophenylmercuric acetate
ATCC	American Type Culture Collection
BAEC	bovine aortic endothelial cells
BAPN	β-aminopropionitrile
BB	bromophenol blue
BHK	baby hamster kidney fibroblasts
BSA	bovine serum albumin
cDNA	complimentary DNA
CHAPS	3-[3-cholamidopropyl-dimethylammonio]-1-propanesulfonate
CIP	calf intestinal phosphatase
CM	carboxymethyl
CMF-PBS	Ca^{2+} and Mg^{2+} free PBS
CNBr	cyanogen bromide
CS	chondroitin sulfate
CsCl	caesium chloride
CsTFA	caesium trifluoracetate
DS	dermatan sulfate
DAB	diaminobenzidine
DEAE	diethylaminoethyl
DEPC	diethylpyrocarbonate
DES	desmosine
DM	Descemet's membrane
DMEM	Dulbecco's modified Eagle's medium
DMSO	dimethyl sulfoxide
DPBS	Dulbecco's phosphate-buffered saline
DP2	disaccharide
DP4	tetrasaccharide
DS	dermatan sulfate
DTT	dithiothreitol
ECM	extracellular matrix
EDTA	ethylenediaminetetraacetic acid
EHS	Engelbreth–Holm–Swarm
ELISA	enzyme-linked immunosorbent assay
EtBr	ethidium bromide
EtOH	ethyl alcohol
FBS	fetal bovine serum

FN	fibronectin
GAG	glycosaminoglycan
Gal	galactose
GalNAc	N-acetylgalactosamine
GAR	goat anti-rabbit
GBM	glomerular basement membrane
Gla	galactose
Glc	glucuronate
GlcA	glucuronic acid
GlcNAc	N-acetylglucosamine
GlcNSO$_3$	N-sulfated glucosamine
Glu	glucose
Gu–HCl	guanidine hydrochloride
HA	hyaluronate
HACC	hydroxyapatite chromatography column
HBSS	Hank's balanced salt solution
HEPES	N-2-hydroxy-ethylpiperazine-N-2-ethanesulfonic acid
Hex A	hexanoic acid
HMW	high molecular weight
HOAc	acetic acid
HPLC	high-pressure liquid chromatography
HRP	horseradish peroxidase
HS	heparan sulfate
IDE	isodesmosine
IdoA	iduronic acid
KOAc	potassium acetate
KS	keratan sulfate
LB	Luria broth
LBM	lens basement membrane
LMW	low molecular weight
LN	laminin
MEM	minimal essential medium
MeOH	methyl alcohol
MOPS	3-[N-morpholino]propane sulfonic acid
NaOAc	sodium acetate
NC1	non-collagenous domain of type IV collagen
NCP	non-collagenous protein
NEM	N-ethylmaleimide
NGS	normal goat serum
NPGB	p-nitrophenyl-p'-guanidobenzoate
PAGE	polyacrylamide gel electrophoresis
PAP	peroxidase–antiperoxidase
PAS	plasminogen activators
PBP	poly-L-proline binding protein

PBS	phosphate-buffered saline
PBS–T	PBS containing 0.5% (v/v) Tween-20
PCR	polymerase chain reaction
PDI	protein disulfide isomerase
pFN	plasma fibronectin
p.f.u.	plaque-forming units
PG	proteoglycan
PIC	protease inhibitor cocktail
PLP	periodate–lysine–paraformaldehyde
PMSF	phenylmethylsulfonyl fluoride
PR	phenol red
RGD	Arg–Gly–Asp
RIA	radioimmunoassay
RT	room temperature
SDS	sodium dodecyl sulfate
SSPE	sodium chloride, sodium phosphate, EDTA buffer
SPARC	secreted protein, acidic and rich in cysteine
TCA	trichloroacetic acid
TBS	Tris-buffered saline
TE	Tris–EDTA
TEMED	N,N,N',N'-tetramethylethylenediamine
TFA	trifluoroacetic acid
TIMPs	tissue inhibitors of metalloproteinases
TPA	tissue type plasminogen activator
TSP	thrombospondin
UPA	urokinase type plasminogen activator

1

The extracellular matrix—an overview

MICHAEL A. HARALSON and JOHN R. HASSELL

1. Introduction

The extracellular matrix (ECM) has been traditionally thought of as the structurally stable material that provides support for cells and tissues. However, a number of discoveries over the past two decades have changed this perspective and established the importance of the ECM in modern biology. First, both biochemical and molecular biological investigations have documented that four major classes of macromolecules—the collagens, proteoglycans, structural glycoproteins, and elastin—collectively comprise the ECM of animal cells (1–6). Furthermore, with the exception of elastin, each class of matrix macromolecules has been found to contain families of related proteins with each member being a unique gene product. Second, individual members of each class and family of ECM molecules were found to exhibit a degree of tissue-specific distribution implicating the matrix in development and tissue function (7–9). Third, specific cell surface receptors for ECM components were identified, which provided a rational basis for linking the ECM with the biology of the cell (10–16). From these discoveries we now appreciate that the extracellular matrix is a complex ordered aggregate composed of a number of different macromolecules whose structural integrity and functional composition are important in maintaining normal tissue architecture, in development, and in tissue specific function (2, 4, 6, 8, 17). Finally, in conjunction with these studies has come the recognition of the importance that dysfunctional matrix components and abnormalities in ECM biosynthesis and catabolism have in both inherited and acquired diseases and in normal wound healing (2, 3, 5, 6, 9, 18).

The purpose of this chapter is not a comprehensive review of the literature concerning the components of the extracellular matrix and their function. Rather, it is to present a brief survey of these topics to orient the reader in the use of the specific protocols contained in this volume and to emphasize unique features of matrix molecules that require special experimental approaches. In keeping with the theme of the *Practical approach* series, we have attempted to limit our citations to secondary sources, such as recent review articles, chapters in books, and books themselves. However, in some instances, we have cited primary sources either for clarity or because they have appeared subsequent to the latest reviews. To those investigators whose discoveries established the facts stated in this chapter and whose articles we have not cited and to those whose current reviews we have not referenced, we apologize and hope they will recognize the necessity of limiting our citations. The reader should also refer to citations in specific chapters in this volume which also contain additional references to a specific topic.

In order to provide the reader with a framework for appreciating the roles of the different families and molecules which comprise the extracellular matrix and their functions in maintaining normal cell function and in pathological processes, this chapter begins with a section on the role of the extracellular matrix as a regulator of the cell's biology. This is followed by sections devoted to each class of matrix components discussed and a brief section on additional aspects of the extracellular matrix requiring the use of distinctive experimental considerations for their study. In each of these sections, we have indicated the chapter in this volume where specific methods for studying the matrix components are presented. Finally, we have included an overview of the current knowledge about the involvement of the extracellular matrix and its components in both inherited and acquired diseases. Hopefully, this overview will enable the reader to orient the protocols in this volume for their studies and suggest new avenues of investigation.

2. The extracellular matrix as a regulator of the cell's biology

Concurrent with the enhanced appreciation of the complexity of the ECM has come the recognition that the ECM profoundly influences the biology of the animal cell (2, 4, 6, 19–22). The ECM and its components have been documented to control the growth, state of differentiation (21, 23), development (7, 8, 24, 25), and metabolic responses of a cell. To date, two major families of cell surface receptors—the integrins (10–13) and the syndecans (14–16)—have been identified as mediating the influence of the ECM on the cell. However, the effects of the extracellular matrix on the cell do not occur in a milieu devoid of other influences that affect the cell's biology. At the same time as studies documented the role of the extracellular matrix in controlling a cell's biology, similar investigations established the importance of growth factors, cytokines, hormones, vitamins and cell-to-cell contact as regulators of a cell's phenotype. From these latter studies, two important concepts have emerged: first, a major effect of growth factors and cytokines (26–28), vitamins and hormones (29), and cell-to-cell contact (30) on the cell is the regulation of extracellular matrix production; and second, many of the biological changes attributed to the effects of growth factors and cytokines on a cell (that is alterations in growth, state of differentiation, and metabolic responses) are similar, if not identical, to the effects of the ECM on cells (2, 4, 19–23, 27, 28). To reconcile these observations, which come from both cell culture and *in vivo* studies, the model, schematically presented in *Figure 1*, has emerged (2, 4, 19–23, 27, 28, 31). This paradigm states that beyond basic genetic programming, four major types of interactions (growth factors/cytokines, hormones/vitamins, cell-to-cell contacts, and the ECM) regulate the growth, shape, state of differentiation, development, and biochemical responses of the cell. Each of these interactions is mediated through specific cell surface or cytoplasmic receptors resulting in altered gene expression. Of additional importance is the fact that each type of interaction illustrated in *Figure 1*, including that with the ECM *per se*, results in altered ECM expression (4, 32)—a phenomenon termed 'mutual reciprocity' (32). The complexity of these interrelationships is further illustrated by the findings documenting that many growth factors and cytokines are specifically bound by matrix components and hence are resident in the ECM (14, 15, 33, 34), and that the extracellular matrix can modulate the expression of receptors for growth factors (35). Thus, it appears that no factor influencing the biology of the animal cell does so without affecting extracellular matrix biosynthesis, and that changes in the extracellular matrix affect the response of the cell to other regulators of the cell's biology. Hence our current knowledge indicates that control of a cell's biological phenotype, both in normal development and function and in pathological responses, represents a complex interplay between

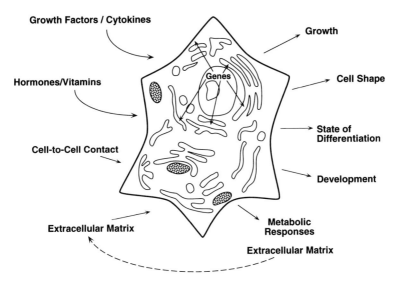

Figure 1. Schematic representation of the interrelationships between growth factors/cytokines, hormones/vitamins, cell-to-cell contact, and the extracellular matrix. Modified and reproduced with permission from reference (31).

soluble and insoluble mediators in which the extracellular matrix plays a pivotal role. It is within this context that studies both on existing and new extracellular matrix components and their biological function must be placed and in which our current knowledge about the macromolecules that form the extracellular matrix is subsequently presented.

3. The collagens

3.1 Introduction

Initially recognized as a distinct molecular entity over a hundred years ago (36–41), collagen is the most abundant protein in the animal kingdom, representing approximately one-third of all protein in tissues (6, 36, 38, 42). Traditionally, collagen was viewed as a protein organized into fibres with the basic monomeric unit structure of the fibre being the collagen molecule (43). These molecules were composed of three subunits (two α1 chains and one α2 chain) assembled into a coiled coil triple helix. This concept of collagen, however, changed approximately 25 years ago with the isolation of collagen molecules containing a collagen α chain having a different primary structure from chick cartilage (44, 45). This new collagen was designated type II collagen, its component subunits designated α1(II) chains, and conventional collagen molecules were designated type I collagen with its respective subunits being termed α1(I) and α2(I). Since the initial description of type II collagen, the collagen family of extracellular matrix macromolecules has expanded to include at least 30 distinct gene products that are present in at least 18 distinct types of collagen (36–40, 42, 46–48). The different types of collagen, their distribution, molecular organization, and function are summarized in *Table 1*. Thus today, the term collagen refers to a family of proteins sharing common biochemical features (see Section 3.2) and forming characteristic histological structures (36, 37), whereas the term 'collagen type' designates proteins with distinct primary structures. The chemistry, biosynthesis, and function of the different collagen types have been the subject of recent books, chapters in books, and numerous reviews in journals (36–42, 47, 49–51) and will not be presented in depth in this chapter. However, features of these molecules important for their isolation and characterization and general information about the different collagen types and their structure will be discussed briefly to orient the reader for use of the protocols in this volume.

3.2 Common features of collagen molecules (Chapters 2 and 6)

The members of the collagen family of extracellular matrix molecules share several common chemical and biological features:

(a) They predominantly reside outside the cell having a structural role in maintaining tissue integrity. It is this feature that partially distinguishes molecules referred to as 'collagen' from other molecules such as the C1q component of complement, acetylcholinesterase, the mannose binding protein, and the apoprotein component of surfactant, which have significant collagenous sequences in their primary structure (37, 38, 40).

(b) The primary structure of the initial gene product (the procollagen chain) contains repeating Gly–X–Y amino acid sequences with X and Y often being proline and hydroxyproline. Glycine represents approximately 33 per cent of the total amino acids in the collagen α chain, occurring in every third position along the chain. This sequence is necessary for the formation of the triple helix (36–40, 42, 49, 50). Proline and 4-hydroxyproline represent about 22 per cent of the total amino acid content in the α chain and alanine about 10–12 per cent of the total (50). Together, this amino acid composition refers in a primary structure containing multiple repeats of the sequence–R–Pro–X–Gly–Pro–R. This repeating sequence confers susceptibility of these molecules to bacterial collagenase (see Chapter 2A), which is useful both for identification of collagenous sequences and for their quantification.

(c) Collagen α chains possess a predominance of imino acids (38, 50) conferring upon them a basic character, which can be utilized for their chromatographic separation and characterization (see Chapter 2B).

(d) The primary gene products form triple helical molecules containing three subunits (α chains) organized in a coiled–coil arrangement. This structure consequently confers upon native collagen molecules resistance to proteases, which can be used for their isolation and purification (see Chapter 2B).

(e) The chains and the subsequent molecules undergo extensive post-translational modifications leading to the formation of both hydroxyproline and hydroxylysine residues and the addition of carbohydrate residues to the hydroxylysine residues. These post-translational modifications are catalysed by the enzymes prolyl hydroxylase (Chapter 6), lysyl hydroxylase, hydroxylysyl galactosyl- and hydroxylysyl glucosyl-transferase (37–39, 49, 50).

(f) The subunits of the molecules and molecules themselves undergo post-translational covalently cross-linking through the hydroxylysine residues (37–39, 50). This cross-linking renders the molecules highly insoluble. While this cross-linking can be inhibited in experimental situations (see Chapter 2B), identification and quantification of collagens from tissues in which the molecules are cross-linked requires special approaches (see Chapter 2D).

Some genetic types of collagen molecules form higher-ordered aggregates, for example, the fibre-forming collagens (Section 3.3.1), which have characteristic histological structures. Other collagen types may interact with these

Table 1. The vertebrate collagens

Collagen type	Constituent chains	Human chromosomal locus	Molecular species	Organization of molecular aggregates	Distribution	Function
I	$\alpha1(I)$ $\alpha2(I)$	17 7	$[\alpha1(I)]_2\alpha2(I)$ $[\alpha1(I)]_3$	Staggered large-diameter banded fibrils	All connective tissues (skin, tendon, ligaments, cornea, lung, bone, etc.) except basement membrane and hyaline cartilage	Supporting fibres; recognized by integrin and syndecan cell surface receptors
II	$\alpha1(II)$	12	$[\alpha1(II)]_3$	Staggered small-diameter banded fibrils	Hyaline cartilage; other cartilagenous tissues, intervertebral discs (annulus fibrosus) and vitreous	Supporting fibres
III	$\alpha1(III)$	2	$[\alpha1(III)]_3$	Staggered small-diameter banded fibrils	Distensible connective tissue, skin, lung, vessels, gingiva, kidney, sclera, spleen, heart muscle, uterus, nerve, and lymph nodes	Small supporting fibres; forms copolymers with type I collagen
IV	$\alpha1(IV)$ $\alpha2(IV)$ $\alpha3(IV)$ $\alpha4(IV)$ $\alpha5(IV)$ $\alpha6(IV)$	13 13 2 2 X X	$[\alpha1(IV)]_2\alpha2(IV)$ $[\alpha1(IV)]_3$	Meshwork sheets	Basement membranes	Meshwork scaffolding; contains multiple cell-binding sites
V	$\alpha1(V)$ $\alpha2(V)$ $\alpha3(V)$	9 2	$[\alpha1(V)]_3$ $[\alpha1(V)]_2\alpha2(V)$ $\alpha1(V)\alpha2(V)\alpha3(V)$	Probably small-diameter staggered fibrils; may form molecules with type XI chains	All connective tissue except BM; abundant in vascularized tissues and in the corneal stroma	Small supporting pericellular fibres; possible core for type I molecules
VI	$\alpha1(VI)$ $\alpha2(VI)$ $\alpha3(VI)$	21 21 2	$\alpha1(VI)\alpha2(VI)\alpha3(VI)$	Beaded filaments	Essentially all (cartilagenous and non-cartilagenous) connective tissue except basement membranes	Possible interface between major collagen fibrils and cells; contains multiple RGD sequences recognized by integrin receptors
VII	$\alpha1(VII)$	3	$[\alpha1(VII)]_3$	Anchoring fibrils	Sub-basal lamina of skin	Anchoring of dermal epithelial cells to underlying stroma

Type	Chains		Molecular form	Tissue distribution	Function
VIII	α1(VIII)[a] α2(VIII)[a]		Unknown	Meshwork sheets	Cultured endothelial cells, sclera, subendothelium of large blood vessels, Descemet's membrane, cartilage growth plate, perichondria, and uncalcified calvaria	Unknown
IX	α1(IX) α2(IX) α3(IX)	6 1	α1(IX)α2(IX)α3(IX)	Fibril associated collagen with interrupted triple helix (FACIT)	Hyaline cartilage; vitreous humour	Association with type II collagen allowing type II collagen fibril association and interaction with cartilage proteoglycans
X	α1(X)	6	[α1(X)]₃	Meshwork sheets	Hypertrophic cartilage	Endochondral bone development; associates with type II collagen fibrils
XI	α1(XI) α2(XI) α3(XI)/α1(II)	6 12	α1(XI)α2(XI)α3(XI)	Small-diameter probably staggered fibrils; can form molecules with type V chains	Hyaline cartilage, vitreous humour	Possible core for type II molecules regulating fibril diameter
XII	α1(XII)	6	[α1(XII)]₃	FACIT	Embryonic tendon and skin; peridontal ligament	Associated with type I collagen allowing association of different ECM components
XIII	α1(XIII)	10	Unknown	Unknown	Product of some cultured tumour cell lines and endothelial cells; cartilage growth plate, epidermis, hair follicles and bone marrow mesenchymal cells	Unknown
XIV	α1(XIV)		[α1(XIV)]₃	Possibly FACIT	Fetal skin and tendon	Possibly similar to type XII
XV	α1(XV)	9	Unknown	Unknown	Fibroblasts, endometrium, HeLa cells	Unknown
XVI	α1(XVI)	1	Unknown	FACIT	Heart, lung, pancreas, skeletal muscle endothelial cells, placenta	Unknown
XVII	α1(XVII)		Unknown	Unknown	Dermal–epidermal junction of skin in hemidesmosomes	Linkage of basal cells to stroma; bullous pemphigoid antigen
XVIII	α1(XVIII)		Unknown	Unknown	Lung, liver, and kidney	Unknown

[a] Designation based on data from full-length cDNA clones established from Descemet's membrane mRNA.

Data compiled and adapted from references 37–42, 49, 51, 53–65, and 90, and references in Section 3 of this chapter.

aggregates, for example the FACIT collagens (Section 3.3.2). The nature of the aggregate differs with the collagen type. Some genetic types of collagen molecules undergo extensive post-translation processing before incorporation into higher-ordered aggregates and deposition in the extracellular matrix. For example type I collagen is initially synthesized as a large precursor (type I procollagen) followed by subsequent cleavage of the amino- and carboxy-propeptides to yield type I tropocollagen, which is the form of type I collagen that exists in tissues (see *Figure 1* in Chapter 2). Other genetic types of collagen form their higher-ordered aggregates and exist in tissues without such processing, for example type IV collagen. These differences in post-translational processing and subsequent structure of the higher-ordered aggregates confer different structures to the molecules and subsequently to the extracellular matrices in which they reside.

3.3 Classification of the collagens based on structure

While sharing common features detailed in Section 3.2 the family of collagen-ous proteins differ greatly in their overall structure. Several schemes based on either the structure of the higher-order aggregates formed by the molecules (36) or the size of the individual chain and the continuity of the helical domains within the chain (42, 52) have been proposed for grouping the differ-ent molecules into different families. The following grouping is a modification of the classification scheme proposed by van der Rest and Garrone (36) with reference to other proposed groupings.

3.3.1 Collagens which form quarter stagger fibrils (Chapters 2 and 3)

Members of this family include the classical collagens—types I, II, III, V, and XI (53–55). These molecules possess long uninterrupted triple helices approximately 300 nm long (52) and share a highly conserved intron–exon structure of their genes (54, 56). The higher-order aggregates arising from each of these five collagen types form banded fibres, and based on similarity in structure have been categorized as Group 1 molecules by Miller (52). Type I collagen is the most abundant of all collagen types and is a heterotrimer composed of two $\alpha 1(I)$ and one $\alpha 2(I)$ chains. It is found in all connective tissues except basement membranes and hyaline cartilage. Its structure, which is representative of the structures of each of the fibril-forming group of molecules, is shown in *Figure 1* in Chapter 2. Type II collagen is the major collagen of cartilagenous tissues and is composed of three $\alpha 1(II)$ chains. Type III collagen contains three $\alpha 1$ (III) chains and has a similar tissue distribution, with the exception of tendon and ligament, with type I collagen. As discussed in the review by van der Rest and Garrone (36), the organization and tissue distribution of type V and type XI molecules is much more complex. Molecules containing different combinations of three chains—designated $\alpha 1(V)$, $\alpha 2(V)$, and α3(V)—have been isolated from tissues containing type I molecules and molecules containing three different chains—designated α1(XI), α2(XI), and α3(XI)—have been isolated from tissues containing type II molecules. However, evidence exists indicating the α1(XI) chain is expressed in some tissues containing type I molecules and that molecules containing a mixture of type V and type XI chains exist. Similar findings indicate that in certain tissues fibrils containing type I molecules also contain type III collagen and that in the cornea, type I fibrils are arranged around a core of type V collagen. Thus, in dealing with most higher-order aggregates of the fibre-forming collagens, fibrils containing a single genetic type of collagen are likely to be the exception rather than the rule. This concept is further borne-out when the FACIT collagens (Section 3.3.2) are considered.

3.3.2 Fibril-associated collagens with interrupted triple helices (Chapter 3)

Members of this family, termed 'FACIT' collagens include types IX, XII, XIV, and XVI molecules. These molecules have significant interruptions in their triple helices and are found on the surface of collagen fibrils. Each FACIT collagen forms homotrimeric molecules with a chain size less than 95 kDa, and hence are classified as group 3 collagens in the scheme of Miller (52). Type IX collagen (57) associates with type II collagen and may possess, but not absolutely, a single chondroitin sulfate proteoglycan chain covalently linked to the NC3 domain of the α2(IX) chain. Type XII and XIV molecules are associated with type I and type III collagen fibrils (58). There are two different sizes of type XII molecules (type XIIA—340 kDa, which contains glycosaminoglycans, and type XIIB, which lacks the glycosaminoglycan chains). Type XIV collagen appears to be a variant of the protein undulin. The FACIT collagens have been postulated to mediate the interactions of the fibrillar collagens with other matrix components.

3.3.3 Collagens which form sheets (Chapters 3, 4, and 5)

Members of this family include type IV collagen (59–61), which is the collagenous component of basement membranes, type VIII molecules (62), which are produced by endothelial cells in culture and are the collagenous constituents of Descemet's membrane in the cornea, and type X collagen (63), which is produced by hypertrophic cartilage. These molecules are characterized by their large size (chain size greater than 95 kDa) and imperfections in their triple helices. Accordingly, these collagen types have been placed in Group 2 molecules (52).

The type IV collagen group of molecules is composed of six unique gene products, whose precise molecular organization remains unresolved. The prototypical structure for the higher-order aggregates of type IV is shown in *Figure 1C*, Chapter 4. The molecular organization of type VIII molecules likewise remains unresolved, with the possibility that type VIII molecules produced by endothelial cells may be different gene products than those found in Descemet's membrane. Type X collagen is produced by hypertrophic collagen and forms homotrimeric molecules. It has a similarity in gene structure and amino acid sequence to type VIII molecules and is found in association with small fibrils containing type II collagen.

3.3.4 Collagens which form beaded filaments

The sole member of this family is type VI collagen (64), which is ubiquitous and found in nearly all non-cartilaginous tissues. Based on chain size and imperfections in the triple helix, type VI molecules are also considered to be members of the Group 2 class. Type VI molecules usually contain three distinct subunits which form heterotrimers. However, the precise molecular composition of the molecules is controversial as there is evidence for alternate splicing and failure of the chains to be synthesized in stoichiometric ratios in all tissues. The molecules possess a number of RGD sequences and appear to associate with the fibre-forming collagens. These findings have led to the speculation that type VI collagen may serve as an interfacing molecule between the cell and the stroma.

3.3.5 Collagens which form anchoring fibrils (Chapter 4)

The only member of this family is type VII collagen, which is the collagenous component of anchoring fibrils in the skin (65–67). Based on chain size and interruption in the triple helix, type VII molecules have been classified as Group 2 molecules. Type VII chains form homotrimers, which interact to form dimers capable of lateral aggregation. These aggregates are capable of binding to the basement membrane and anchoring plaques at their ends, and hence contribute to the structural adherence of the epithelium to the underlying stroma.

3.3.6 Collagens with unknown structure

To date, four additional genetic types of collagen have been reported based on the sequence of cDNA clones. These are: types XIII, XV, XVII, and XVIII. While their function and structure remains obscure, recent data, reviewed by Mayne and Brewton (47), does provide some insights into their structure and function. Type XIII collagen has some similarity with type IX collagen and type XIII transcripts are especially abundant in epidermal tissues. Of interest is the fact that the type XIII gene can undergo alternative splicing resulting in several transcripts. Type XV collagen has been found to contain a number of short collagenous domains interrupted by non-collagenous domains but shows no similarity to any of the other collagen types. Type XVII was originally isolated from hemidesmosomes, based upon it being one of the autoantigens in the blistering disease bullous pemphigoid (48, 69, 70). Its sequence bears no resemblance to any other collagen type, and it remains to be demonstrated that the initial gene product can form triple-helical molecules.

Type XVIII molecules show sequence similarities with type XV but have a different tissue distribution. To date, the structure, molecular composition, and function of both collagen types remains unresolved.

3.4 Future perspectives

To date over 30 distinct structural genes coding for the collagens found in the extracellular matrix have been described. Undoubtedly more such genes will be found, with one candidate being the recently described gene for the $\alpha1(Y)$ chain (47). However, information about the chemistry of many of the newly recognized collagen types remains incomplete. The newly recognized collagen types, as well as those that may yet be discovered, are likely to represent minor constituents of specialized extracellular matrices and will exhibit a limited tissue distribution. Thus, the future challenges in collagen biology are to define the structure of the new molecules and their function. Such information should provide insights into both the role of the collagens in maintaining normal tissue architecture and function, as well as into the evolutionary origin and purpose of these multiple genes.

4. The proteoglycans (Chapters 9 and 10)

4.1 Introduction

Proteoglycans consist of a protein core and one or more glycosaminoglycan side chains (GAG) covalently bound to the core protein (for reviews see references 71, 72, and 73). Proteoglycans also often have N- and O-linked oligosaccharides attached to their core proteins. There are basically three types of sulfated GAGs: (1) chondroitin/dermatan sulfate, a polymer consisting of alternating galactosamine and glucuronic/iduronic acid units; (2) heparin and heparan sulfate, a polymer consisting of alternating glucosamine and glucuronic/iduronic acid units; and (3) keratan sulfate, a polymer consisting of alternating glucosamine and galactose units. All these polymers are sulfated in various positions. Thus, for each GAG, at least two sugar transferases are required to polymerize the repeating disaccharide and two sulfotransferases are required to add the sulfate esters on to the resulting polymer. Hyaluronan is not sulfated and is not covalently attached to protein. Consequently, it is not considered as part of proteoglycan structure. The structure of all the GAGs and hyaluronan is discussed in detail in Chapter 8.

GAGs were discovered and characterized before core proteins were identified and distinguished from one another. Furthermore, some tissues were found to be a particular rich source of certain types of GAGs. Consequently, many proteoglycans have been named according to the type GAG attached to its core protein or its tissue source. In 1985, however, the first proteoglycan was cloned (74) and since then the core protein structure of 18 proteoglycans have been deduced from cDNA clones. As a result, proteoglycan nomenclature has changed to naming cloned proteoglycans according to the gene product or core protein rather than the type of GAG or tissue source. These new names often reflect some feature, activity, or property of the core protein.

4.2 Proteoglycan core proteins

The structural characterization of the core proteins, as deduced from cDNA clones, has radically changed our perception of proteoglycan structure and function as well as their relationship to each other and to other matrix components. Unlike collagens, which all have a Gly–X–Y repeat, the core proteins of proteoglycans do not all have a single structural feature in common. Core proteins of most proteoglycans are multidomained and, in many cases, the domains are homologous to other matrix structural glycoproteins and collagens. Like the structural glycoproteins and collagens, many domains of proteoglycan core proteins interact with distinct structural elements present on cells or in extracellular matrices. This indicates that proteoglycans are essentially a class of collagens and structural glycoproteins made special by the addition of GAG side chains.

Traditionally, proteoglycans were found to be on the surface of cells and in the extracellular matrix, and the sequences of the core proteins deduced from cDNA clones indicate they contain structural features that place them in one of these two locations (*Table 2*). The core proteins of cell surface proteoglycans have a transmembrane segment, a small cytoplasmic domain and a

---

Table 2. Proteoglycan gene families

Cell surface	Extracellular matrix
Syndecan family	*Leucine-rich family*
Syndecan I	Decorin
Syndecan II (fibroglycan)	Biglycan
Syndecan III	Fibromodulin
Syndecan IV (ryudocan)	Lumican
	PG-Lb
Orphans	*Hyaluronate binding family*
NG2	Aggrecan
Betaglycan	Versican
CD44	Neurocan
Glypican	CD44
	Collagenous family
	Collagen IX
	Collagen XII
	Orphans
	Perlecan
	Serglycin

Table 3. The vertebrate proteoglycans

Name	Core size (kDa)	GAG type	Location	Function
Syndecan 1	32	HS + C/DS	Cell surfaces of	Binds matrix constituents and growth factors via GAG side chains
Syndecan 2	23	C/DS	many cell types	
Syndecan 3		C/DS		
Syndecan 4	22	C/DS		
Betaglycan	92	C/DS + HS		Binds TGF-β
NG2	251	C/DS	Neural tissues	Binds collagen VI
CD44		C/DS		
Glypican	62	HS	Liver, Schwann cells	
Decorin	36	C/DS		
Biglycan	38	C/DS		Associates with fibrillar collagens
Fibromodulin	39	KS		
Lumican	36	KS		
PG-Lb	32	C/DS		
Aggrecan	221	CS + KS	Cartilage	Resists compression
Versican		C/DS		
Neurocan		C/DS		
Collagen IX	Trimer	C/DS		
Collagen XII	Trimer	C/DS		
Perlecan	396	HS	Basement membrane	Filtration, cell attachment
Serglycin	19	H + CS	Mast cells	GAG synthesis

larger, although somewhat featureless, extracellular domain. The core pro-teins of extracellular matrix proteoglycans lack a transmembrane segment but have multidomain core proteins that often have homology to the structural glycoproteins of the cell surface and extracellular matrix (*Table 2*). Some of the proteoglycans in each of these subdivisions have homologous core pro-teins, and therefore can be grouped into families (*Table 2*). The structural features and known functions of the proteoglycans in most of these families are listed in *Table 3*.

4.3 Cell surface proteoglycans

The cell surface proteoglycan subdivision contains one major family—the syndecans. This family contains four proteoglycans (designated 1, 2, 3, and 4) each with transmembrane segments and cytoplasmic domains that are nearly identical (75–77). The sizes of their core proteins range from 22 kDa to 40 kDa and they have two or three heparan sulfate side chains per core. There are also four other proteoglycans with transmembrane segments; NG2 (78), betaglycan (79), and CD44 (68). These have no homology to each other or to the syndecan family and are, for the present, termed orphans. Glycipan does not have a transmembrane segment but maintains its association with cell surfaces via a phosphoinositol linkage (80).

4.4 Extracellular matrix proteoglycans

The extracellular matrix subdivision of proteoglycans contains three major families. The first family, the leucine-rich family, consists of five members: decorin (81), biglycan (82), fibromodulin (83), lumican (84), and PG-Lb (85). Members of this family are characterized by relatively small core proteins (32–39 kDa) and a centrally located leucine-rich domain that comprises about 60 per cent of the total sequence. Lumican and fibromodulin have keratan sulfate side chains while the rest of the members of this family usually have chondroitin/dermatin sulfate side chains. The second major family of this subdivision, the hyaluronate-binding family, is characterized by the presence of a hylauronate-binding domain in the core protein. This is homologous to the hyaluronate-binding domain that is in link protein. This family consists of aggrecan (86), versican (87), neurocan (88), and CD44 (68). Members of this family usually have chondroitin or dermatan sulfate side chains but aggrecan is also often produced with keratan sulfate side chains. The core protein sizes of these proteoglycans are large (approximately 100–220 kDa), with the exception of CD44 which is smaller. CD44 can be alternately spliced with a transmembrane segment and is therefore also placed in the cell surface subdivision. The third family, collagenous family, consists of two collagens, types IX and XII, that bear chondroitin/dermatan sulfate side chains. There are two proteoglycans, perlecan (89) and serglycin (74), in this subdivision that have no homology with any other proteoglycan and are therefore in the orphan category. Two domains of perlecan, however, are highly homologous to the N- and C-terminal domains of the A chain of laminin. Perlecan has an exceptionally large core protein (about 396 kDa) and is most often produced with heparan sulfate side chains.

It should also be recognized that the core proteins of some proteoglycans are not always produced with GAG side chains. Lumican, the major keratan sulfate containing proteoglycan of the corneal stroma is produced as a glycoprotein in the aorta. Likewise, collagens IX and XII as well as CD44 are not always produced with their C/DS side chain. Perlecan is made without heparan sulfate side chains when produced by chondrocytes. These observations suggest that GAG addition is more cell type-specific than core protein-specific. The diversity of proteoglycan structures is illustrated in *Figure 2*. Methods to separate different proteoglycans from each other and from glycoproteins are discussed in Chapter 10.

5. The structural glycoproteins

5.1 Introduction

Just as the collagen and proteoglycan classes of extracellular matrix molecules have been found over the last two decades to consist of many members, so has the class of structural glycoproteins. While members of this class are hetero-

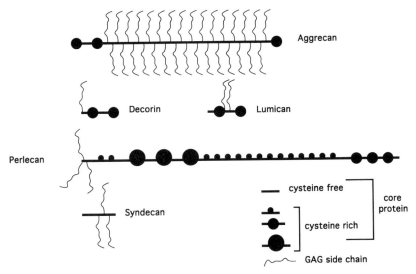

Figure 2. Structural models of different proteoglycans. The core protein of five different proteoglycans are diagrammaticly drawn to scale. The relative proportion of cysteines in that region of the core are represented by the size of the black circle. The location of the GAG side chains on the core are indicated. However, the size, type, and number (in the core of aggrecan) of GAG varies considerably between these proteoglycans, and this is not indicated in these models. Lumican and decorin have similar core protein models because of their high homology. Additional information is contained in Section 4 of this chapter and in references 71, 72, and 73.

geneous in size, structure, and tissue distribution, as indicated in the review by von der Mark and Goodman (90), these matrix macromolecules exhibit several common features. First, their primary structures have revealed them to be multidomain, both structural and functional, proteins. Second, they possess independent binding sites for cells and for other matrix components. Arg–Gly–Asp (RGD) is the major cell recognition sequence, which allows specific interaction of the structural glycoproteins and some of the collagens with cells through the integrin superfamily of cell surface receptors (10–13, 90, 91). Finally, many of the structural glycoproteins consist of multiple subunits. A summary of the major structural glycoproteins, their tissue distribution and function is presented in *Table 4*. Many times the same protein was discovered independently in different laboratories, and thus has multiple names in the literature. We have tried to list these other names under the accepted name for the same protein and in some instances have listed the names of other members of the same family of structural glycoproteins in *Table 4*. This section will summarize the features of the major extracellular matrix structural glycoproteins referring the reader to recent reviews for primary sources.

Table 4. The structural glycoproteins of the extracellular matrix

Protein	Size	Location	Function
Fibronectin LETS Cold insoluble globulin	Two 220 kDa S–S bonded chains	Plasma (produced by hepatocytes); cellular fibronectin produced by fibroblasts, epithelial and endothelial cells, chondrocytes, macrophages, and platelets	Cell attachment; binds collagen types I and III, heparin and fibrin; involved in cell differentiation, early developmental processes and wound repair; opsonization of blood debris; thrombosis; cytoskeletal organization
Laminin[a] α β γ	850 kDa α chain—400 kDa β chain—230 kDa γ chain—210 kDa	Basement membranes	Interaction with type IV collagen, entactin, and heparan sulfate proteoglycan; cell binding to the ECM; involved in development and cell differentiation, wound healing, angiogenesis, and nerve regeneration
Entactin Nidogen	150 kDa	Basement membranes	Binds to laminin, type IV collagen and fibronectin; contains RGD sequence and EGF-like domains
Tenascin[b] Cytotactin Hexabrachion Brachionectin J1 Myotendinous Ag Restrictin	~ 1240 kDa S–S bonded trimers that connect to form a hexamer; 3 chains —230,200,190 kDa	Transient expression during embryonic development, in tumours and healing wounds; present in skin, bone, cartilage, tendon, adult brain, and myotendinous junction	Fibronectin, heparin, and chondroitin sulfate proteoglycan interactions; EGF and fibrinogen-like domains; contains integrin specific cell binding domain but is strongly linked to cell rounding and detachment

Protein	Molecular properties	Occurrence	Function
Thrombospondin	420 kDa; 3 identical chains ~ 150 kDa each	Platelet α-granules; produced by endothelial, glial, smooth muscle, and bone cells; fibroblasts, monocytes, macrophages, type II pneumocytes, and keratinocytes	Cell attachment; platelet aggregation, proliferation, and migration of smooth muscle cells; binding sites for thrombin, fibrinogen, fibronectin, laminin, tPA, and plasminogen
SPARC/Osteonectin BM-40 43 kDa culture-shock protein	32 kDa	Bone, basement membrane, and tissues undergoing remodelling or morphogenesis; produced by endothelial and smooth muscle cells, fibroblasts, and osteoblasts; present in platelets	Binds collagen types I, II, and V, calcium, thrombospondin and cells; possible regulator of cell responses during injury and development by inhibiting cell binding and spreading
Vitronectin Serum-spreading factor S protein	75 kDa single chain; 65 kDa and 10 kDa S–S bonded proteolytic fragments	Present in liver, platelets, and macrophages; present in plasma at ~250 μg/ml and platelets	RGD-mediated attachment of fibroblasts and endothelial cells; associated with elastic fibres; binds C5b-7, heparin and plasminogen activator inhibitor-1; prevents C9 polymerization that inhibits membrane attack complex; binds thrombin–antithrombin III complex and inhibits its heparin-catalysed formation
Link protein	43–49 kDa; different forms vary in CHO content and ± a small peptide via alternative splicing	Cartilage	Binds to PG aggrecan and hyaluronate, stabilizing their interaction
Fibrillin	350 kDa	Skin, aorta, lung, muscle, cornea, placenta, and ciliary zone	A constituent of the microfibrillar component of elastic fibres

Information primarily compiled from references, 73, 90, 92–100, 105, 107, 109–112, and references in Section 5 of this chapter.

[a] Members of the laminin family have been described in the literature under several names, including merosin, S-laminin, Kalinin, *etc*. (see references 61, 90, and 97). Recently, a systematic nomenclature for the laminins and their constituent chains has been adopted (69) and is used in this table.

[b] Tenascin is used in this context to denote a family of proteins. Tenascin-C corresponds to cytotactin. Other members of the tenascin family are tenascin-R (restrictin) and tenascin-X. For additional information see reference 110 and references in Section 5 of this chapter.

5.2 Fibronectin (Chapter 7)

The initial structural glycoprotein, based on its immunohistochemical and cell-adhesion promoting activity, defined as an extracellular matrix component was fibronectin (FN). The structure, isolation, properties, biosynthesis, and biological activities of this molecule have been summarized in several recent reviews (90, 92–95). Fibronectin contains two subunits, each with a molecular mass of approximately 250 kDa, that are disulfide-bonded at their carboxy-terminus. The molecule is a product of most mesenchymal and epithelial cells and is present in both the extracellular matrix and in plasma. There is a single fibronectin gene located on human chromosome 2; however, FNs isolated from different sources contain chains that differ in size, indicating a number of different slicing variants—a concept proven by the sequencing of cDNA clones. To date the family of fibronectins contains at least eight splice variants. Depending on the cellular origin and splicing variant, 4–10 per cent of the mass of a FN molecule can be O- and N-linked carbohydrates. Fibronectin contains binding domains for cells, gelatin (collagen) heparin, some proteoglycans, DNA, hyaluronic acid, and fibrin. These structural domains provide the chemical basis for the diverse biological functions of fibronectin, including the regulation of growth, differentiation, cell shape, and migration, as well as its interactions with other matrix components.

5.3 Laminin (Chapter 4)

Laminin (LN) is the major structural glycoprotein of basement membranes which separate epithelial, endothelial, muscle, and nerve cells from the stromal matrix (90, 93, 96–100). Laminin was initially identified as a product of a cultured mouse carcinoma line (101). Subsequently, this structural glycoprotein was isolated from the transplantable mouse Englebreth–Holm–Swarm (EHS) sarcoma (102), which produces significant quantities of basement membrane (see Chapter 13). The prototypical laminin molecule, the extract from the EHS sarcoma, is a disulfide-bonded heterotrimer containing three non-identical chains designated A ($M_r = 440\,000$), B1, and B2, each with an $M_r = 220\,000$. The three chains form a cruciform structure with globular domains at the end of each arm (see *Figure 1C*, Chapter 4). Over the past decade, evidence has emerged indicating that the EHS laminin represents one member of a family of proteins containing a number of isoforms composed of at least nine genetically distinct subunits (61, 90, 96, 97). The chain organization, structural features, and proposed nomenclature system for the laminin isoforms are summarized in the reviews of Tryggvason (97) and Engel (96). Among the major laminin isoforms are merosin, which is a component of the basement membrane surrounding striated muscle and Schwann cells, S-laminin found in the basement membrane of the glomerulus and the synaptic basement membrane, and K-laminin or kalinin. Like the fibronectins, a variety of biological activities have been ascribed to the laminins. These include: promotion of cell adhesion and neurite outgrowth, the regulation of cell shape, proliferation and differentiation, and establishment of the polarity of a variety of cell types.

5.4 Entactin/nidogen (Chapter 4)

Extraction under gentle conditions revealed a fourth molecule closely associated with laminin. This molecule was originally named entactin (103) and later a fragment of entactin, which formed nest-like structures, was termed nidogen (104). Entactin/nidogen is a single polypeptide chain, which exhibits both type IV collagen and laminin-binding activities, and contains RGD sequences, which are cell attachment sites (105, 106). The location of entactin/nidogen in basement membranes is schematically presented in *Figure 1C*, Chapter 4.

5.5 Tenascin (Chapter 8)

The family of tenascin (TN) molecules consists of large ($M_r > 1.2 \times 10^6$), hexameric structural glycoproteins containing two groups of three nonidentical subunits (90, 107–110). Using the nomenclature system given in the review by Erickson (110), tenascin will be used to refer to tenascin-C. The two three-arm units are covalently linked at the centre of the molecule resulting in a molecule which, by rotary shadowing, displays a six-arm structure. The tenascins are multidomain proteins containing multiple EGF and FN type III domains. Tenascin has been identified in a number of tissues, tumours, and healing wounds. However, its precise biological function remains to be elucidated.

5.6 Thrombospondin (Chapter 5)

Thrombospondin was initially identified as a 450 kDa molecule released from the α-granules of platelets upon activation, and was subsequently identified as a product secreted and incorporated into the extracellular matrix by growing cells including smooth muscle cells, endothelial cells in culture, and fibroblasts (90, 111, 112). Thrombospondin consists of three identical subunits each with multiple binding domains allowing its interaction with a number of matrix molecules (fibronectin, collagen, heparin, etc.), Ca^{2+}, and components involved in thrombosis and homeostasis (for example thrombin/plasminogen, fibrinogen, fibrin). Consequently, thrombospondin has been implicated as a regulator of many biological processes including cell growth, adhesion and migration, platelet aggregation, and fibrin deposition and lysis.

5.7 SPARC/Osteonectin (Chapter 5)

Originally identified as a 43 kDa bone phosphoprotein, osteonectin subsequently isolated from other tissues has been termed SPARC (secreted protein, acidic and rich in cysteine, or from basement membranes termed BM-40 (73)). In addition to its presence in bone, it is produced by cultured cells, is found in the kidney, and has been shown to be present in other tissues. Osteonectin/SPARC binds calcium and hence its suggested role in mineralization. It also contains separate domains capable of binding cells and extracellular matrix components and has thus been proposed as a regulator of the interaction of cells with the extracellular matrix. It probably has other functions, as suggested by its structural homology to heat-shock protein and its production in response to cellular injury, but additional information is required to precisely resolve these roles.

5.8 Other structural glycoproteins

Many additional structural glycoproteins have been described in the literature. It is beyond the scope of this chapter to address each of these in detail. A few of these molecules, which are listed in *Table 4*, will be discussed because of their relationships to molecules covered in this book. For additional details, the reader should refer to the recent reviews (73, 90) and the additional citations.

Vitronectin is a 75 kDa serum and matrix protein with a similar tissue distribution to fibronectin and is involved in cell attachment, regulation of blood coagulation, and immune responses (113). Vitronectin interaction with cells can be inhibited by RGD peptides, and its integrin receptor also recognizes fibronectin (13, 113). *Link protein* was originally isolated from cartilage, and studies have established that it stabilizes the interaction of hyaluronate with the proteoglycan aggrecan. *Fibrillin* is found in association with elastic fibres. Additionally, there are many structural glycoproteins associated with bone and cartilage, for example osteocalcin, osteopontin, bone sialoprotein, anchorin, etc. The reader is referred to the chapter by Heinegård and Oldberg (73) for recent information regarding these molecules. Undoubtedly, additional structural glycoproteins of the extracellular matrix will be identified in the future, and their function, as well as resolution of the function of many of the known structural glycoproteins, will provide further insights into the organization and biological actions of the extracellular matrix.

6. Elastin (Chapter 11)

The elastic fibre, which confers flexibility and distensibility to all vertebrate tissues, is a complex of elastin with several non-elastic glycoproteins termed fibrillar components, the most notable of which are fibrillin (114–116) and lysyl oxidase (117). In contrast to the other families of extracellular matrix components, the elastin family is composed only of a single gene product (tropoelastin) that is coded for by a single copy gene located on human chromosome 7 (116–118). The elastin molecules in elastic fibres are highly insoluble, reflecting the presence of the unique intermolecular cross-links desmosine and isodesmosine that arise from hydroxylysine. These cross-links can be used to quantitate the elastin content of tissues (Chapter 11). However, purification of elastin from tissues and from the microfibrillar components of the elastin fibre requires special approaches that are presented in Chapter 11. Elastin has a similar but non-identical amino acid content to collagen, that is about 33 per cent glycine, 10–13 per cent proline, and about 24 per cent alanine, but is distinguished by the lack of hydroxylysine, methionine, and other amino acids (see *Table 1*, Chapter 11).

7. Other unique features of the extracellular matrix requiring special techniques

In contrast to more traditional proteins, the structure and organization outside the cell of the matrix proteins requires additional considerations in their studies. In addition to the special approaches necessary for studying the individual matrix components included in the chapters previously referenced, some additional considerations involving other aspects of the matrix are necessary for their study.

The extracellular matrix is a highly organized superstructure containing many members of the families of matrix components. Catabolism of the extracellular matrix is an important aspect both of normal tissue development and in the initiation and progression of disease and is the sum of a balance between the activities of the matrix metalloproteinases and their inhibitors (119–121). Because of its organization and the primary, secondary, and tertiary structures of the components, degradation of the extracellular matrix involves families of specialized enzymes to facilitate its catabolism (119–121). The methods detailed in Chapter 12 provide strategies for studying matrix degradation. Similarly, the unique organization and chemical properties of the matrix require, in some instances, special technical considerations beyond standard immunohistochemical approaches for properly identifying the tissue location of matrix macromolecules. These methodologies are detailed in Chapter 14.

Most of the extracellular matrix molecules and their components are large. Consequently, special considerations in approach (detailed in Chapter 15) are required for isolating the mRNAs which code for them. Elucidating the regulation of matrix gene expression is crucial not only for the understanding of normal cell and tissue function, but also for understanding the molecular basis of many inherited and acquired diseases (see Sections 8.2 and 8.3). The size of most matrix components and hence their mRNAs are large. Consequently, the size and complexity of the genes coding for extracellular matrix components likewise are very large (56, 122, 123). Thus, the study of the regulation of intact genes requires special approaches for their isolation, and these methods are detailed in Chapter 16.

Finally, because the extracellular matrix affects the biology of the cell (see Section 2), it has been used as a substrate for cell culture both for promoting cell survival and growth and for altering the biology of the cell in culture. Techniques for using extracellular matrix and its components in culture studies are presented in Chapter 13.

8. Involvement of the extracellular matrix in disease

8.1 Introduction

For over two centuries, the medical community has appreciated the widespread and significant consequences that heritable changes in connective tissue cause (124). As our awareness of the number of molecules found in the extracellular matrix has increased exponentially over the past two decades, so has our recognition not only of how specific mutations in extracellular matrix genes result in heritable diseases, but also of the critical roles that changes in these molecules play in the development and progression of a variety of acquired pathological states (for reviews see references 1–3, 5, 8, 125 and 41). We now also recognize the major roles that changes in ECM synthesis, accumulation, and catabolism play in wound healing and fibrosis (5, 18). The following is a brief summary of the major inherited and acquired diseases involving the extracellular matrix. While by no means complete, we hope this section will orient the reader to diseases involving those matrix components for which methods are presented in this volume.

8.2 Heritable connective tissue diseases reflect mutations in matrix genes

In the past decade, cloning of the extracellular matrix genes and establishment of their DNA sequences have led to the acquisition of a large body of evidence documenting mutations in these genes as underlying numerous inherited pathological conditions (3, 125–127). Mutations in the type I collagen ($\alpha1(I)$ and $\alpha2(I)$) genes cause osteogenesis imperfecta (115, 128, 129), at least some of the chondrodysplasias (130) reflect mutations in the type II collagen gene, and mutations in the type III collagen gene have been linked with the lethal Ehlers–Danlos syndrome (127, 131). The Alport syndrome, a heritable kidney disorder, has been connected with mutations in the collagen $\alpha5(IV)$ gene (132), and epidermolysis bullosa has been linked to mutations in the collagen $\alpha1(VII)$ gene (66, 67, 133). Additionally, other inherited diseases, such as progeria, similarly reflect mutations in collagen genes (134).

Fewer mutations involving the proteoglycans have been described. Macular corneal dystrophy (MCD) is an inherited human disease resulting from a failure to synthesize keratan sulfate (135). Lumican is the major keratan sulfate-containing proteoglycan of the corneal stroma, and its production without sulfate esters, as is the case in MCD, results in corneal opacities. Two animal models carrying mutations resulting in a defective aggrecan core protein have been described: nanomelia in the chicken (136) and cartilage matrix deficiency in the mouse (137). Both of these mutations result in shortened long bones, as well as other abnormalities of cartilagenous tissue. While not involving the biosynthesis of proteoglycans, the mucopolysaccharidoses are a group of inherited diseases (both X-linked and autosomal recessive) resulting from the failure to produce an enzyme (α-L-iduronidase) responsible for one step in the degradation of three different GAGs (138, 139). The accumulation of these incompletely degraded GAGs results in a number of clinical presentations, such as the Hurler and Hunter syndromes.

In contrast to the collagen and proteoglycan superfamilies, with one exception no inherited diseases resulting from mutations in the structural glycoprotein genes have been described. This paucity of clinical manifestations may reflect the importance of the majority of the molecules in this family in embryonic development—a speculation consistent with the recent report that homozygotic transgenic mice carrying an inactive fibronectin gene display neural tube abnormalities and defects in other mesodermally derived tissues resulting in early embryonic death (140). The one notable exception to this concept arises from the recent evidence that has identified mutations in the fibrillin genes (114, 141, 142). These mutations have been linked with the Marfan syndrome, ectopia lentis, and congenital contractural arachnodactyly (143). Additionally cutis laxa, pseudoxanthoma elasticum, and others, as well as premature ageing syndromes are associated with histological alterations in elastic fibres (134, 144). As we delineate more of the DNA sequences coding for both the collagenous and non-collagenous extracellular matrix components, as well as their biosynthesis and assembly, there is a high likelihood that additional diseases whose molecular basis reflects mutations in matrix genes will be identified.

8.3 Involvement of changes in the extracellular matrix in acquired diseases

Changes in the extracellular matrix have likewise come to be recognized as critical components in the initiation and progression of a variety of acquired diseases. Among these are atherosclerosis (145), liver fibrosis (146), glomerulonephritis and glomerulosclerosis (147), scleroderma (148, 149), pulmonary fibrosis (150, 151), and the secondary consequences of diabetes (152–154), to name a few.

Extracellular matrix components have also been implicated as the autoimmune antigen in several diseases, including the Goodpasture syndrome (155), which reflects an autoimmune response against the NC1 domain of the α3(IV) collagen chain, and bullous pemphigoid (48, 69, 70) that represents an autoimmune response against type XVII molecules present in hemidesmosomes. Finally, changes in extracellular matrix synthesis, deposition, metabolism of matrix molecules and matrix receptors are important components in cancer development and metastasis (156–158). Clearly, as we identify new components, and understand more of the chemistry and function of existing matrix components, more diseases having matrix derangements as a major component will be recognized and their roles in the pathogenesis of these conditions will be elucidated.

9. Summary

We now recognize that the extracellular matrix is a complex ordered structure containing multiple members of families (collagens, proteoglycans, structural glycoproteins, and elastin) of macromolecules. It functions not only as the structural supporting element for cells and tissues but plays a major role in modulating the biology of the cell and the response of the cell to growth factors/cytokines and vitamins/hormones. Individual members of each matrix family can interact with the cell through specific cell surface receptors and/or with other matrix macromolecules through specific binding domains. The composition of the extracellular matrix is not, however, static, but changes during both normal development and in tissue repair and regeneration. Mutations in matrix genes result in a variety of pathological consequences, and changes in matrix metabolism (both synthesis and catabolism) characterize many acquired diseases. Thus the extracellular matrix is a dynamic complex intimately involved in both normal biological function and response to injury.

Acknowledgements

The authors would like to thank Dr Gregory Sephel for his advice and guidance in compiling the data on the structural glycoproteins presented in *Table 4*, and Ms Jean McClure for her assistance in the preparation of this chapter. This work was supported in part by USPHS NIH Grants HL-14214, DK-39261, EY-08104, and GM-45380.

References

1. Piez, K. A. and Reddi, A. H. (1984). *Extracellular matrix biochemistry*. Elsevier, New York.
2. Hay, E. D. (1991). *Cell biology of extracellular matrix* (2nd edn). Plenum Press, New York.
3. Royce, P. M. and Steinmann, B. (1993). *Connective tissue and its heritable disorders. Molecular, genetic and medical aspects*. Wiley-Liss, Inc., New York.
4. Labat-Robert, J., Bihari-Varga, M., and Robert, L. (1990). *FEBS Letters*, **268**, 386–93.
5. Cohen, I. K., Diegelmann, R. F., and Lindblad, W. J. (1992). *Wound healing. Biochemical and clinical aspects*. W. B. Saunders, Philadelphia.

6. Zern, M. A. and Reid, L. M. (1993). *Extracellular matrix: chemistry, biology and pathobiology with emphasis on the liver*. Marcel Decker, Inc., New York.
7. Hay, E. D. (1991). In *Cell biology of extracellular matrix* (2nd edn) (ed. E. D. Hay), pp. 419–62. Plenum Press, New York.
8. Trelstad, R. L. (1984). *The role of the extracellular matrix in development*. Alan R. Liss, New York.
9. Rohrbach, D. H. and Murrah, V. A. (1993). In *Molecular and cellular aspects of basement membranes* (ed. D. R. Rohrbach and R. Timpl), pp. 385–419. Academic Press, New York.
10. Juliano, R. L. and Haskill, S. (1993). *J. Cell. Biol.*, **120**, 577–85.
11. Ruoslahti, E. (1991). *J. Clin. Invest.*, **87**, 1–5.
12. Hynes, R. O. (1987). *Cell*, **48**, 549–52.
13. Kramer, R. H., Enenstein, J., and Waleh, N. S. (1993). In *Molecular and cellular aspects of basement membranes* (ed. D. R. Rohrbach and R. Timpl), pp. 239–65. Academic Press, San Diego.
14. Rapraeger, A. C. (1993). In *Molecular and cellular aspects of basement membranes* (ed. D. R. Rohrbach and R. Timpl), pp. 267–88. Academic Press, San Diego.
15. Bernfield, M. R. and Sanderson, R. D. (1990). *Phil. Trans. R. Soc. London (Ser. B).*, **327** 171–86.
16. Rapraeger, A. C. (1993). *Curr. Opin. Cell Biol.*, **5**, 844–53.
17. Birk, D. E., Silver, F. H., and Trelstad, R. L. (1991). In *Cell biology of extracellular matrix* (2nd edn) (ed. E. D. Hay), pp. 221–54. Plenum Press, New York.
18. Martinez-Hernandez, A. (1988). In *Pathology* (ed. E. Rubin and J. L. Farber), pp. 66–95. J. B. Lippincott, Philadelphia.
19. Madri, J. A. and Basson, M. D. (1993). *Lab Invest.*, **66**, 519–21.
20. Schnaper, H. W. and Kleinman, H. K. (1993). *Pediatr. Nephrol.*, **7**, 96–104.
21. Lin, C. Q. and Bissell, M. J. (1993). *FASEB J.*, **7**, 737–43.
22. Teti, A. (1992). *J. Am. Soc. Nephrol.*, **2**, S83–S87.
23. Kleinman, H. K., Kibbey, M. C., Schnaper, H. W., Hadley, M. A. Dym, M., and Grant, D. S. (1993). In *Molecular and cellular aspects of basement membranes* (ed. D. R. Rohrbach and R. Timpl), pp. 309–26. Academic Press, New York.
24. Ekblom, P. (1993). In *Molecular and cellular aspects of basement membranes* (ed. D. R. Rohrbach and R. Timpl), pp. 359–83. Academic Press, New York.
25. Ekblom, P. (1989). *FASEB J.*, **3**, 2141–50.
26. Esposito, S. D. and Zern, M. A. (1993). In *Extracellular matrix: chemistry, biology and pathobiology with emphasis on the liver* (ed. M. A. Zern and L. M. Reid), pp. 331–49. Marcel Decker, Inc., New York.
27. Roberts, A. B., Heine, U. I., Flanders, K. C., and Sporn, M. B. (1990). *Ann. NY Acad. Sci.*, **580**, 225–32.
28. Roberts, A. B., McCune, B. K., and Sporn, M. B. (1992). *Kidney Int.*, **41**, 557–9.
29. Cutroneo, K. R. (1993). In *Extracellular matrix: chemistry, biology and pathobiology with emphasis on the liver* (ed. M. A. Zern and L. M. Reid), pp. 351–68. Marcel Decker, Inc., New York.
30. Gumbiner, B. M. and Yamada, K. M. (1993). *Curr. Opin. Cell Biol.*, **5**, 769–71.
31. Haralson, M. A. (1993). *Lab. Invest.*, **69**, 369–72.
32. Sage, E. H. and Bornstein, P. (1991). *J. Biol. Chem.*, **266**, 14831–4.

33. Vlodavsky, I., Bar-Shavit, R., Korner, G., and Fuks, Z. (1993). In *Molecular and cellular aspects of basement membranes* (ed. D. R. Rohrbach, and R. Timpl), pp. 327–43. Academic Press, New York.
34. Vlodavsky, I., Korner, G., Eldor, A., Bar-Shavit, R., Klagsbrun, M., and Fuks, Z. (1993). In *Extracellular matrix: chemistry, biology and pathobiology with emphasis on the liver* (ed. M. A. Zern and L. M. Reid), Marcel Decker, Inc., New York.
35. Marx, M., Daniel, T. O., Kashgarian, M., and Madri, J. A. (1993). *Kidney Int.*, **43**, 1027–41.
36. van der Rest, M. and Garrone, R. (1991). *FASEB J.*, **5**, 2814–23.
37. Kielty, C. M., Hopkinson, I., and Grant, M. E. (1993). In *Connective tissue and its heritable disorders. Molecular, genetic and medical aspects* (ed. P. M. Royce and B. Steinmann), pp. 103–47. Wiley-Liss, New York.
38. Miller, E. J. and Gay, S. (1992). In *Wound healing: biochemical and clinical aspects* (ed. I. K. Cohen, R. F. Diegelmann, and W. J. Lindblad), pp. 130–51. W. B. Saunders, Philadelphia.
39. Nimni, M. E. (1993). In *Extracellular matrix: chemistry, biology and pathobiology with emphasis on the liver* (ed. M. A. Zern and L. M. Reid), pp. 121–48. Marcel Decker, Inc., New York.
40. van der Rest, M. and Garrone, R. (1990). *Biochimie*, **72**, 473–84.
41. Kucharz, E. J. (1992). *The collagens: biochemistry and pathophysiology.* Springer–Verlag, Berlin.
42. Burgeson, R. E. and Nimni, M. E. (1992). *Clin. Orth. Rel. Res.*, **282**, 250–72.
43. Martin, G. R. and Piez, K. A. (1993). In *Molecular and cellular aspects of basement membranes* (ed. D. R. Rohrbach and R. Timpl), pp. 3–18. Academic Press, San Diego.
44. Miller, E. J. and Matukas, V. J. (1969). *Proc. Natl. Acad. Sci. USA*, **64**, 1264–8.
45. Miller, E. J. (1971). *Biochemistry*, **10**, 1652–9.
46. Ramirez, F. and Di Liberto, M. (1990). *FASEB J.*, **4**, 1616–23.
47. Mayne, R. and Brewton, R. G. (1993). *Curr. Opin. Cell Biol.*, **5**, 883–90.
48. Li, K., Tamai, K., Tan, E. M. L., and Uitto, J. (1993). *J. Biol. Chem.*, **268**, 8825–34.
49. Linsenmayer, T. F. (1991). In *Cell biology of extracellular matrix* (2nd edn) (ed. E. D. Hay), pp. 7–44. Plenum Press, New York.
50. Miller, E. J. (1984). In *Extracellular matrix biochemistry* (ed. K. A. Piez and A. H. Reddi), pp. 41–81. Elsevier, New York.
51. Mayne, R. and Burgeson, R. E. (1987). *Structure and function of collagen types.* Academic Press, Orlando.
52. Miller, E. J. (1985). *Ann. NY Acad. Sci.*, **460**, 1–13.
53. Fessler, J. H. and Fessler, L. I. (1987). In *Structure and function of collagen types* (ed. R. Mayne and R. E. Burgeson), pp. 81–103. Academic Press, Orlando.
54. Künn, K. (1987). In *Structure and function of collagen types* (ed. R. Mayne and R. E. Burgeson), pp. 1–42. Academic Press, Orlando.
55. Eyre, D. R. and Wu, J.-J. (1987). In *Structure and function of collagen types* (ed. R. Mayne and R. E. Burgeson), pp. 261–81. Academic Press, Orlando.
56. Vuorio, E. and De Crombrugghe, B. (1990). *Annu. Rev. Biochem.*, **59**, 837–72.
57. Van der Rest, M. and Mayne, R. (1987). In *Structure and function of collagen types* (ed. R. Mayne and R. E. Burgeson), pp. 195–221. Academic Press, Orlando.
58. Fine, L. G. and Norman, J. T. (1992). *J. Am. Soc. Nephrol.*, **2**, S206–S211.

59. Hudson, B. G., Reeders, S. T., and Tryggvason, K. (1993). *J. Biol. Chem.*, **268**, 26033–6.
60. Glanville, R. W. (1987). In *Structure and function of collagen types* (ed. R. Mayne and R. E. Burgeson), pp. 43–79. Academic Press, Orlando.
61. Paulsson, M. (1993). In *Molecular and cellular aspects of basement membranes* (ed. D. R. Rohrbach and R. Timpl), pp. 177–87. Academic Press, New York.
62. Sage, H. and Bornstein, P. (1987). In *Structure and function of collagen types* (ed R. Mayne and R. E. Burgeson), pp. 173–94. Academic Press, Orlando.
63. Schmid, T. M. and Linsenmayer, T. F. (1987). In *Structure and function of collagen types* (ed. R. Mayne and R. E. Burgeson), pp. 223–59. Academic Press, Orlando.
64. Timpl, R. and Engel, J. (1987). In *Structure and function of collagen types* (ed. R. Mayne and R. E. Burgeson), pp. 105–43. Academic Press, Orlando.
65. Burgeson, R. E. (1987). In *Structure and function of collagen types* (ed. R. Mayne and R. E. Burgeson), pp. 145–72. Academic Press, Orlando.
66. Burgeson, R. E. (1993). *J. Invest. Dermatol.*, **101**, 252–5.
67. Burgeson, R. E. (1993). In *Molecular and cellular aspects of basement membranes* (ed. D. R. Rohrbach and R. Timpl), pp. 49–66. Academic Press, San Diego.
68. Goldstein, L. A., Zhou, D. F. H., Picker, L. J., Minty, C. N., Bargatze, R. F., Ding, J. F., and Butcher, E. C. (1989). *Cell*, **56**, 1063–72.
69. Burgeson, R. E., Chiquet, M., Deutzmann, R., Ekblom, P., Engel, J., Kleinman, H., Martin, G. R., Meneguzzi, G., Paulsson, M., Sanes, J., Timpl, R., Tryggvason, K., Yamada, Y., and Yurchenco, P. D. (1994) *Matrix Biol.*, **14**, 209–11.
70. Giudice, G. J., Emery, D. J., and Diaz, L. A. (1992). *J. Invest. Dermatol.*, **99**, 243–50.
71. Hassell, J. R., Blochberger, T. C., Rada, J. A., Chakravarti, S., and Noonan, D. (1993). *Adv. Molec. Cell Biol.*, **6**, 69–113.
72. Kjellén, L. and Lindahl, U. (1991). *Annu. Rev. Biochem.*, **60**, 443–75.
73. Heinegård, D. and Oldberg, Å. (1993). In *Connective tissue and its heritable disorders. Molecular, genetic and medical aspects* (ed. P. M. Royce and B. Steinmann), pp. 189–209. Wiley-Liss, New York.
74. Bourdon, M. A., Shiga, M., and Ruoslahti, E. (1986). *J. Biol. Chem.*, **261**, 12534–7.
75. Sanders, S., Jalkanen, M., O'Farrell, S., and Bernfield, M. (1989). *J. Cell Biol.*, **108**, 1547–56.
76. Marynen, P., Zhang, J., Cassiman, J.-J., van den Berghe, H., and David, G. (1989). *J. Biol. Chem.*, **264**, 7017–24.
77. Kojima, T., Shworak, N. W., and Rosenberg, R. D. (1992). *J. Biol. Chem.*, **267**, 4870–7.
78. Nishiyama, A., Dahlin, K. J., Prince, J. T., Johnstone, S. R., and Stallcup, W. B. (1990). *J. Cell Biol.*, **114**, 359–71.
79. López-Casillas, F., Cheifetz, S., Doody, J., Andres, J. L., Lane, W. S., and Massagué, J. (1991). *Cell*, **67**, 785–95.
80. David, G., Lories, V., Decock, B., Marynen, P., Cassiman, J. J., and van den Berghe, H. (1990). *J. Cell Biol.*, **111**, 3165–7.
81. Krusius, T. and Ruoslahti, E. (1986). *Proc. Natl. Acad. Sci. USA*, **83**, 7683–7.
82. Fisher, L. W., Termine, J. D., and Young, M. F. (1989). *J. Biol. Chem.*, **264**, 4571–6.

83. Oldberg, Å., Antonsson, P., Lindblom, K., and Heinegååd, D. (1989). *EMBO J.*, **8**, 2601–4.
84. Blochberger, T. C., Vergnes, J. P., Hempel, J., and Hassell, J. R. (1992). *J. Biol. Chem.*, **267**, 347–52.
85. Shinomura, T. and Kimata, K. (1992). *J. Biol. Chem.*, **267**, 1265–70.
86. Doege, K., Sasaki, M., Horigan, E., Hassell, J. R., and Yamada, Y. (1987). *J. Biol. Chem.*, **262**, 17757–67.
87. Zimmermann, D. R. and Ruoslahti, E. (1989). *EMBO J.*, **8**, 2975–81.
88. Rauch, U., Karthikeyan, L., Maurel, P., Margolis, R. U., and Margolis, R. K. (1992). *J. Biol. Chem.*, **267**, 19536–47.
89. Noonan, D. M., Fulle, A., Valente, P., Cai, S., Horigan, E., Sasaki, M., Yamada, Y., and Hassell, J. R. (1991). *J. Biol. Chem.*, **266**, 22939–47.
90. Von der Mark, K. and Goodman, S. (1993). In *Connective tissue and its heritable disorders. Molecular, genetic and medical aspects* (ed. P. M. Royce and B. Steinmann), pp. 211–36. Wiley-Liss, New York.
91. Hynes, R. O. (1992). *Cell*, **69**, 11–25.
92. Mosher, D. F. (1989). *Fibronectin*. Academic Press, New York.
93. Yamada, K. M. (1991). In *Cell biology of extracellular matrix* (2nd edn) (ed. E. D. Hay), pp. 111–46. Plenum Press, New York.
94. Ruoslahti, E. (1988). *Annu. Rev. Biochem.*, **57**, 375–413.
95. Paolella, G., Barone, M. V., and Baralle, F. E. (1993). In *Extracellular matrix: chemistry, biology and pathobiology with emphasis on the liver* (ed. M. A. Zern and L. M. Reid), pp. 3–24. Marcel Decker, Inc., New York.
96. Engel, J. (1993). In *Molecular and cellular aspects of basement membranes* (ed. D. R. Rohrbach and R. Timpl), pp. 147–76. Academic Press, San Diego.
97. Tryggvason, K. (1993). *Curr. Opin. Cell Biol.*, **5**, 877–82.
98. Martinez-Hernandez, A. and Amenta, P. (1983). *Lab. Invest.*, **48**, 656–77.
99. Martin, G. R., Timpl, R., and Künn, K. (1988). *Adv. Prot. Chem.*, **39**, 1–50.
100. Chung, A. E. (1993). In *Extracellular matrix: chemistry, biology and pathobiology with emphasis on the liver* (ed. M. A. Zern and L. M. Reid), pp. 25–48. Marcel Decker, Inc., New York.
101. Chung, A. E., Jaffe, R., Freeman, I. L., Vergnes, J. P., Braginski, J. E., and Carlin, B. (1979). *Cell*, **16**, 277–87.
102. Timpl, R., Rhode, H., Robey, P. G., Rennard, S. I., Foidart, J., and Martin, G. R. (1979). *J. Biol. Chem.*, **254**, 9933–7.
103. Carlin, B., Jaffe, R., Bender, B., and Chung, A. E. (1981). *J. Biol. Chem.*, **256**, 5209–14.
104. Timpl, R., Dziadek, M., Fujiwara, S., Nowack, H., and Wick, G. (1983). *Eur. J. Biochem.*, **137**, 455–65.
105. Paulsson, M. (1992). *Crit. Rev. Biochem. Mol. Biol.*, **27**, 93–127.
106. Timpl, R. (1989). *Eur. J. Biochem.*, **180**, 487–502.
107. Chiquet, M. (1992). *Kidney Int.*, **41**, 629–31.
108. Venstrom, K. A. and Reichardt, L. F. (1993). *FASEB J.*, **7**, 996–1003.
109. Erickson, H. P. and Bourdon, M. A. (1989). *Annu. Rev. Cell Biol.*, **5**, 71–92.
110. Erickson, H. P. (1993). *Curr. Opin. Cell Biol.*, **5**, 869–76.
111. Timpl, R. and Aumailley, M. (1993). In *Molecular and cellular aspects of basement membranes* (ed. D. R. Rohrbach, and R. Timpl). pp. 211–35. Academic Press, San Diego.

112. Mosher, D. F. (1990). *Annu. Rev. Med.*, **41**, 85–97.
113. Felding-Habermann, B. and Cheresh, D. A. (1993). *Curr. Opin. Cell Biol.*, **5**, 864–8.
114. Sakai, L. Y., Keene, D. R., and Engvall, E. (1986). *J. Cell Biol.*, **103**, 2499–509.
115. Byers, P. H. (1993). In *Connective tissue and its heritable disorders. Molecular, genetic, and medical aspects* (ed. P. M. Royce and B. Steinmann), pp. 317–50.
116. Rosenbloom, J. (1993). In *Connective tissue and its heritable disorders. Molecular, genetic and medical aspects* (ed. P. M. Royce and B. Steinmann), pp. 167–88. Wiley-Liss, New York.
117. Mecham, R. P. and Heuser, J. E. (1991). In *Cell biology of extracellular matrix* (2nd edn) (ed. E. D. Hay), pp. 79–109. Plenum Press, New York.
118. Rosenbloom, J., Abrams, W. R., and Mecham, R. (1993). *FASEB J.*, **7**, 1208–18.
119. Birkedal-Hansen, H., Moore, W. G. I., Bodden, M. K., Windsor, L. J., Birkedal-Hansen, B., DeCarlo, A., and Engler, J. A. (1993). *Crit. Rev. Oral Biol. Med.*, **4**, 197–250.
120. Murphy, G. and Reynolds, J. J. (1993). In *Connective tissue and its heritable disorders. Molecular, genetic and medical aspects* (ed. P. M. Royce and B. Steinmann), pp. 287–316. Wiley-Liss, New York.
121. Kleiner, D. E., Jr and Stetler-Stevenson, W. G. (1993). *Curr. Opin. Cell Biol.*, **5**, 891–7.
122. Chu, M.-L. and Prockop, D. J. (1993). In *Connective tissue and its heritable disorders. Molecular, genetic and medical aspects* (ed. P. M. Royce and B. Steinmann), pp. 149–65. Wiley-Liss, New York.
123. Sandell, L. J. and Boyd, C. D. (1990). *Extracellular matrix genes*. Academic Press, San Diego.
124. McKusick, V. A. (1972). *Heritable disorders of connective tissue*. C. V. Mosby, Saint Louis.
125. Byers, P. H. (1990). In *Extracellular matrix genes* (ed. L. J. Sandell, and C. D. Boyd), pp. 251–63. Academic Press, San Diego.
126. Kuivaniemi, H., Tromp, G., and Prockop, D. J. (1991). *FASEB J.*, **5**, 2052–60.
127. Prockop, D. J. (1992). *N. Engl. J. Med.*, **326**, 540–6.
128. Prockop, D. J., Colige, A., Helminen, H., Khillan, J. S., Pereira, R., and Vandenberg, P. (1993). *J. Bone Miner. Res.*, **8**, S489–S492.
129. Byers, P. H. and Steiner, R. D. (1992). *Annu. Rev. Med.*, **43**, 269–82.
130. Horton, W. A. and Hecht, J. T. (1993). In *Connective tissue and its heritable disorders. Molecular, genetic and medical aspects* (ed. P. M. Royce and B. Steinmann), pp. 641–75. Wiley-Liss, New York.
131. Steinmann, B., Royce, P. M., and Superti-Furga, A. (1993). In *Connective tissue and its heritable disorders. Molecular, genetic, and medical aspects* (ed. P. M. Royce and B. Steinmann), pp. 351–407. Wiley-Liss, New York.
132. Zhou, J., Barker, D. F., Hostikka, S. L., Gregory, M. C., Atkin, C. L., and Tryggvason, K. (1991). *Genomics*, **9**, 10–18.
133. Bruckner-Tuderman, L. (1993). In *Connective tissue and its heritable disorders. Molecular, genetic, and medical aspects* (ed. P. M. Royce and B. Steinmann), pp. 507–31. Wiley-Liss, New York.
134. Uitto, J., Fazio, M. J., and Christiano, A. M. (1993). In *Connective tissue and its heritable disorders. Molecular, genetic, and medical aspects* (ed. P. M. Royce and B. Steinmann), pp. 409–23. Wiley-Liss, New York.

135. Hassell, J. R., Newsome, D. A., Krachmer, J. H., and Rodrigues, M. M. (1980). *Proc. Natl. Acad. Sci. USA*, **77**, 3705–9.

136. Argraves, W. S., McKeown-Longo, P. J., and Goetinck, P. F. (1981). *FEBS Lett.*, **131**, 265–8.

137. Kimata, K., Barrach, H. J., Brown, K. S., and Pennypacker, J. P. (1981). *J. Biol. Chem.*, **256**, 6961–8.

138. McKusick, V. A. (1972). In *Heritable disorders of connective tissue* (4th edn) (ed. Anonymous), pp. 521–686. C. V. Mosby, Saint Louis.

139. Leroy, J. G. and Wiesmann, U. (1993). In *Connective tissue and its heritable disease. Molecular, genetic, and medical aspects* (ed. P. M. Royce and B. Steinmann), pp. 613–39. Wiley-Liss, New York.

140. George, E. L., Georges-Labouesse, E. N., Patel-King, R., Rayburn, H., and Hynes, R. O. (1993). *Development*, **119**, 1079–91.

141. Milewicz, D. M., Pyeritz, R. E., Crawford, E. S., and Byers, P. H. (1992). *J. Clin. Invest.*, **89**, 79–86.

142. Pyeritz, R. E. (1993). In *Connective tissue and its heritable disorders. Molecular, genetic, and medical aspects* (ed. P. M. Royce and B. Steinmann), pp. 437–68. Wiley-Liss, New York.

143. Tsipouras, P., Mastro, R. D., Sarfarazi, M., Lee, B., Vitale, E., Child, A., et al. (1992). *N. Engl. J. Med.*, **326**, 905–9.

144. Nelder, K. H. (1993). In *Connective tissue and its heritable disorders. Molecular, genetic, and medical aspects* (ed. P. M. Royce and B. Steinmann), pp. 425–36. Wiley-Liss, New York.

145. Badimon, J. J., Fuster, V., Chesebro, J., and Badimon, L. (1993). *Circulation*, **87**, II3–II16.

146. Gressner, A. M. (1991). *Eur. J. Clin. Chem. Clin. Biochem.*, **29**, 293–311.

147. Sterzel, R. B., Schulze-Lohoff, E., Weber, E., and Goodman, S. L. (1992). *J. Am. Soc. Nephrol.*, **2**, S126–S131.

148. Smith, E. A. (1992). *Curr. Opin. Rheumatol.*, **4**, 869–77.

149. LeRoy, E. C. (1989). *Clin. Exp. Rheumatol.*, **7**, S135–S137.

150. Khalil, N. and Greenberg, A. H. (1991). *Ciba Foundation Symposium*, **157**, 194–207.

151. Pelton, R. W. and Moses, H. L. (1990). *Am. Rev. Respir. Dis.*, **142** (Suppl.), S31–S35.

152. Wiedemann, P. (1992). *Surv. Ophthalmol.*, **36**, 373–84.

153. Brownlee, M. (1992). *Diabetes Care*, **15**, 1835–43.

154. Ziyadeh, F. N., Goldfarb, S., and Kern, E. F. O. (1989). In *The kidney in diabetes mellitus* (ed. B. M. Brenner and J. H. Stein), pp. 87–113. Churchill Livingstone, New York.

155. Hudson, B. G., Wisdom, B. J., Jr, Gunwar, S., and Noelken, M. E. (1991). In *Collagen* (ed. A. Kang) CRC Press, Boca Raton.

156. Zvibel, I. and Kraft, A. (1993). In *Extracellular matrix: chemistry, biology and pathobiology with emphasis on the liver* (ed. M. A. Zern and L. M. Reid), pp. 559–80. Marcel Decker, Inc., New York.

157. Stetler-Stevenson, W. G., Liotta, L. A., and Kleiner, D. E., Jr. (1993). *FASEB J.*, **7**, 1434–41.

158. Juliano, R. L. and Varner, J. A. (1993). *Curr. Opin. Cell Biol.*, **5**, 812–18.

2

Types I, III, and V collagen and total collagen from cultured cells and tissues

BEVERLY PETERKOFSKY, MICHAEL A. HARALSON,
SAMUEL J. DIMARI, and EDWARD J. MILLER

1. Introduction

The fibre-forming collagens—types I, III, and V—constitute the majority of collagens in non-cartilagenous tissues. The common structure of these molecules is typified by the type I molecule, which is illustrated in *Figure 1* (1). The general and specific features of these collagen types have been the subject of many reviews (2–10) and will not be exhaustively discussed in this chapter. However, several unique structural aspects of these molecules that are exploited for their isolation, identification, and quantification are briefly reviewed for purposes of orientation. First, the molecules possess long, non-interrupted triple-helical regions. This feature confers upon the molecules resistance to proteolytic digestion, particularly to the enzyme pepsin. Second, these molecules, as do the other collagen types, contain multiple repeats of the sequence −R−Pro−X−Gly−Pro−R−. This sequence is recognized by the protease produced by *Clostridium histolyticum*, which cleaves the molecules

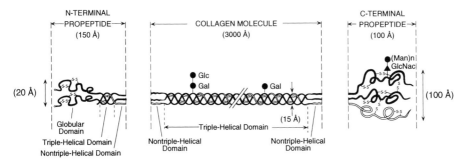

Figure 1. Schematic representation of the structure of type I procollagen. (Reproduced with permission from reference 1.)

Table 1. Methionine contents of human types I, III, and V collagen chains

Collagen chain	Residues/1000 amino acid residues
α1(I)	7
α2(I)	5
α1(III)	8
α1(V)	9
α2(V)	11
α3(V)	8

Data from reference 4 with permission.

into small molecular weight peptides. Third, the chains that compose these molecules have a limited number of methionyl residues (*Table 1*). This feature has been utilized to identify the chains by the unique sizes and chemical properties of the limited number of peptides generated by digestion of the molecules with CNBr.

This chapter contains several approaches for identifying and quantitating types I, III, and V molecules, as well as for determining total collagen content. In Parts A, B, and C, methods are described for estimating total collagen (Part A) and identifying types I, III, and V molecules (Parts A, B, and C) produced by cultured cells and tissue or organ culture. These approaches use radioactive proline to isotopically label the collagens, which offers several experimental opportunities, including:

(a) Ascorbic acid, which stimulates collagen expression, can be added to the culture medium (11–13).

(b) The effects of soluble substances, such as growth factors, hormones, and other pharmacological agents, can be directly assessed by including them in the growth medium.

(c) The effects of different collagen-containing substrates (see Chapter 14) on collagen production can be evaluated.

(d) The amounts of type I-heterotrimers ($[\alpha1(I)]_2\alpha2(I)$) to type I-homo-trimers ($[\alpha1(I)]_3$) and of the different molecular species of type V molecules (see *Table 1* in Chapter 1) can be determined.

(e) Covalent cross-links between the subunits and molecules, which preclude quantification of types I, III, and V molecules from tissues, can be inhibited by adding β-aminopropionitrile (BAPN), a specific inhibitor of lysyl oxidase (14).

The approach described in Part B relies on the resolution of pepsin-derived collagen chains by ion-exchange chromatography to identify and quantitate types I, III, and V collagens, and the methods detailed in Part C are based on identification of specific CNBr- and V8 protease-derived peptides in these molecules. In contrast to these methods, which measure isotopically labelled molecules, Part D describes an approach for the direct assessment of the amounts of type I, III, and V molecules in tissues. This method is based upon HPLC separation of CNBr-derived peptides not involved in either inter- or intrachain covalent cross-links. Together, these approaches, both alone or in combination, offer the investigator the capability to assess the production of total collagen, in general, and types I, III, and V molecules, in particular, under a variety of experimental conditions.

2A. ESTIMATION OF TOTAL COLLAGEN PRODUCTION BY COLLAGENASE DIGESTION

BEVERLY PETERKOFSKY

2. Introduction

In this procedure, cells or tissues are labelled with a radioactive amino acid, and the free amino acid is removed by acid precipitation. The collagen molecules are then degraded by bacterial collagenase to acid-soluble peptides that can be separated from the non-collagenous proteins (NCP). A highly purified preparation of collagenase free of other protease activities (15) must be used, and N-ethylmaleimide (NEM) is added to ensure complete inhibition of trace amounts of clostripain, the major contaminant in collagenase.

3. Estimation of total collagen production by cultured cells using collagenase digestion

---

Protocol 1. Estimation of total collagen production by cultured cells using collagenase digestion

Reagents

- 0.11 M NaCl/0.05 M Tris–HCl, pH 7.4
- 1 M Hepes buffer, pH 7.2: dissolve 23.8 g Hepes in 80 ml of H_2O; add 10 M NaOH until pH reaches 7.2; adjust volume to 100 ml with H_2O. Store at $-20\,°C$.
- 50% (w/v) Trichloroacetic acid (TCA)/50 mM proline. Store in the refrigerator.
- BSA: 10 mg/ml. Store at $-20\,°C$.
- 1 mM BAPN

- 10 mM ascorbic acid (prepared fresh)
- 0.2 M NaOH (dilute from a standardized 1 M solution)
- 0.15 M HCl (dilute from a standardized 1 M solution)
- 62.5 mM NEM. Adjust pH to ~6.0 by adding solid $NaHCO_3$. Store at $-20\,°C$.
- 25 mM $CaCl_2$

Beverly Peterkofsky

Protocol 1. *Continued*

- 10% TCA/0.5% tannic acid. Store at 4 °C for no longer than 1 week.
- *Clostridium histolyticum* collagenase: either purified by chromatography on S200 (15) or purchased from Advanced Biofactures, Lynbrook, NY. Store at −20 °C.
- Elution buffer (0.05 M Tris–HCl, pH 7.6, 5 mM CaCl₂) or a solution equivalent to the buffer in which the enzyme is dissolved.
- 13 × 100 mm screw-cap glass tubes

- Standard radioactive collagen:
  (a) Label fibroblasts in several 100 mm dishes in serum-free medium with [5-³H]proline for 6–24 h as described in Part A, steps 1–11, below.
  (b) Precipitate collagen from the medium with ammonium sulfate as described in *Protocol 12*, steps 8–12.
  (c) Determine the amount of collagen in the solution by digestion of aliquots with collagenase, as detailed in Part C, steps 5–20, and store in small aliquots at −70 °C or in liquid nitrogen.

A. *Preparation of samples*

1. When cells reach the desired density, remove the growth medium and rinse the cell layer twice with warm (37 °C) serum-free medium. Cells in logarithmic phase generally have a higher level of incorporation than confluent cells, but the stage at which cells are used will depend on the objective of the experiment.

2. For cells in 60 mm dishes, add 0.9 ml serum-free medium containing 0.1 mM BAPN and 10 μl of freshly prepared 10 mM ascorbate.

3. Incubate for 15 min at 37 °C under 95% air/5% CO₂.

4. Add 0.1 ml of [5−³H]proline (100 μCi/ml, 1 mM Sp. Act = 100μ Ci/μmol) in serum-free medium and continue the incubation at 37 °C for an additional 2–4 h. Incorporation should be in the linear phase of a time-course. For 100 mm dishes, use 2.4 ml of serum-free medium and 0.1 ml of radioactive proline. For 24-well plates, use 0.4 ml serum-free medium and 0.1 ml of the radioactive solution.

5. Chill dishes on ice and immediately remove the medium.

6. Add 1 ml per dish of a solution containing 0.11 M NaCl + 0.05 M Tris–HCl (pH 7.4) + 10 mM proline. Scrape off cells with a Cell-Lifter (Corning-Costar). For 24-well plates, a plastic transfer pipette can be used. Remove resulting suspension, and place in a 15 ml centrifuge tube. The volumes for 24-well plates and 100 mm dishes should be adjusted accordingly.

7. Rinse dishes with 1 ml of saline per dish and add the rinse to the suspension.

8. Centrifuge cells at 1000 *g*, 5 min at 4 °C.

9. Add the supernatant fraction to the medium.

10. Add 0.5 ml of 0.4 M NaCl/0.1 M Hepes buffer, pH 7.2/10 mM proline to the cell pellet.

11. If the assay is not going to be performed immediately, store samples at −20 °C.

B. *Precipitation of proteins with TCA*

1. Sonicate the cell suspension twice with the needle probe of a sonicator for 20 sec at 20% of maximum voltage each time, keeping the tube in a salt/ice-water bath at $-5\,°C$.

2. Precipitate proteins in the sonicate, medium, and standard collagen preparation with TCA at $0–4\,°C$ as detailed in *Table 2*.

3. Mix the samples and place at $0\,°C$ for 5 min.

4. Centrifuge at 1000 g for 5 min at $4\,°C$.

5. Remove the supernatant fraction.

6. Suspend the pellet in 1 ml of a 5% TCA/10 mM proline solution and centrifuge.

7. Repeat step 6 three times.

C. *Collagenase digestion*

1. Dissolve the pellets in 0.6 ml of 0.2 M NaOH.

2. Transfer two 0.2 ml aliquots of each sample to 5 ml conical polystyrene tubes.

3. Neutralize by adding 0.26 ml of a solution containing 0.16 ml 0.15 M HCl + 0.10 ml 1 M Hepes buffer, pH 7.2.

4. Check the pH of a few representative samples using pH paper by removing 1 µl with a microcapillary tube. The pH should be between 6.8–7.4.

5. Add 40 µl of the digestion cocktail minus $(-)$ or plus $(+)$ purified bacterial collagenase (*Table 3*) to each 0.2 ml aliquot.

6. Place the rack of tubes in a water bath at $37\,°C$ and incubate for 90 min.

7. Place the rack of tubes in ice.

8. Add 0.5 ml of a 10% TCA/0.5% tannic acid solution, mix, and let stand at $0\,°C$ for 5 min.

9. Centrifuge at 1000 g for 5 min $(4\,°C)$.

10. Transfer the supernatant fractions to scintillation vials.

11. Suspend the pellets in 0.5 ml of a 5% TCA/0.25% tannic acid solution and centrifuge as in step 9.

12. Add the supernatant fraction from each tube to the initial supernatant in the scintillation vial.

13. Add 9 ml of scintillation fluid and measure the radioactivity in a scintillation counter.

14. Suspend the pellets from the $(+)$ samples (NCP) in 0.5 ml 6 M HCl.

15. Transfer the suspensions to Teflon-lined, screw-capped 13 × 100 mm glass tubes (hydrolysis tubes).

Protocol 1. *Continued*

16. Rinse the original tubes with 0.5 ml 6 M HCl and transfer to the hydrolysis tubes.

17. Cap the tubes tightly and place in a heating block at 150 °C for 30 min.

18. Cool the hydrolysates to room temperature and transfer 0.2 ml to a scintillation vial.

19. Add 8 ml of scintillation fluid and measure the radioactivity. If there is insufficient radioactivity, add additional sample and proportionately more scintillation fluid.

20. Prepare counting standards to determine counting efficiency by adding approximately 5000 d.p.m. of $[^3H]H_2O$ to mixtures duplicating the two different systems.
    (a) *For digests*: 1.5 ml of a 5% TCA/0.25% tannic acid solution plus 9 ml of scintillation fluid.
    (b) *For hydrolysates*: 0.2 ml 6 M HCl plus 8 ml of scintillation fluid.
    (c) *For background samples*: do not add radioactivity.

Table 2. Precipitation of proteins with TCA

(a) *Cell sonicate*[a]	0.50 ml
20% TCA/20 mM proline	0.50 ml
(b) *Medium*	3.00 ml
BSA: 10 mg/ml	0.15 ml
50% TCA/50 mM proline	0.79 ml
(c) *Standard [^3H]proline labelled collagen*	0.05 ml
BSA: 10 mg/ml	0.15 ml
20% TCA/20 mM proline	0.20 ml

[a] For cell sonicates from 24-well plates, BSA (bovine serum albumin) may be needed to give a visible precipitate.

Table 3. Preparation of bacterial collagenase digestion cocktails

	(−) µl per tube	(+) µl per tube	Final concentration
NEM: 62.5 mM	20	20	2.5 mM
CaCl₂: 25 mM	10	10	0.5 mM
Elution buffer	10	none	
Purified collagenase: 600–800 µg/ml	none	10	12–16 µg/ml

4. Calculation of collagen synthesis

Protocol 2. Calculation of collagen synthesis

1. Subtract the radioactivity (c.p.m.) in the supernatant fraction of the minus collagenase blanks from the radioactivity in the digests (+) to obtain the radioactivity incorporated into collagen. To calculate total incorporation for the sample, multiply by 3 since only a 0.2 ml aliquot was analysed from the original 0.6 ml NaOH solution.

2. Subtract background c.p.m. from the radioactivity in the hydrolysate. For total incorporation, multiply by 15 to correct for the 0.2 ml aliquot initially used and for the fact that a 0.2 ml portion out of the 1 ml hydrolysate was analysed.

3. Convert c.p.m. to d.p.m. using the counting efficiencies determined using tritiated water.

4. Multiply the d.p.m. in the hydrolysate by 5.4 in order to correct for the large proportion of imino acids in collagen compared to other proteins.

5. Calculate the relative rate or percentage of collagen produced relative to total proteins by the formula:

$$\frac{\text{collagen d.p.m.}}{\text{NCP d.p.m.} + \text{collagen d.p.m.}} \times 100. \qquad [1]$$

6. If short labelling periods are used for cell cultures, so that very little of the procollagen is processed, the following formula can be used to calculate the percentage of procollagen synthesized:

$$\frac{1.13\,(\text{collagen d.p.m.})}{4.3\,(\text{NCP d.p.m.} - 0.13 \times \text{collagen d.p.m.}) + 1.13\,(\text{collagen d.p.m.})} \times 100. \qquad [2]$$

---

5. Example of typical results using collagenase digestion

Table 4 shows the results obtained with this procedure using baby hamster kidney (BHK) fibroblasts. Incorporation of radioactive proline into collagen specifically and into NCP were determined (*Protocol 1*). These results were used to calculate the percentage of collagen secreted into the medium and the relative rate of collagen synthesis (*Protocol 2*). Applications of the procedure to other cell types and tissues have been published (16, 17).

Table 4. Determination of collagen synthesis and secretion in BHK fibroblasts[a]

Fraction	[³H]Proline incorporated (d.p.m. × 10⁻³)[b]	
	Collagen	NCP
Cell	37.2	302.6
Medium	34.1	203.1
Total	71.3	505.7
$T_{NCP} \times 5.4$		2731

$$\text{Collagen secretion}^c = \frac{34.1 \times 10^3}{71.3 \times 10^3} \times 100 = 45.4\%$$

$$\text{Collagen synthesis}^d = \frac{71.3 \times 10^3}{2731 \times 10^3 + 71.3 \times 10^3} \times 100 = 2.54\%$$

[a] Cells were incubated for 2.5 h with [5-³H]proline in 60 mm dishes as detailed in *Protocol 1*.
[b] Results were calculated from the radioactivity in the collagenase digests and NCP hydrolysates as detailed in steps 1–4 of *Protocol 2*.
[c] Secretion is the percentage of total collagen in the medium.
[d] Results were calculated using the formulae (Equations 1 and 2) detailed in *Protocol 2*.

6. Estimation of total collagen production by tissues, using collagenase digestion

Tissues labelled in organ culture are treated similarly to cell cultures except for tissue disruption.

Protocol 3. Estimation of total collagen production by tissues using collagenase digestion

1. Remove the medium and rinse the tissue twice with buffered saline containing 10 mM proline.
2. Mince the tissue, suspend the mince in 0.05 M Tris–HCl/10 mM proline and homogenize. In the case of hard tissues, use a Polytron (Brinkmann) homogenizer or a similar tissue disintegrator. A glass homogenizer may be used for soft tissues.
3. Add one-fourth volume of a solution containing 50% TCA and 50 mM proline to the homogenate and Vortex well.
4. Follow the procedure described above (*Protocol 2C* onward), except that for hard tissues it may be necessary to heat at 100 °C for 5 min in order to dissolve the TCA-precipitated protein in NaOH.

2B. IDENTIFICATION AND QUANTIFICATION OF COLLAGEN TYPES I, III, AND V PRODUCED BY CELLS IN CULTURE

MICHAEL HARALSON and SAMUEL DIMARI

7. Introduction

In general, sufficient chemical quantities of collagen to permit character-ization cannot be isolated from cultured cells. To circumvent this problem, the collagens synthesized by cells in culture are isotopically labelled. The approach described in this section utilizes differential salt fractionation and ion-exchange chromatography to identify and quantitate radioactive types I, III, and V molecules based upon the behaviour of the pepsin-derived subunits from these molecules. It should be noted that in most instances cells pro-ducing types I, III, and V collagen also synthesize type IV molecules, which are recovered with type V collagen. Thus, this approach represents an alter-nate procedure to that described in Chapter 3 for evaluating type IV collagen. This approach has been used for identifying and quantitating types I, III, IV, and V molecules from a variety of cultured cells of both epithelial and non-epithelial origin (18–24), and for evaluating the effects of growth factors, hormones, and mechanical forces on collagen production (25–31).

8. Isotopic labelling of collagens

8.1 Introduction

Cells are cultured under their standard conditions. To obtain sufficient quanti-ties of material to permit analysis beyond SDS–PAGE, cells should be grown in dishes at least 60 mm in diameter. If the collagen types to be studied are minor products of the cells, use 150 mm diameter dishes. Dishes, rather than flasks, are preferred because of the ease of separating the culture medium from the attached cells, as well as for removing the cells and deposited matrix from the dish. In general, the following volumes of labelling medium have been used in this laboratory:

- 60 mm dish   5 ml
- 100 mm dish   10 ml
- 150 mm dish   25 ml

The number of dishes employed depends upon the collagen biosynthetic activity of the cells. Generally, at least 50 000 c.p.m. of pepsin-resistant material is required for quantitative analysis. As a rule, begin with five dishes and adjust this number in subsequent experiments based on the amount of radioactivity recovered in pepsin-resistant molecules.

Protocol 4. Preparation of labelling medium and labelling of cells

Equipment and reagents

- BSA
- BAPN
- Ascorbic acid

- [2,3,4,5-^{3}H]proline (Sp.Act. ~100 Ci/mmol)
- 0.2 μm filters

Method

1. Dissolve the following components at the indicated concentrations in the appropriate amount of serum-free culture medium:
   - BSA   1 mg/ml
   - BAPN   100 μg/ml
   - Ascorbic acid   50 μg/ml
2. Filter into a sterile container using a 0.2 μm filter.
3. Add [^{3}H]proline to a final concentration of 20 μCi/ml.
4. Remove medium from cells and replace with labelling medium.
5. Incubate cells in labelling medium for the desired amount of time (generally 12–18 h).

8.2 Special considerations when labelling cells

(a) If the incubation time with the cells exceeds 24 hours, the labelling medium should be supplemented with fresh ascorbic acid each day.

(b) If growth factors or hormones are to be included in the labelling medium, they should be added after filtering the other components (step 2, *Protocol 4*).

(c) [^{3}H]proline labelled in the 2, 3, and 5 position is the preferred label because it is generally not metabolized directly to other products and because it and its hydroxylated derivatives represent about 20% of the total amino acids in pepsin-derived collagen molecules. [^{3}H]proline labelled in the 2, 3, 4, and 5 positions is commonly used even though the label in the 4 position will be lost due to hydroxylation.

9. Preparation of secreted and cell-associated culture fractions

9.1 Introduction

Several studies have documented a disparity in both the types and relative amounts of the collagens secreted into the medium by cultured cells and those deposited in the cell matrix. Additionally, growth factors may alter the culture distribution of the collagens. Thus, accurately assessing the collagen phenotype of a cell and completely determining the effects of mediators of collagen production, require separate evaluation of each culture compartment.

9.2 Secreted collagens

Protocol 5. Preparation of secreted collagen fraction

Reagents

- NEM
- Phenylmethylsulfonyl fluoride (PMSF)
- 0.5 M Acetic acid (HOAc)

- Non-radioactive type I collagen (2 mg/ml in 0.5 M HOAc) either purchased commercially (Sigma) or prepared from human placenta as previously detailed (32).

Method

1. Carefully remove the culture medium using a pipette and a hand-held pipetter. Do not aspirate medium because this will denature the proteins.

2. Place the medium in chilled centrifuge tubes and centrifuge at 2000 *g* for 15 min at 4°C. Save any visible pellet for inclusion in the cell-associated fraction (step 6, *Protocol 6*).

3. Combine the supernatant fractions and adjust to a final concentration of 10 mM NEM and 0.5 mM PMSF.

4. Add 5 mg of non-radioactive type I collagen.

5. The preparation can be immediately used for isolation of pepsin-resistant molecules (*Protocol 7*) or can be stored at −70°C.

9.3 Cell-associated collagens

Protocol 6. Preparation of cell-associated collagen fraction

Reagents

- NEM
- PMSF
- 0.5 M HOAc

- Non-radioactive type I collagen (2 mg/ml in 0.5 M HOAc) either purchased commercially (Sigma) or prepared from human placenta as previously detailed (32).

Method

1. Add 0.5 M HOAc (2 ml for 60 mm dishes and 4 ml for 100 mm and 150 mm dishes) to each dish from which the labelling medium has been removed.

Protocol 6. *Continued*

2. Scrape the cell layers, using a cell lifter (Costar Corporation), into the acid.

3. Remove the suspension with a pipette and place in a chilled container (a fleaker or flat-bottom centrifuge bottle is excellent for this purpose).

4. Add an additional 2–4 ml aliquot of 0.5 M HOAc to each dish and repeat the scraping.

5. Remove the suspension and combine with previous suspension.

6. If any visible pellet was obtained upon centrifugation of the culture medium (step 2, *Protocol 5*), suspend it in 1–2 ml 0.5 M HOAc and combine with the suspension obtained from the dishes.

7. Adjust the combined suspension to a final concentration of 10 mM NEM and 0.5 mM PMSF based upon a volume of 4 ml/dish for 60 mm dishes and 8 ml/dish for 100 and 150 mm dishes + the volume of the suspended pellets from the culture medium.

8. Add 5 mg of non-radioactive type I collagen.

9. Stir the suspension at 4 °C for 24 h on a magnetic stirrer.

10. Centrifuge the suspension at 15 000 *g* for 30 min at 4 °C and decant the supernatant fraction.

11. This supernatant fraction can be used immediately for isolating pepsin-resistant molecules (*Protocol 7*) or stored at −70 °C.

10. Isolation of secreted and cell-associated collagens synthesized by cells in culture

10.1 Introduction

While a number of different approaches have been described for isolating types I, III, and V collagen, the approach described herein and schematically represented in *Figure 2* employs limited pepsin digestion and differential salt fractionation. While this approach results in the preparation of collagens substantially free of non-collagenous proteins, its applicability primarily applies only to the preparation of those genetic types of collagens containing significant triple-helical domains, that is types I, II, III, IV, and V molecules. Approaches for isolating other collagen types are detailed in Chapters 3, 4 and 5.

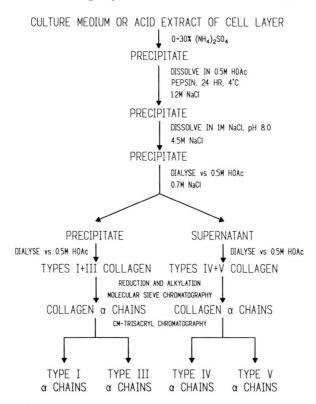

Figure 2. Schematic representation of the protocol for the isolation of collagen molecules produced by cultured cells. For the preparation and partial purification of secreted and cell-associated pepsin-derived collagen molecules synthesized by cultured cells, the sequence of steps indicated is employed. Details are presented in *Protocol 7*. (Reproduced with permission from reference 27.)

Protocol 7. Isolation of collagens from the secreted and cell-associated fractions

Equipment and reagents

- Pepsin
- Dialysis tubing
- Solid NaCl and $(NH_4)_2SO_4$

- 1 M NaCl dissolved in 0.05 M Tris–HCl, pH 8.0
- 0.5 M HOAc

Method

N.B. *Perform all procedures at 4 °C.*

1. Count a 25 μl aliquot from each newly generated solution in the protocol to monitor and quantitate the efficiency of each step.

43

Protocol 7. *Continued*

2. Adjust either the clarified culture medium or the acid extract of the cell layer to 30% (w/v) $(NH_4)_2SO_4$ by the slow addition of solid $(NH_4)_2SO_4$ while stirring in an ice bath, and continue stirring for 4–6 h.

3. Collect the precipitate by centrifugation at 15 000 *g* for 1 h.

4. Dissolve the precipitate in ~30 ml 0.5 M HOAc, and dialyse the solution for 24 h against 2 litres 0.5 M HOAc.

5. Measure the volume of the solution and adjust to 0.1 mg/ml pepsin by the addition of solid enzyme.

6. Stir the solution for 24 h.

7. Slowly add, with stirring, solid NaCl to a final concentration of 7% (w/v) and continue stirring for 6 h.

8. Collect precipitate by centrifugation at 15 000 *g* for 1 h.

9. Dissolve the precipitate by trituration in ~30 ml 1 M NaCl/0.05 M Tris–HCl, pH 8.0.

10. Slowly add, with stirring, solid NaCl to a final concentration of 4.5 M (204 mg/ml) and stir the suspension overnight.

11. Collect the precipitate by centrifugation at 15 000 *g* for 1 h.

12. Dissolve the precipitate by trituration in ~10 ml 0.5 M HOAc and dialyse the solution for ~4 h against 2 litres 0.5 M HOAc.

13. Dialyse the solution for 24 h against 2 litres of 0.7 M NaCl/0.5 M HOAc.

14. Collect the precipitate (types I + III molecules—see *Figure 2*) by centrifuging at 15 000 *g* for 1 h. Save the supernatant fraction which contains types IV + V molecules (see *Figure 2*).

15. Dissolve the precipitate in ~10 ml 0.5 M HOAc and dialyse for 2–4 h against 2 litres 0.5 M HOAc.

16. Determine the amount of radioactivity in a 25 μl aliquot.

17. Freeze and lyophilize the majority of the sample in a 20 ml liquid scintillation vial. This facilitates future manipulation of the sample during the isolation of the collagen chains. Before freezing, small aliquots containing ~20 000 c.p.m. can be removed for analysis by SDS–PAGE if desired. These should likewise be frozen and lyophilized.

18. Dialyse the supernatant fraction obtained in step 14 against two 2 litre portions of 0.5 M HOAc (total dialysis time 4–6 h). The dialysed preparation contains types IV and V molecules and should be processed as described in step 17 of this protocol.

10.2 Additional considerations in the isolation of collagens

(a) The amounts of radioactivity in steps 16 and 17 can be useful indicators of the relative amounts of type I + III molecules and type IV + V molecules.

(b) Type I-homotrimers are partially soluble at 0.7 M NaCl under acidic conditions. If further analysis (SDS–PAGE and/or ion-exchange chromatography) indicate this to be the case, perform the differential salt fractionation (step 13) at 0.9 M NaCl.

(c) If subsequent analysis indicates that the majority of the collagen produced consists of either type IV or type V molecules, omit the differential salt fractionation (step 13) and proceed directly to step 15.

(d) Only fully assembled triple-helical molecules will be recovered by this protocol. Hence, the collagens isolated from the cell layer represent those molecules deposited in the matrix and any fully assembled intracellular procollagen; newly synthesized chains and incompletely assembled molecules inside the cell will be digested by pepsin.

(e) This approach can be used for the isolation of collagens from tissues (32). However, only a qualitative approximation of the collagen types present will be realized. Higher-order collagen aggregates (β- and γ-components as well as higher molecular weight insoluble aggregates) resulting from the Hyl–Lys covalent cross-links will be recovered precluding quantitative evaluation. If a quantitative assessment of types I, III, and V molecules from tissues is desired, the protocol described in Part D should be used.

11. Isolation of collagen chains by molecular-sieve chromatography

11.1 Introduction

Before quantitative evaluation of collagen types by ion-exchange chromatography, it is necessary to isolate the subunits (α chains) of the collagen molecules. This is achieved by reducing and alkylating the sample followed by chromatography of the preparation on Agarose A-5m. The arrangement of equipment for performing this procedure is shown in *Figure 3*.

11.2 Preparation of samples for molecular-sieve chromatography

Types III and IV collagen contain interchain disulfide bonds. Therefore, it is necessary to reduce these bonds and to alkylate the cysteine residues to prevent their reformation. Each sample should contain approximately 5 mg of non-radioactive type I collagen. Samples containing types I + III molecules contain sufficient carrier (added at step 4, *Protocol 5* and step 8, *Protocol 6*).

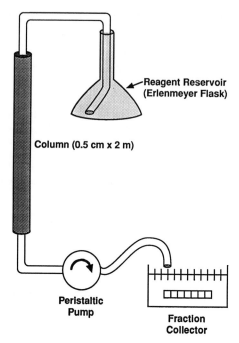

Figure 3. Schematic representation of equipment arrangement for isolation of collagen chains by molecular-sieve chromatography.

However, it is necessary to add carrier to each sample containing types IV + V molecules; that is preparations soluble at 0.7 M NaCl (step 13, *Protocol 7*).

Protocol 8. Reduction and alkylation of collagen molecules

Equipment and reagents

- 2-Mercaptoethanol
- 42 °C Water bath on top of a magnetic stirrer
- Solid iodoacetic acid and Trizma base (Sigma)
- 5 M Urea dissolved in 0.05 M Tris–HCl, pH 8.0
- Small magnetic stirring bar
- Non-radioactive type I collagen either purchased commercially (Sigma) or prepared from human placenta (32).

Method

1. To the lyophilized sample in the glass scintillation vial, add 3 ml 5 M urea/0.05 M Tris–HCl, pH 8.0, and 20 µl 2-mercaptoethanol.
2. Place a small stirring bar in the vial, cap the vial, and place in a 42 °C water bath on top of a magnetic stirrer.
3. Stir the solution for 30 min.
4. Periodically during the stirring, remove the vial from the water bath, tip and rotate it to wash any undissolved material from the sides.

---

5. After the incubation, add 125 mg of solid iodoacetic acid in five equal aliquots. Simultaneously add equal amounts of Trizma base.

6. After the additions, remove a small aliquot of the sample using a Pasteur pipette and measure the pH using pH paper.

7. If the sample is acidic, add additional Trizma base until the pH is approximately 8.

8. Cap the vial and place it in the dark for 10 min.

11.3 Molecular-sieve chromatography of reduced and alkylated collagen chains

Protocol 9. Molecular-sieve chromatography

Equipment and reagents

- Agarose A-5m (Bio-Rad)
- Fraction collector
- Peristaltic pump
- UV spectrophotometer (optional)
- 2 M Guanidine hydrochloride dissolved in 0.05 M Tris–HCl, pH 7.5
- Glass chromatography column (0.5 cm × 200 cm)
- Aquasol
- 0.5 M Acetic acid

A. *Column preparation*

1. Equilibrate about 400 ml of Agarose A-5m in 200–300 ml 2 M guanidine/0.05 M Tris–HCl, pH 7.5, at 42 °C under vacuum for approximately 3 h. Swirl occasionally to allow air bubbles to surface. Cool to room temperature under vacuum.

2. Slowly add the resin to the column.

3. To eliminate trapped air bubbles, gently tap the sides of the column with a rubber stopper while the resin settles.

4. Connect a peristaltic pump to the bottom of the column and begin pumping at a flow rate of 9 ml/h. When the buffer level reaches the top of the column, add an additional 2 ml of buffer and continue pumping until the buffer again reaches the top of the resin.

5. Fill the column above the resin with buffer, attach a siphon-flow buffer reservoir to the top as illustrated in *Figure 3*.

6. Pump 350–400 ml of 2 M guanidine/0.05 M Tris–HCl, pH 7.5, through the column at a flow rate of 7 ml/h. Collect the effluent in a beaker or flask.

7. The final bed volume of the column should be between 350 and 375 ml.

Protocol 9. *Continued*

8. Connect the pump to a fraction collector. Alternatively, the effluent can be monitored at 226 nm by connecting a spectrophotometer between the pump and the fraction collector.

9. One column can be repeatedly used for periods as long as 6 months to a year.

B. *Method*

1. Remove buffer from the column until the top of the resin is exposed.

2. Gently layer the reduced and alkylated sample, prepared as described in *Protocol 8*, on top of the resin.

3. Wash the vial with 0.5 ml 2 M guanidine/0.05 M Tris–HCl, pH 7.5, and mix this solution with the sample on top of the resin.

4. Turn on the peristaltic pump and collect fractions every 30 min at a flow rate of 7 ml/h.

5. When the sample has completely entered the resin, wash the sides of the column with about 2 ml of 2 M guanidine/0.05 M Tris–HCl, pH 7.5.

6. When the wash has entered the resin, fill the void at the top of the column with 2 M guanidine/0.05 M Tris–HCl, pH 7.5, and connect a siphon-flow reservoir containing the buffer.

7. Continue elution until 80 fractions (3.5 ml) are collected.

8. Monitor the effluent by counting between 25 and 100 μl of each fraction in 5 ml of Aquasol. Combine the fractions containing the radioactive collagen α chains, dialyse against repeated changes of 2 litre aliquots of 0.5 M HOAc to ensure <10 μg of guanidine remains in the dialysed solution. Freeze and lyophilize the sample.

9. Monitor the positions of elution of the components derived from the non-radioactive type I collagen carrier by measuring the absorbance of each fraction at 226 nm.

10. Pump an additional 100 ml of buffer through the column to ensure complete removal of small molecular weight compounds.

11.3.1 Molecular-sieve chromatographic profile of types I + III collagen chains

A typical elution profile of chains recovered in the types I + III fraction is shown in *Figure 4A*. Only the fractions containing radioactivity that co-elutes with the non-radioactive collagen α chains (dashed line, *Figure 4A*) should be combined and further processed. This area of the chromatogram contains the $\alpha 1(I)$, $\alpha 2(I)$, and $\alpha 1(III)$ collagen chains.

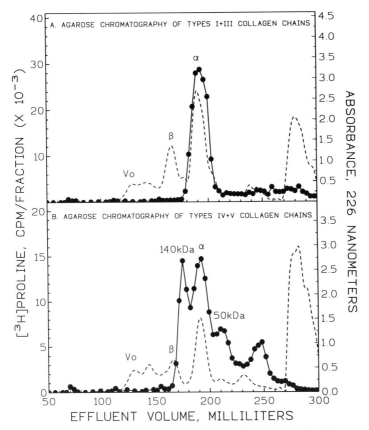

Figure 4. Molecular-sieve chromatography of collagens. Collagen molecules recovered in either the types I + III fraction (Panel A) or the types IV + V fraction (Panel B) were chromatographed after reduction and alkylation on Agarose A-5 m as detailed in *Protocol 9*. The elution of the radioactive collagen chains or fragments (—●—●—●—) was determined by liquid scintillation counting and of the non-radioactive carrier collagen (–––––) by measuring the absorbance at 226 nm. In Panel A, ∼ 275 000 c.p.m. of material isolated in the types I + III fraction (see Protocol 7) from the products synthesized by rat kidney mesangial cells (18) were chromatographed. In Panel B, ∼ 550 000 c.p.m. of material isolated in the types IV + V fraction (*Protocol 7*) from the products synthesized by fetal rat lung epithelial cells (23) were chromatographed. Radioactivity was determined by counting a 50 μl aliquot of each fraction in Aquasol.

11.3.2 Molecular-sieve chromatographic profile of types IV + V collagen chains

A typical elution profile of chains recovered in the types IV + V fraction is shown in *Figure 4B*. This profile is more complex than that obtained upon chromatography of the I + III fraction (*Figure 4A*) reflecting the fact that treatment of type IV molecules with pepsin results in multiple cleavage products. Thus, molecular-sieve chromatography of this fraction may contain, depending upon the cell of origin of the products, three distinct areas. The material co-eluting with the non-radioactive collagen α chains contains the three type V collagen chains and may contain the α2(IV) as well as the 100 kDa pepsin-derived α1(IV) chain. The material eluting between the non-radioactive collagen α chains and β components (designated 140 kDa on *Figure 4B*) contains components corresponding to only the largest pepsin-derived fragment of the α1(IV) chain. The material eluting in the region of the chromatogram designated 50 kDa contains pepsin-derived 50 kDa fragments derived from the α1(IV) and α2(IV) chains. If the cells produce significant amounts of type IV, it is recommended that each region of the chromatogram be separately combined for further processing. These fractions can be successfully chromatographed on CM–Trisacryl (*Protocol 10*) to quantitate type IV synthesis. If type IV molecules represent a minor amount of the total radioactive collagen recovered in this fraction, the α chain peak will contain a shoulder, and all of the radioactivity in this region should be combined for further processing.

12. Identification and quantification of collagen chains by carboxymethyl–Trisacryl chromatography

12.1 Introduction

Each collagen chain differs in its primary structure and hence possesses a unique charge (4, 5, 8, 33). Consequently, each chain elutes in a unique position upon ion-exchange chromatography. A common method used to separate and identify the individual collagen chains is chromatography on carboxymethyl (CM)–Trisacryl or CM–cellulose.

12.2 CM–Trisacryl chromatography of collagen chains

Table 5. Preparation of CM–Trisacryl chromatography buffers

	Components (amount/1000 ml)		
	1 M NaOAc, pH 4.8	Urea	NaCl
Column buffer	20 ml	60 g	–
1st Gradient buffer	20 ml	60 g	0.88 g
2nd Gradient buffer	20 ml	60 g	5.26 g
Wash buffer	20 ml	60 g	58.44 g

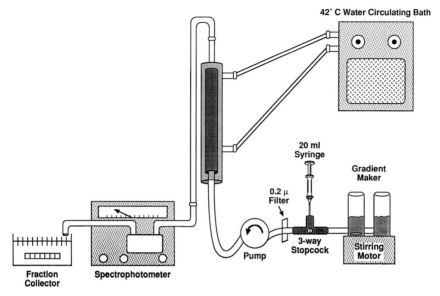

Figure 5. Schematic representation of equipment arrangement for CM–Trisacryl chromatography of collagen chains.

Protocol 10. CM–Trisacryl chromatography of collagen chains

Equipment and reagents

- CM–Trisacryl M, bead size: 40–80 μm (IBF)
- Fraction collector
- Peristaltic pump
- 42 °C Circulating water bath
- Jacketed glass chromatography column (0.9 × 9 cm) with adjustable end plungers
- Urea—electrophoresis grade
- 1 M NaOAc (HPLC grade), pH 4.8 buffer

A. *Preparation of buffers*

The composition of the various buffers is given in *Table 5*. Make these up as follows:

1. Weigh the NaCl and urea out separately for each buffer.

2. After combining the components, make the solution up to approximately 900 ml with glass-distilled water at room temperature.

3. Titrate each solution to a final pH of 4.8.

4. Adjust the volume of each solution to 1 litre in a volumetric flask.

5. Immediately before use, degas each solution at 42 °C with stirring for at least 20 min.

Protocol 10. *Continued*

B. *Preparation of resin*

1. Place approximately 20 g resin in a 250 ml side-arm flask.

2. Add approximately 100 ml H_2O, swirl, and let settle for 15 min.

3. Decant the supernatant, generously pouring off fines.

4. Repeat steps 2 and 3 twice.

5. Add about 100 ml wash buffer, swirl, and let settle.

6. Decant the supernatant, generously pouring-off fines.

7. Repeat steps 5 and 6 until the volume of the packed resin is about a half of the original.

8. Place the side-arm flask in a 42 °C water bath, connect to a vacuum line and degas the suspension with gentle swirling for 30 min.

C. *Column preparation*

1. Slowly add the resin to a water-jacketed column equilibrated at 42 °C until the bed height is about 10 cm.

2. Connect a peristaltic pump to the bottom of the column and begin an upward pumping of approximately 40 ml of column buffer at a flow rate of 60–70 ml/min.

3. When the column is packed, pump about 100 ml of column buffer through the column at a flow rate of 120–150 ml/h. **NOTE:** The pH of the column **must** be 4.8. If it is not, pass approximately 10 column volumes of 1 M NaOAc, pH 4.8/1 M urea through the column followed by 150 ml of column buffer (*Table 5*).

4. Connect the pump to a gradient maker and the column output to either a spectrophotometer or a fraction collector (see *Figure 5*).

5. One column can be repeatedly used for as long as 6 months.

D. *Method*

1. Prepare the gradient by adding 150 ml of Gradient Buffer 1 in the 1st (outlet) chamber and 150 ml of Gradient Buffer 2 in the 2nd chamber (*Table 5*).

2. Dissolve the lyophilized sample of collagen α chains in 15 ml of column buffer by warming for 30 min at 42 °C.

3. Place the sample in the syringe, and apply the sample to the column by pumping at a flow rate of 100–120 ml/h.

4. Begin collecting approximately 3 ml fractions.

5. Wash the sample vial with 10 ml of column buffer and apply this solution to the sample reservoir.

6. When a total of 10 fractions have been collected, start the gradient.

7. Elution is continued until 110 fractions are collected.

8. Add 50 ml of wash buffer (*Table 5*) to the first chamber in the gradient maker and collect an additional 10 fractions.

9. Monitor the effluent by counting between 25 and 100 μl aliquots of each fraction in 5 ml of Aquasol. Combine the fractions containing the radioactive collagen α chains. Dialyse combined fractions against repeated changes of 2 litre aliquots of 0.5 M HOAc until urea in the dialysed solution is reduced to <10 μg. Count an aliquot of the dialysed sample. If further analysis (CNBr mapping or SDS–PAGE) of the fractions is desired, freeze and lyophilize.

10. Monitor the positions of elution of the components derived from the non-radioactive type I collagen carrier by measuring the absorbance of each fraction at 226 nm.

11. Store the column equilibrated in wash buffer (*Table 5*).

12.2.1 Profile of types I + III collagen chains obtained by CM–Trisacryl chromatography

A typical profile obtained upon CM–Trisacryl chromatography of a mixture of chains derived from types I and III molecules is shown in *Figure 6A*. The identification of the radioactive peaks corresponding to the α1(I) and α2(I) chains is established by their co-elution with the non-radioactive type I chains included in the sample. Type III chains, which are more basic than the α1(I) chains but more acidic than the α2(I) chains, elute in an intermediate position. These positions of elution are characteristic of the three chains and generally suffice to establish their identity. If additional evidence is required to initially identify an eluted peak, the radioactivity eluting in the peak can be separately combined and dialysed exhaustively against 0.5 M HOAc. After lyophilization, the preparation is suitable for analysis by either CNBr or V8 protease cleavage as described in *Protocols 14* and *15*.

Theoretically, the ratio of radioactivity recovered in the α1(I) peak to that in the α2(I) (*Figure 6A*) peak should be approximately 2:1. However, in some instances, an excess of α1(I) to α2(I) chains is observed. This finding reflects the recovery of a mixture of type I-heterotrimers and type I-homotrimers among the products of the cells (reviewed in reference 21). Furthermore, a major consequence of the action of growth factors and hormones on collagen production is a change in this ratio (27–30). The relative amounts of type I-heterotrimers and type I-homotrimers can be determined by the calculations detailed in *Protocol 11*.

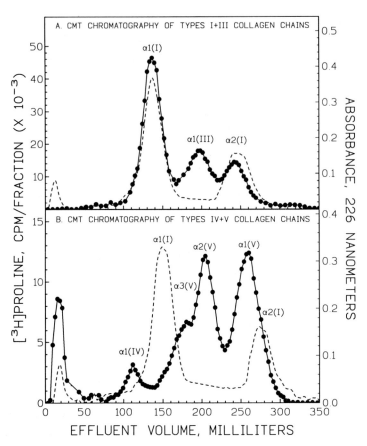

Figure 6. CM–Trisacryl chromatography of collagen chains. Reduced and alkylated collagen chains prepared as described in *Protocol 9* were chromatographed on CM–Trisacryl as detailed in *Protocol 10*. Panel A, Chromatographic profile of types I + III chains (~ 1 200 000 c.p.m.) recovered in the types I + III fraction derived from the products synthesized by fetal rat lung epithelial cells (23). Panel B, Chromatographic profile of the types IV + V chains (~ 400 000 c.p.m.) derived from the products synthesized by normal rat kidney epithelial cells (22). The elution of the radioactive collagen chains (—●—●—●—) was determined by liquid scintillation counting and of the non-radioactive α1(I) and α2(I) collagen chains (–––––) by measuring the absorbance at 226 nm.

Protocol 11. Calculation of the relative amounts of type I-heterotrimers to type I-homotrimers

1. Determine the amount of radioactivity recovered in the peaks corresponding to the α1(I) and α2(I) chains (*Figure 6A*).

2. Multiply the amount of radioactivity recovered as α2(I) chains by 0.112.

This value reflects the reduction of prolyl and 3- and 4-hydroxyprolyl residue content of the α2(I) relative to the α1(I) chain (4).

3. Add the above calculated amount of radioactivity to the amount of radioactivity recovered as α2(I) chains.

4. Multiply the sum by 2. This value represents the amount of radioactivity in α1(I) chains that was recovered in type I-heterotrimers.

5. Subtract this amount of radioactivity from the total amount of radio-activity recovered as α1(I) chains. The resulting value represents the amount of radioactivity recovered as type I-homotrimers.

6. The amount of radioactivity recovered as type I-heterotrimers is the sum of the amount of radioactivity recovered as α2(I) chains plus the amount of radioactivity calculated in step 4 for α1(I) chains.

12.2.2 Profile of types IV + V collagen chains obtained by CM–Trisacryl chromatography

A typical profile obtained upon CM–Trisacryl chromatography of a mixture of chains derived from types IV and V molecules is shown in *Figure 6B*. Pepsin-derived type V chains elute between the α1(I) and α2(I) non-radioactive components. In many instances, the resolution of the α2(V) and α3(V) chains will be incomplete allowing quantification of only their sum. It should be noted that the position of elution of the α1(V) chain overlaps that of the α2(I) chain and the position of elution of the α2(V) chain overlaps that of the α1(III) chain (compare *Figures 6A* and *6B*). This fact demonstrates the necessity of separating types I and III molecules from types IV and V in the initial isolation of the collagens (step 13 in *Protocol 7*) if the cells produce all four collagen types.

The three major components (140 kDa, 100 kDa, and 50 kDa) derived with pepsin from the α1(IV) chain are the most acidic of the collagen components and elute slightly before the α1(I) chain. In contrast, pepsin-derived α2(IV) components elute as more basic molecules slightly behind the α1(I) chain and co-incident with the α3(V) chain. If evidence by SDS-PAGE suggests the co-production of both of the type IV components and type V molecules, additional salt fractionation, as described in reference 4, should be performed before isolation of the chains and ion-exchange chromatography. To date, no information exists regarding the behaviour of the more recently described α3(IV), α4(IV), α5(IV), and α6(IV) chains during differential salt fractionation or their elution from ion-exchange resins.

2C. IDENTIFICATION OF RADIOACTIVE COLLAGENS BY PEPTIDE MAPPING ON SDS–PAGE

BEVERLY PETERKOFSKY

13. Introduction

As previously mentioned, the collagen phenotype of cultured cells generally consists of more than one genetic type of collagen. This section describes an alternative approach to that described in Part B for identifying the collagens synthesized by cells in culture. This approach is based on the same structural features of the fibrillar collagens, but is geared to small-scale analysis using SDS–PAGE. Protocols are provided for distinguishing collagenous from non-collagenous proteins by digestion with bacterial collagenase prior to SDS–PAGE (*Protocol 13*), for converting procollagen to collagen using pepsin digestion (*Protocol 14*), and for identifying the collagen α chains excised from gels by cleavage with either cyanogen bromide (CNBr) or V8 protease to yield unique peptide patterns on SDS–PAGE (*Protocols 16* and *17*).

14. Isolation of procollagen

Protocol 12. Isolation of procollagen produced by cultured cells

Reagents

- L-Proline: 500 mM (store at −20 °C)
- NEM: 100 mM. Adjust pH to approximately 6 with solid NaHCO₃. Store at −20 °C.
- EDTA: 200 mM. Adjust pH to 7 and store at −20 °C.
- PMSF: 50 mM in dimethyl sulfoxide. Store at −20 °C.
- Saturated ammonium sulfate (3.9 M). Store at room temperature and chill a portion on ice just before use.

Method

1. Label procollagens as described in *Protocol 1* except [^{14}C]proline is used since the efficiency for tritium in gels is low.
2. Add 0.1 ml of [^{14}C]proline (100 μCi/ml, 2.5 mM, Sp. Act = 40 μCi/μmol) to 2.4 ml of serum-free medium without phenol red in a 100 mm culture dish.
3. Harvest the cell and medium fractions as detailed in *Protocols 5* and *6*. Perform all subsequent steps at 0–4 °C.
4. Sonicate the cell pellet in a volume of 0.4 M NaCl/0.1 M Hepes buffer, pH 7.2, equal to one-tenth the volume of the medium fraction.

---

5. Add proline and protease inhibitors to cell and medium fractions to give the following concentrations: proline, 10 mM; NEM, 1 mM; EDTA, 2 mM; PMSF, 0.2 mM.

6. Remove about 10% of each fraction; precipitate the proteins with TCA and measure incorporation into collagen using bacterial collagenase as described in *Protocol 1*. Store the remainder of the fractions at $-20\,°C$.

7. Thaw the remainder of the fractions.

8. Adjust the concentration of the medium to 0.4 M NaCl and 0.1 M Hepes buffer, pH 7.2, taking into consideration that the medium contains approximately 0.15 M NaCl. Measure the total volume.

9. Add one-half volume of saturated ammonium sulfate to the sonicate and medium with mixing and continue mixing for 30 min. The $(NH_4)_2SO_4$ will be 33% saturated (P33 fraction). Most procollagens and collagens precipitate at 33% $(NH_4)_2SO_4$, but some non-helical procollagens require 50% saturation for precipitation (34). For initial characterization of a cell's collagen phenotype, add a volume of $(NH_4)_2SO_4$ equal to that measured in step 8 and follow the same procedure (P50 fraction).

10. Centrifuge at 20 000 *g* for 10 min at 4°C.

11. Remove the supernatant fractions, drain the tubes by inverting, and wipe the sides of the tubes.

12. Dissolve the precipitates (P33 or P50 fractions) in a small volume of 0.2 M NaCl/0.05 M Hepes buffer, pH 7.2, so that the solution contains 100–400 d.p.m./μl of collagen, based on the initial assay.

15. SDS–PAGE analysis of procollagens, collagens, and peptides derived from collagen α chains

Typical results for peptide mapping of collagens produced by a fibroblast culture are shown in *Figure 7*. The pepsin-derived collagen was applied to a 5% SDS–polyacrylamide gel (*Figure 7A*) and the α chains of type I [α1(I) and α2(I)] and type V [α1(V) and α2(V)] collagens were observed. The positions of migration of the α chains give some indication of their identity. However, for further identification, the α1(I) and α2(I) bands were separately excised from the dried gel, as described in *Protocol 15*, and subjected to digestion with CNBr (*Figure 7B*) or V8 protease (*Figure 7C*), as detailed in *Protocols 16* and *17*. In *Figure 7B* (CNBr-derived peptides) the peptides were electrophoresed on an 8% polyacrylamide gel, and the V8 protease-generated peptides (*Figure 7C*) were resolved by electrophoresis on a 10% gel.

Beverly Peterkofsky

PEPTIDE MAPPING OF COLLAGEN α - CHAINS ON SDS-PAGE

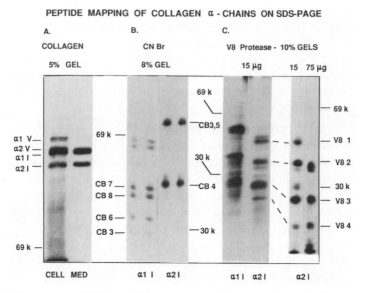

Figure 7. Fluorogram of CNBr and V8 protease peptide maps of type I collagen α chains. P33 fractions of cell and medium fractions from Syrian hamster embryo fibroblasts were treated with pepsin and peptides generated as described in *Protocols 14–17*. The cell (6200 d.p.m.) and medium (5500 d.p.m.) collagens were electrophoresed on a 5% gel and a fluorogram was prepared (Panel A). For the cyanogen bromide (CB) peptide maps (Panel B), α1(I) and α2(I) chains from the medium fraction were excised from a 5% gel that had 8000 d.p.m. of collagen applied. The positions of the CNBr peptides are indicated at the sides. For the V8 protease digestion (Panel C), the α chains were excised from a 5% gel that had 23 000 d.p.m. of collagen applied. Digestion with 15 μg of enzyme partially digests the chains and yields distinct peptide patterns (Panel C, lanes 1–3). For α2(I), the positions of four major peptides (V8 1–4) are indicated. The final digest maps obtained with 75 μg of enzyme also are distinct, but only the α2(I) pattern is shown here (Panel C, lane 4). The positions of globular markers (30 and 69 kDa) are indicated at the sides of each fluorogram. The CB peptide gel was exposed for 1 week and the V8 protease peptide gel was exposed for 2 weeks.

15.1 Preparation of SDS–polyacrylamide gels

Prepare gels with a 3.5% stacking gel, pH 6.8, and 5%, 8%, or 10% running gels (pH 8.8) according to standard recipes (15, 35). Use Tris–glycine, pH 8.3, reservoir buffer. Protocols below are designed for 14 × 16 × 0.15 cm gels.

15.2 Identification of collagenous proteins on gels by bacterial collagenase digestion

Procollagens and collagens can be identified on gels as bands that disappear when samples are treated with purified bacterial collagenase prior to electrophoresis. The reagents used for the digestion are described in *Protocol 1*. However, the digestion cocktail differs (*Table 6*).

58

Protocol 13. Identification of collagenous proteins on gels by bacterial collagenase digestion

Reagents

- 2 × Denaturing solution (DS), for samples already in buffer:

0.04% (w/v) Bromophenol blue	1.0 ml
40% (w/v) sucrose	5.0 ml
10% (w/v) SDS	4.0 ml

  (a) Dissolve 600 mg of urea in the solution and store at −20°C.
  (b) SDS will precipitate during freezing but will dissolve upon warming the thawed solution.

  (c) For 2 × DS + dithiothreitol (DTT) (20 mM), add 15.4 mg of DTT to 5 ml of 2 × DS.
- 2 × Sample buffer, for samples that do not contain any buffer: add 1 μl of 1 M Hepes, pH 7.2, per 10 μl of 2 × DS + urea

Method

1. Mix 20 μl aliquots of a ^{14}C-labelled protein solution with 10 μl of (−) or (+) cocktails (see *Table 6*) and incubate for 20 min at 37°C.

2. Add 30 μl of 2 × DS + DTT to each tube and heat for 5 min at 100°C.

3. Cool the sample, apply to the wells of a 5% SDS–polyacrylamide gel and start electrophoresis at 20 mA per gel. Increase power to 40 mA when the tracking dye enters the resolving gel and continue electrophoresis until tracking dye reaches the bottom of gel.

4. After stopping electrophoresis, remove the gel and immerse it in an enhancer solution such as FluoroHance (RPI) and shake on a platform shaker for 20 min.

5. Remove the gel from the enhancer and place it on a piece of filter paper slightly bigger than the gel.

6. Cover the gel with plastic wrap and dry using a gel dryer.

7. In the darkroom, place the dried gel, gel side down, on X-ray film and enclose in an X-ray cassette.

8. Develop the film after overnight exposure. If necessary, the gel can be exposed again for longer periods to visualize minor bands.

Table 6. Cocktail for digestion of procollagens by collagenase

	Volume (μl)	
	(−)	(+)
NEM: 62.5 mM	1.2	1.2
CaCl$_2$: 25 mM	1.2	1.2
H$_2$O	5.2	5.2
Elution buffer	2.4	—
Purified collagenase: 700–800 μg/ml	—	2.4

16. Pepsin digestion of procollagens for peptide mapping

This protocol is designed to prepare sufficient radioactive collagen to perform cyanogen bromide (CNBr) and V8 protease mapping. It can be scaled down to 10% of the indicated volumes if only the pepsin-digested products are required.

Table 7. Cocktails for pepsin digestion of procollagen

	Volume (μl)	
	Medium P33	**Cell P33**
Sample	100	100
	(80 000 d.p.m. collagen)	(37 000 d.p.m. collagen)
	(184 μg protein)	(750 μg protein)
Acetic acid: 1 M	186	224
NEM: 100 mM	25	25
H₂O	60	98
Pepsin: 100 μg/ml	129	—
Pepsin: 1 mg/ml	—	53

Mix ingredients at 0 °C.

Protocol 14. Pepsin digestion of procollagens for peptide mapping

Reagents

- Acetic acid: 0.5 M and 1.0 M
- Pepsin: 1 mg/ml freshly prepared, in cold 0.5 M acetic acid and diluted as needed in 0.5 M acetic acid
- NEM: 100 mM

- Protein assay reagent: the Bio-Rad Coomassie Blue reagent or Pierce BCA (bicinchoninic acid) reagent are suitable. Use an immunoglobin solution as a standard

Method

1. Measure the protein content of the procollagen solutions.

2. Set up the pepsin digestion in a final volume of 500 μl at 0 °C, using a portion of a P33 fraction (step 12, *Protocol 12*) containing at least 20 000 d.p.m. of collagen. Calculate the amount of protein in this portion and multiply by 0.07 to determine the amount of pepsin required. An example of treatment of fractions from fibroblast cultures is shown in *Table 7*.

3. Incubate tubes at 4 °C for 2 h.

4. Immediately add 250 μl of 1 M NH_4HCO_3 and mix vigorously to disperse the CO_2 bubbles. The pH should be approximately 7.

5. Freeze the samples in dry ice and lyophilize to remove the ammonium acetate that formed upon neutralization.

6. Dissolve the residue in 200 μl of water, mix vigorously, freeze in dry ice, and lyophilize again.

7. Dissolve the samples in a volume of cold water equal to the volume of the original sample (in the example (*Table 7*), 100 μl). For a small-scale experiment to analyse the pepsin products, the entire sample can be dissolved in 30 μl of 1 × sample buffer and electrophoresed directly or dissolved in water and treated with collagenase as described in *Protocol 1*.

17. Electrophoresis of pepsin-derived collagens on 5% gels and excision of α chains

Protocol 15. Electrophoresis of pepsin-derived collagens and excision of α chains

1. Electrophorese radioactive collagen samples on a 5% SDS–polyacrylamide gel using a 10-well comb.

2. Prepare a fluorogram as described in *Protocol 13*.

3. Mark the corners of the dried gel with a radioactive or fluorescent solution before exposing it to an X-ray film, so that the gel and resulting fluorogram can be aligned.

4. Make a template from the exposed X-ray film by cutting out the desired bands with a single-edged razor blade or scalpel. The area excised should be 11 mm wide and 3–4 mm high.

5. Align the X-ray film with the orienting marks on the gel and, using a scalpel, cut out gel strips within the areas cut from the film.

6. Try to have very little blotting paper adhering to the gel. Half of a strip fits into the well formed using a 15-well comb.

18. Cyanogen bromide and V8 protease cleavage of collagen α chains

The following two protocols describe procedures for cleaving the pepsin-derived collagen α chains (*Protocol 14*). Each collagen chain exhibits a unique set of cleavage products (*Figure 7*) that can be used in its identification.

Protocol 16. Cyanogen bromide (CNBr) cleavage of collagen
α chains

Equipment and reagents

- CNBr: 100 mg/ml solution in 70% formic acid
- Sample buffer for SDS–PAGE (*Protocol 13*)
- A 15-well, 8% SDS–polyacrylamide gel

Method

1. Place half of a dry gel strip in a small flat-bottomed vial or tube.

2. Add 0.5 ml of water, stopper tightly, and shake at room temperature for 15 min on a rotary platform shaker.

3. Remove the water by aspiration.

4. In a fume hood, add 0.25 ml of the CNBr solution to the washed, hydrated gel strips and stopper tightly.

5. Shake for 2 h at room temperature.

6. Add 1 ml of water to each vial, mix, and remove the solution.

7. Add 1 ml of water to the gel strips and shake for 10 min.

8. Remove the water and repeat the washing twice.

9. Add 0.25 ml of the sample buffer, stopper, and shake for a few minutes. The solution will turn yellow if formic acid is still present.

10. Remove the sample buffer, add another 0.25 ml, and shake as in step 9. Repeat this step until the dye remains blue.

11. Place the gel strip in a well of an 8% gel using a flat spatula.

12. Add 25 µl of sample buffer to cover the gel strip. Be sure the bottom of the gel strip is in close contact with the bottom of the well.

13. Add reservoir buffer to the wells with a transfer pipette, fill the reservoir, and start electrophoresis.

14. After electrophoresis, prepare a fluorogram as described in *Protocol 13*.

Protocol 17. V8 protease cleavage of collagen α chains

Equipment and reagents

- V8 protease (endoproteinase Glu–C): 500 or 750 µg/ml in sample buffer + DTT. The solutions are stable for months at −20 °C
- A 15-well, 10% SDS–polyacrylamide gel

Method

1. Prepare a fluorogram from a 5%, 10-well gel and cut out strips containing collagen α chains, as described in *Protocol 15*.

2. Place half of a gel strip in the well of the 10% SDS–polyacrylamide gel.

3. Add 30 μl of a V8 protease solution.

4. Incubate for 45 min, during which time the gel hydrates and digestion occurs.

5. Add reservoir buffer and electrophorese for 45 min at 15 mA.

6. Increase the current to 40 mA.

7. Stop electrophoresis when the tracking dye reaches the end of the gel.

8. Prepare a fluorogram as described in *Protocol 13*.

2D. DETERMINATION OF THE RELATIVE AMOUNTS OF TYPE I, III, AND V COLLAGENS BY RESOLUTION AND QUANTITATION OF LOW MOLECULAR WEIGHT MARKER CYANOGEN BROMIDE PEPTIDES FOR EACH COLLAGEN

EDWARD J. MILLER

19. Introduction

This section presents still another approach for determining the proportions of type I, III, and V collagens. The method is based on a recently described procedure (36) involving ion-exchange chromatography followed by size-exclusion chromatography to resolve and quantitate marker CNBr-derived peptides for each of the indicated collagens. The peptides used for this purpose are α1(I)-CB2, α1(III)-CB2, and α1(V)-CB1 or α2(V)-CB5. Quantitation of these peptides is achieved as a function of UV absorbance observed as the peptides are eluted from the size-exclusion column. The unique advantages of this approach are:

(a) The quantitative recovery of the collagens as soluble peptide fragments from virtually any tissue source.

(b) The resolution of the marker peptides using relatively simple chromatographic procedures.

(c) The quantitation of the marker peptides in the second chromatographic step by UV absorbance.

(d) The sensitivity of the method is such that it can be used for studies on the collagen composition of a standard needle-biopsy specimen.

Additionally, this approach could also be used to evaluate the collagen phenotype of cultured cells, provided the collagen produced contains radioactively-labelled amino acids and suitable carrier collagens are added at the time of CNBr cleavage.

20. Preparation of tissues

Protocol 18. Preparation of tissue samples for optimum cleavage at methionyl residues with CNBr

Reagents

- 2-Mercaptoethanol
- Ammonium bicarbonate: 0.2 M. Adjust pH to 7.0

- NaCl: 1 M. Adjust pH to 7.5 with 50 mM Tris

Method

1. Divide tissue mechanically.

2. Extract tissue in several volumes of 1 M NaCl, pH 7.5, at 4 °C to remove blood, lyse cells, and eliminate easily-solubilized proteins.

3. Incubate tissue residue in shaking water bath at 45 °C for 18 h in 10 vols of 0.2 M ammonium bicarbonate, pH 7.5, containing 25% 2-mercapto-ethanol.

4. Recover tissue by centrifugation at room temperature, wash with water, and lyophilize.

---

21. CNBr cleavage of proteins

Protocol 19. CNBr cleavage of collagen sample or whole tissue sample

Reagents

- Formic acid (70%)
- CNBr (crystalline)
- Nitrogen (N₂ gas)

- Lyophilized collagen sample
- Lyophilized tissue sample
- HPLC-grade water

Method

1. Dissolve collagen in 70% formic acid at 5–10 mg/ml or add 5 vol. of 70% formic acid to insoluble tissue samples.

2. Gas the solution (solvent) with a steady stream of N_2 for 30 min.

3. For each 10 mg of collagen or tissue, add 15 mg of crystalline CNBr. Stopper tightly and swirl gently to dissolve crystals.

4. Incubate stoppered flasks in a shaking water bath at 30 °C for 4 h.

5. Following incubation, dilute reaction mixture 10-fold with water and lyophilize.

6. Suspend lyophilized reaction mixture in 5–10 ml of HPLC-grade water, warm to 30 °C for 5 min.

7. Centrifuge suspension at 2500 *g* at room temperature for 15 min using a table-top centrifuge and lyophilize the clear supernatant solution.

22. Ion-exchange chromatography

Protocol 20. Ion-exchange chromatography of freeze-dried CNBr peptides

Equipment and reagents

- Buffers
  (a) Starting: 20 mM (Na$^+$) formate, pH 3.8
  (b) Limit: starting buffer containing 200 mM NaCl
- Column: Mono S HR5/5 (Pharmacia)
- Liquid chromatography system equipped with at least two programmable pumps for gradient elution

- Variable wavelength UV detector plus recorder allowing full-scale deflection for a range of absorbance levels as well as integration of resulting peaks
- Fraction collector

Method

1. Set system parameters: flow rate, 1 ml/min; UV detection, 225 nm; recorder (full-scale deflection), 0–0.2 absorbance units.

2. Equilibrate the column with starting buffer.

3. Dissolve 1–6 mg of CNBr-derived peptides in 0.5 ml of starting buffer and inject for delivery to column.

4. Initiate elution using starting buffer for 10 min.

5. Start gradient after 10 min delay, increasing the proportion of limit buffer at a rate of 0.5% per min.

6. After 50 min (total elution time), apply limit buffer for 15 min to elute the more basic peptides.

Protocol 20. *Continued*

7. When the baseline returns to its initial level equilibrate the column with starting buffer for 15 min.

Results obtained using the above procedure with samples of type I, III, and V collagens are presented in *Figure 8*.

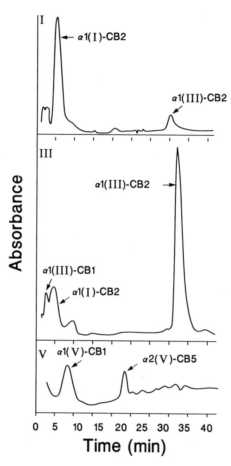

Figure 8. Elution patterns for CNBr peptides derived from type I, III, and V collagens when chromatographed on a Mono S column (Pharmacia) as described in the text. Note that there is little or no overlap of the marker peptides for each collagen. It is also apparent that many preparations of type I collagen (top portion of the figure) contain about 5% type III collagen. Similarly, preparations of type III collagen (middle portion of the figure) invariably contain about 10% type I collagen. Amount of peptides applied: 5 mg, type I collagen; 6 mg, type III collagen; 2 mg, type V collagen.

23. Size-exclusion chromatography

Protocol 21. Size-exclusion chromatography of peptides eluted from ion-exchange column

Equipment and reagents

- Buffer: 0.5 M guanidine–HCl, 50 mM Tris–HCl, pH 7.0
- Columns: two, tandemly arranged, 300 × 7.8 mm, Bio-Sil SEC 125 (Bio-Rad Laboratories)
- Apparatus: same as that described in *Protocol 20* for ion-exchange chromatography

Method

1. Freeze dry the effluent from the ion-exchange column containing peaks of marker peptides (*Protocol 20*).

2. Dissolve lyophilized material in 100 ml of buffer.

3. Inject the sample and elute isocratically at a flow rate of 0.6 ml/min using a wavelength of 220 nm for detection. An alternative wavelength for detection may be used depending on system parameters. The main concern here is that peaks be large enough to allow accurate and reproducible integration for peak area.

Since the marker peptides commonly used in these determinations have unique molecular weights, they exhibit characteristic elution intervals during size-exclusion chromatography (*Figure 9*). This serves to confirm the identity of the peptide as well as to provide data for calculation of the proportions of the collagens present in the sample.

---

24. Calculations
24.1 General considerations and formula

In calculating the results of these determinations, the integration area (A) observed for each marker peptide eluted from size-exclusion columns must be normalized for size and the proportion of the molecule it represents (36). It is convenient to normalize for size to $\alpha1(I)$-CB2, the marker peptide for type I collagen. However, the peak area for this peptide is multiplied by 1.5 since the peptide is present in only two of three chains of type I collagen. This factor is not necessary for $\alpha1(III)$-CB2 since all three chains of the type III molecule contribute to the marker peptide. Nevertheless, $\alpha1(III)$-CB2 and $\alpha1(I)$-CB2 contain 40 and 36 residues, respectively, requiring that the peak area for $\alpha1(III)$-CB2 be multiplied by 0.9 to normalize its area to that of the type I marker. If one utilizes $\alpha1(V)$-CB1 as the marker peptide for type V collagen, no correction factor is needed since the applicable normalization

67

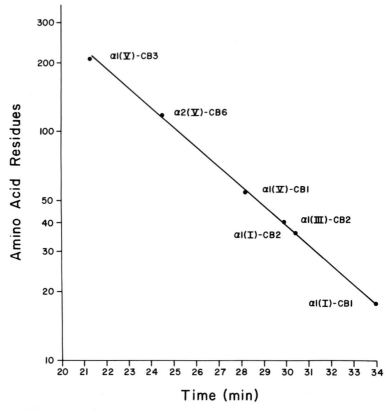

Figure 9. The elution time for different CNBr peptides from size-exclusion columns shows the conventional relationship between elution time and log of molecular weight or peptide size.

factors are equivalent to unity. If α2(V)-CB5 is used as the type V marker, the normalization factor is 3.6 (3 × 1.2) since the α2(V) chain is one-third of the type V molecule and α2(V)-CB5 contains only 30 amino acids (37). Since α1(V)-CB1 elutes during ion-exchange chromatography in a position close to α1(I)-CB2 (*Figure 7*), it may be advantageous to use the more strategically located α2(V)-CB5 peptide (*Figure 7*) for determinations on type V collagen.

The proportion of a given collagen in the sample is then calculated as a function of the normalized peak area for its marker peptide, divided by the sum of normalized peak areas for each marker peptide. The equation for type I collagen would be:

$$\% \text{ type I} = \frac{\text{Normalized } A \text{ for } \alpha1(I)CB2}{\Sigma \text{ normalized } A \text{ for I, III, V markers}} \times 100.$$

24.2 Application of method and example of results

This method has been most extensively applied to an evaluation of collagens in the wall of major human arteries. Representative analytical results from these studies on media–intima preparations are presented in *Table 8*. In all samples, the amount of type V collagen accounts for less than 2% of the total collagen. The data are therefore presented in terms of type I/III collagen ratios.

Table 8. Type I/III collagen ratios in different segments of human arteries at various ages

Age (years)	Thoracic aorta	Abdominal aorta
15–19	2.2 ± 0.8 (13)[a]	2.0 ± 0.7 (10)
30–34	1.6 ± 0.7 (12)	1.8 ± 1.0 (10)

[a] Results expressed as mean ± 1 standard deviation. The number of samples analysed is indicated in parentheses.

The data reveal that the type I and III ratios in thoracic and abdominal aorta are essentially equal and that the value of the ratio tends to decrease with age in the thoracic aorta. The relative increase in type III content in the thoracic aorta indicates a dramatic change in vessel wall chemistry over the indicated age span. The relatively large standard deviations observed in these determinations indicate a surprisingly high degree of individual variability in vessel wall chemistry.

Acknowledgements

This work was supported, in part, by USPHS NIH Grants HL-14214 and DK-39261.

References

1. Prockop, D. J., Kivirikko, K. I., Tuderman, L., and Guzman, N. A. (1979). *N. Engl. J. Med.*, **301**, 13–23.
2. van der Rest, M. and Garrone, R. (1990). *Biochimie*, **72**, 473–84.
3. van der Rest, M. and Garrone, R. (1991). *FASEB J.*, **5**, 2814–23.
4. Miller, E. J. (1984). In *Extracellular matrix biochemistry* (ed. K. A. Piez and A. H. Reddi), pp. 41–81. Elsevier, New York.
5. Miller, E. J. and Gay, S. (1982). In *Methods in enzymology* (2nd edn) (ed. L. W. Cunningham and D. W. Frederiksen), pp. 3–32. Academic Press, New York.
6. Linsenmayer, T. F. (1991). In *Cell biology of extracellular matrix* (2nd edn) (ed. E. D. Hay), pp. 7–44. Plenum Press, New York.

7. Kucharz, E. J. (1992). In *The collagens: biochemistry and pathophysiology*, pp. 5–29. Springer–Verlag, Berlin.
8. Miller, E. J. and Gay, S. (1992). In *Wound healing: biochemical and clinical aspects* (ed. I. K. Cohen, R. F. Diegelmann, and W. J. Lindblad), pp. 130–51. W. B. Saunders, Philadelphia.
9. Kuhn, K. (1987). In *Structure and function of collagen types* (ed. R. Mayne and R. E. Burgeson), pp. 1–42. Academic Press, Orlando.
10. Fessler, J. H. and Fessler, L. I. (1987). In *Structure and function of collagen types* (ed. R. Mayne and R. E. Burgeson), pp. 81–103. Academic Press, Orlando.
11. Geesin, J., Murad, S., and Pinnell, S. R. (1986). *Biochim. Biophys. Acta*, **886**, 272–4.
12. Anderson, S. M. L., McLean, W. H. I., and Elliott, R. J. (1991). *Biochem. Soc. Trans.*, **19**, 48S.
13. Geesin, J. C., Hendricks, L. J., Falkenstein, P. A., Gordon, J. S., and Berg, R. A. (1991). *Arch. Biochem. Biophys.*, **290**, 127–32.
14. Fuller, G. C. and Cutroneo, K. R. (1992). In *Wound healing. Biochemical and clinical aspects* (ed. I. K. Cohen, R. F. Diegelmann, and W. J. Lindblad), pp. 305–15. W. B. Saunders, Philadelphia.
15. Peterkofsky, B. (1982). In *Methods in enzymology* (82nd edn) (ed. L. W. Cunningham and D. W. Frederiksen), pp. 453–71. Academic Press, New York.
16. Peterkofsky, B. and Prather, W. (1992). *J. Biol. Chem.*, **267**, 5388–95.
17. Chojkier, M., Spanheimer, R., and Peterkofsky, B. (1983). *J. Clin. Invest.*, **72**, 826–35.
18. Haralson, M. A., Jacobson, H. R., and Hoover, R. L. (1987). *Lab. Invest.*, **57**, 513–23.
19. Haralson, M. A., Mitchell, W. M., Rhodes, R. K., and Miller, E. J. (1984). *Arch. Biochem. Biophys.*, **229**, 509–18.
20. Haralson, M. A., Federspiel, S. J., Martinez-Hernandez, A., Rhodes, R. K., and Miller, E. J. (1985). *Biochemistry*, **24**, 5792–7.
21. Rupard, H. J., DiMari, S. J., Damjanov, I., and Haralson, M. A. (1988). *Am. J. Pathol.*, **133**, 316–26.
22. Creely, J. J., Commers, P. A., and Haralson, M. A. (1988). *Connect. tissue Res.*, **18**, 107–22.
23. Leheup, B. P., Federspiel, S. J., Guerry-Force, M. L., Wetherall, N. T., Commers, P. A., DiMari, S. J., and Haralson, M. A. (1989). *Lab. Invest.*, **60**, 791–807.
24. Haralson, M. A., Gibbs, S. R., and DiMari, S. J. (1990). *Ann. N.Y. Acad. Sci.*, **580**, 498–500.
25. Creely, J. J., DiMari, S. J., Howe, A. M., and Haralson, M. A. (1992). *Am. J. Pathol.*, **140**, 45–55.
26. Creely, J. J., DiMari, S. J., Howe, A. M., Hyde, C. P., and Haralson, M. A. (1990). *Am. J. Pathol.*, **136**, 1247–57.
27. Federspiel, S. J., DiMari, S. J., Guerry-Force, M. L., and Haralson, M. A. (1990). *Lab. Invest.*, **63**, 455–66.
28. Federspiel, S. J., DiMari, S. J., Howe, A. M., Guerry-Force, M. L., and Haralson, M. A. (1991). *Lab. Invest.*, **64**, 463–73.
29. Federspiel, S. J., DiMari, S. J., Howe, A. M., Guerry-Force, M. L., and Haralson, M. A. (1991). *Lab. Invest.*, **65**, 441–50.
30. DiMari, S. J., Howe, A. M., and Haralson, M. A. (1991). *Am. J. Respir. Cell Molec. Biol.*, **4**, 455–62.

31. Harris, R. C., Haralson, M. A., and Badr, K. F. (1992). *Lab. Invest.*, **66**, 548–54.
32. Klasson, S. C., Klein, L. N., DiMari, S. J., and Haralson, M. A. (1986). *Collagen Relat. Res.*, **12**, 397–408.
33. Miller, E. J. and Rhodes, R. K. (1982). In *Methods in enzymology* (82nd edn) (ed. L. W. Cunningham, and D. W. Frederiksen), pp. 33–64. Academic Press, New York.
34. Peterkofsky, B. and Prather, W. (1992). *J. Biol. Chem.*, **267**, 5388–95.
35. Hames, B. D. (1990). In *Gel electrophoresis of proteins* (2nd edn) (ed. B. D. Hames and D. Rickwood), pp. 1–147. IRL Press, Oxford.
36. Miller, E. J., Furuto D. K., and Narkates, A. J. (1991). *Anal. Biochem.*, **196**, 54–60.
37. Rhodes, R. K., Gibson, K. D., and Miller, E. J. (1981). *Biochemistry*, **20**, 3117–21.

3

The collagens of cartilage (types II, IX, X, and XI) and the type IX-related collagens of other tissues (types XII and XIV)

RICHARD MAYNE, MICHEL VAN DER REST, PETER BRUCKNER, and THOMAS M. SCHMID

1. Introduction

Cartilage is probably the most extensively studied of all connective tissues, and the functions (reviewed in references 1–3) of the different collagen types within the matrix are beginning to be understood. Collagen fibrils of cartilage are assembled from type II and type XI collagen with type IX collagen being located on the surface of the fibrils (see *Figure 1* for the structure of cartilage collagens and procollagens). Type X collagen is synthesized only by hypertrophic cartilage and becomes associated with the collagen fibrils during development (4). Its function is unknown. Recently, two other collagens (designated type XII and type XIV) which are structurally related to type IX collagen and that are present in several connective tissues have been identified (2, 3). These two collagens are also present in articular cartilage but perhaps not in other cartilages (5). The methods used for isolating and characterizing type XII and type XIV collagens are similar to those used for type IX collagen, and these collagens have therefore been included in this chapter.

2. Sources

Cartilage collagens are isolated from a variety of species and anatomical locations, with the choice of tissue usually being determined by its availability in large quantity and by relevance to clinical conditions such as osteoarthritis. The sterna of adult chickens are often used (6) and usually can be obtained in quantity from a local processing plant. Successful isolation of cartilage collagens (types II, IX, and XI) can also be accomplished from bovine, porcine, and human articular cartilages and bovine nasal septum. Preferably tissues from fetal or young animals should be used as, with increasing age, the

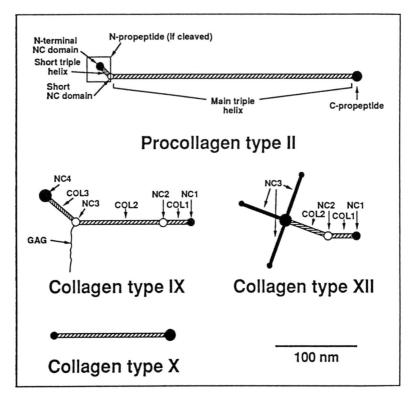

Figure 1. Diagram showing the structural organization of cartilage collagens. For type II procollagen specific cleavages both of the N- and C-propeptides give rise to type II collagen. Type XI procollagen is not included in the diagram although it is generally similar to type II procollagen (36–38). Tissue processing of type XI procollagen is still poorly understood and it is possible that the N-propeptide is not completely removed from the tissue form of the molecule. Note the location of the collagenous and non-collagenous domains of type IX and type XII collagen.

collagen becomes essentially insoluble even after extensive pepsin digestion. Another important source is the transplantable (Swarm) chondrosarcoma of rats (7), and the procedures used to passage and harvest the tumour are briefly described in Section 2.3.

2.1 Lathyrism

To increase the solubility of cartilage collagens, it is often useful to induce lathyrism during a period of rapid growth by the administration of BAPN, which inhibits lysyl oxidase activity, the first reaction required for cross-link formation. For chickens, one week after hatching, BAPN is usually administered for 2–3 weeks at a concentration of 0.02% in the drinking water. Within about one week, chickens will show evidence of weakened leg joints and eventually will have difficulty in supporting their weight. However, the dosage of BAPN required will depend on the growth rate of the chickens, and chickens bred for very fast growth will require proportionately less BAPN. It is also possible to administer BAPN *in ovo* and obtain extensive lathyrism of the developing sternum. It is recommended that a solution (0.1 ml) of BAPN be injected into the egg consecutively on days 14 (1 mg), 15 (1 mg), and 16 (2 mg), and the embryonic sterna dissected out from the surviving embryos on day 17. For rats, it is necessary to include BAPN in the diet. To induce lathyrism in the transplantable chondrosarcoma, a dosage of 0.3–0.5% (w/w) is recommended for 2–4 weeks during tumour growth.

2.2 Cultured cells

Embryonic chick sterna or vertebra are useful sources of chondrocytes that will grow well in culture and retain their phenotypic properties. Chondrocytes must first be dissociated from their matrix using a mixture of trypsin and collagenase. The methodology for chick chondrocyte isolation and culture remains essentially unchanged after more than two decades and is described in detail elsewhere (8). The only improvements over the original method are attempts to select against contaminating fibroblasts. This can be achieved by:

(a) collecting 'floaters' in which the cells are first preplated for a few days without feeding; in these conditions small clusters of the most chondrocytic cells float off into the medium and are collected by centrifugation, briefly treated with trypsin and replated;

(b) removal of the perichondrium by predigestion for 15 min in trypsin/collagenase.

Chick chondrocytes, if grown properly in culture, will maintain their phenotypic properties for many generations (that is they will continue to synthesize type II collagen plus cartilage-specific proteoglycan and to accumulate a matrix). It is also possible, using a similar methodology, to culture chondrocytes from adult rabbit articular cartilage (9) and adult pig articular cartilage (10). A key development in chondrocyte culture was the recognition that mammalian cells will retain their phenotypic properties if grown in suspension either in agarose or in a collagen gel. This technique has been successfully used both for rabbit and pig articular cartilage cells (9, 11) and also to obtain long-term cultures of the rat chondrosarcoma (12), although the latter cells only grow slowly. Permanent cell lines of chondrocytic cells were recently selected from a line of cells derived from fetal rat calvaria (13) and by transformation of fetal rat costal chondrocytes (14). Also, human epiphyseal chondrocytes were recently successfully grown in suspension on the surface of an agarose gel making it possible to investigate directly potential mutations of the various collagen types that give rise to the human chondrodystrophies (15).

Another recent method for the successful long-term culture of mammalian chondrocytes is growth within alginate beads (45). This method has the advantage that chondrocytes can be released easily from the matrix by chelation of Ca^{2+}.

2.3 Transplantable rat chondrosarcoma

Small amounts of matrix are shaved from the surface of the tumour, diced, and injected subcutaneously using a cancer implant needle (Model 7927, Popper and Sons, New Hyde Park, NY 11040). Within three to four weeks the tumour grows to the size of a small walnut. However, histologic analysis shows that the centre of the tumour is often necrotic and is best avoided. Large amounts of cartilage can be cut, however, from the surface of the tumour with a scalpel, and the host connective tissue subsequently dissected away. It is also possible to store the isolated chondrocytes for extended periods in liquid nitrogen. The following procedure provided by Dr Jeffrey Stevens (Division of Clinical Immunology and Rheumatology, School of Medicine, University of Alabama at Birmingham, USA) is used for transplantation.

Protocol 1. Transplantation of rat chondrosarcoma

Equipment and reagents

- Digestion buffer: Ca^{2+}- and Mg^{2+}-free Earle's balanced salt solution containing 0.25% trypsin and 0.01% collagenase
- Lens paper
- Minimal essential medium (MEM) + 10% horse serum
- Serum-free MEM
- Freezing medium: MEM + 10% fetal bovine serum + 10% dimethyl sulfoxide (DMSO)

Method

1. Incubate small cubes of tumour in digestion buffer for 2 h at 32°C with shaking.

2. Repeatedly pipette the digested tumour up and down in a 5 ml pipette to dissociate the cells, and subsequently filter through lens paper to give a single cell suspension.

3. Lightly centrifuge the cells at 800 g for 10 min at 4°C. Wash once in Minimal Essential Medium (MEM) plus 10% horse serum to inactivate the trypsin and then wash 3 × with MEM but without serum.

4. Suspend the cells in freezing medium (MEM + 10% fetal bovine serum + 10% dimethyl sulfoxide) to a final concentration of 1.0×10^7 cells/ml.

5. Freeze the cells at −70°C, transfer to liquid nitrogen, and store.

6. Cells can be recovered by washing with MEM + 10% fetal bovine serum and reinjected into a host to give a tumour.

---

3. Isolation and separation of cartilage collagens (types II, IX, X, and XI)

The collagen of adult cartilages is usually extensively cross-linked and cannot be solubilized except by pepsin digestion which cleaves away the cross-linking sites located in the telopeptides.

3.1.1 Solubilization by pepsin digestion

The following protocol is suitable for all hyaline cartilages and is adapted from earlier work (6, 16).

Protocol 2. Solubilization of cartilage collagens by pepsin digestion

Equipment and reagents

- Extraction buffer: 4 M guanidine–HCl, 0.05 M Tris–HCl, pH 7.5
- Miniblender of household Osterizer
- 0.2 M NaCl, 0.5 M HOAc
- Pepsin (p-7012 Sigma)
- 5 M NaOH
- 250 ml centrifuge bottles
- 0.5 M HOAc
- Dialysis solutions
  - —0.7 M NaCl, 0.5 M HOAc
  - —0.9 M NaCl, 0.5 M HOAc
  - —1.2 M NaCl, 0.5 M HOAc
  - —2.0 M NaCl, 0.5 M HOAc
  - —0.1 M HOAc

Method

1. Remove the perichondrium from 500 g wet wt of cartilage and chop the cartilage into 1 mm cubes with a scalpel.
2. Wash small cubes of cartilage 3 times with deionized H_2O, homogenize in extraction buffer, using the miniblender of a household Osterizer (this is far more effective than a Brinkmann tissue homogenizer) and stir at 4°C overnight. Another method which can be used is to homogenize the tissue with a freezer mill (see Section 6.1 in this chapter).
3. Centrifuge the extract for 1 h at 14 000 g and re-extract the pellet with a further 1500 ml of extraction buffer, overnight at 4°C.
4. Wash the pellet 4 times with cold H_2O, resuspend in 0.2 M NaCl, 0.5 M HOAc to give a total volume of 1800 ml, and digest with pepsin (1 mg/ml) with stirring overnight at 4°C.
5. Centrifuge the suspension at 14 000 g for 30 min at 4°C and adjust the supernatant to pH 8.0 by the *slow* addition of 5 M NaOH with rapid stirring in order to inactivate the pepsin. (It is also possible to perform a second pepsin digestion but the product is usually slightly more degraded than that obtained in the first extraction when examined by SDS-PAGE.)
6. Dialyse the solution extensively for 3 days at 4°C against three changes of 0.7 M NaCl, 0.5 M HOAc to selectively precipitate *type II collagen*.

Protocol 2. *Continued*

7. Collect the precipitate by centrifugation at 14000 *g* for 30 min at 4°C and dialyse the solution against 1.2 M NaCl, 0.5 M HOAc for several days at 4°C.

8. Collect the precipitate by centrifugation, and dialyse the solution against 2.0 M NaCl, 0.5 M HOAc for several days to precipitate *type IX collagen*.

9. To obtain *type XI collagen*, dissolve the precipitate at 1.2 M NaCl in 200 ml of 0.5 M HOAc and subsequently dialyse first against 0.9 M NaCl, 0.5 M HOAc and then against 1.2 M NaCl, 0.5 M HOAc. The precipitate obtained between 0.9 M NaCl and 1.2 M NaCl should contain a highly purified fraction of type XI collagen on analysis by SDS-PAGE (see *Figure 2*).

10. Dissolve all precipitates in 0.5 M HOAc. Dialyse extensively at 4°C for 4 days against 0.1 M HOAc, and then lyophilize.

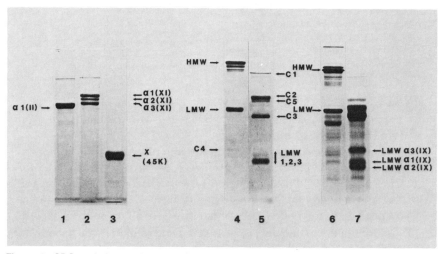

Figure 2. SDS–gel electrophoresis of cartilage collagens solubilized by pepsin digestion. Lane 1: α1(II) chain of chicken type II collagen. Lane 2: chicken type XI collagen [note that the α3(XI) chains migrate slightly slower than the α1(II) chain]. Lane 3: chicken type X collagen after pepsin digestion. Lane 4: chicken type IX collagen after pepsin digestion without reduction. Note two major bands designated HMW (includes the COL2 and COL3 domains, see *Figure 1*) and LMW (includes the COL1 domain) so that pepsin cleavage occurs across the molecule at NC2. The C4 fragment arises from the COL3 domain and is not disulfide-bonded to this domain (17). Lane 5: chicken type IX after reduction. The location of these fragments (C1, C2, C3, C5, and LMW 1, 2, and 3) within type IX collagen is known (17). Lane 6: type IX collagen from bovine nasal septum. Note that bovine HMW migrates faster than chicken HMW and additional bands are present below LMW. Lane 7: after reduction, the fragments of bovine HMW are not fully resolved whereas LMW gives rise to three bands whose identity is known from N-terminal amino acid sequencing (M. van der Rest and D. Herbage, unpublished results).

Comments
(a) Reprecipitation between 0.9 M NaCl, 0.5 M HOAc and 1.2 M NaCl, 0.5 M HOAc at a higher concentration of collagen is a highly effective means of obtaining a preparation of type XI collagen free of type II collagen and type IX collagen.

(b) For some strains of rapidly growing chickens, we have not obtained type XI collagen from adult sternal cartilage after pepsin digestion.

(c) If hypertrophic cartilage is used, type X collagen will be present in the precipitate obtained at 2.0 M NaCl, 0.5 M HOAc. The molecular weight of the pepsin-resistant fragment of type X collagen is 45 000, and this fragment can be easily separated in large amounts from the other collagen chains by gel filtration using a column of Bio-Gel A–1.5 m, 200–400 mesh (Bio-Rad Laboratories) and elution of the column with 1 M CaCl$_2$, pH 7.4 (6).

3.1.2 Characterization of pepsin-solubilized collagen

Usually the molecules are characterized initially by SDS-PAGE (illustrated in *Figure 2* for a selection of chicken and bovine collagens). Several important observations include:

(a) The type II collagen fraction should not contain a band migrating slightly faster than α1(II) which would indicate the presence of an α2(I) chain and hence contamination with type I collagen. (This usually arises from a failure to completely dissect away the perichondrium during initial tissue preparation.)

(b) For type XI collagen, the 1α, 2α, and 3α bands should be in a 1:1:1 proportion. If the 3α is in excess of the other two bands, this indicates the presence of some type II collagen, and the differential salt precipitation step between 0.9 M and 1.2 M NaCl, 0.5 M HOAc should be repeated.

(c) For chicken type IX collagen two bands are obtained (designated HMW and LMW) after pepsin digestion and similar major bands are obtained for mammalian type IX collagen. However, chicken type IX gives a series of well-defined peptides after reduction (called C1, C2, C3, C4, C5) whose organization within intact type IX collagen is known (17). In contrast, mammalian type IX does not give a discrete set of fragments on reduction and further additional cleavages by pepsin occur at NC3 so that fragments of COL3 are differentially cleaved from HMW (18,19; see *Figure 1*).

(d) The same LMW fraction is obtained from chicken and mammals, and after reduction of bovine LMW three bands are obtained whose identity is known from N-terminal amino acid sequencing (M. van der Rest and D. Herbage, unpublished results).

3.2 Solubilization and purification of collagens without pepsin digestion

Some collagen can usually be extracted either from fetal cartilages or from lathyritic cartilage using 0.5–1.0 M NaCl, 0.05 M Tris–HCl, pH 7.4 (7, 20). However, some proteoglycan and type IX collagen is also extracted and must be removed by passage over DEAE–cellulose to which type II collagen does not bind (20).

3.2.1 Type II collagen

Highly purified type II collagen with intact telopeptides is required for studies of fibril formation *in vitro* (7) and for the production of monoclonal antibodies that uniquely recognize lathyritic but not pepsin-digested type II collagen (*Table 1*). The procedure described in detail in reference 7 is recommended for lathyritic cartilage either from rat chondrosarcoma or chicken sternum. We have found that the precipitate obtained between 4% and 5% NaCl, 0.5 M HOAc gives the most highly purified preparation of type II collagen as judged by SDS-PAGE (*Figure 3*).

3.2.2 Type IX collagen

Intact type IX collagen can be prepared in limited amounts from embryonic chick cartilage (21, 22), vitreous of the eye (23), or from lathyritic chicken cartilage (24). More recent methods involve the use of cell cultures of chondrocytes from which intact type IX collagen can be isolated in good yield from the culture medium (25).

Purification of native type IX collagen in high yield represents somewhat of a biochemical challenge. In normal cartilages, the molecule is extensively cross-linked to type II collagen and is difficult to extract unless cross-linking is inhibited by BAPN. Intact type IX collagen also has a single chondroitin sulfate chain of variable length located at NC3 (see *Figure 1*) which makes the molecule highly anionic. Extraction of type IX collagen from cartilage of lathyritic chickens results in a preparation which also contains some type II collagen probably due to incomplete inhibition of cross-linking by BAPN (24). Better preparations can be obtained either from organ cultures of small pieces of cartilage kept in the presence of BAPN or from chick embryo sternal chondrocytes kept in agarose gel culture for two weeks in the presence of BAPN. However, even in agarose cultures cross-linking cannot entirely be prevented, presumably due to the formation of cross-links not derived from allysyl residues. Further cross-linking may occur to a large extent when agarose cultures are frozen for several weeks to months.

Table 1. Monoclonal antibodies to chicken cartilage collagens

	Reference	Comments
Type II collagen		
II$_6$B$_3$	39	Cross-reacts with mammalian species including human. Reacts by immunoblotting with the TC^A fragment after cleavage with mammalian collagenase. Also recognizes CB-11 on immunoblotting of a CNBr digest of type II collagen after SDS-PAGE (46).
5B2	46	Cross-reacts with human and mouse. Reacts with lathyritic but not pepsin-digested type II collagen and recognizes the TC^A fragment on immunoblotting after cleavage with mammalian collagenase. Recognizes the 3α chain of type XI collagen on immunoblotting.
2B1	40	Cross-reacts with human and mouse. Recognizes an epitope in the TC^B fragment on immunoblotting after cleavage with mammalian collagenase. Does not recognize the 3α chain of type XI collagen on immunoblotting (46).
CII-D3	41	Recognizes an epitope present in chick, mouse, rat, bovine, and human type II collagen. It also recognizes both α1(II) and the 3α chain of type XI collagen by immunoblotting.
CII-C1	42	Recognizes an epitope present in chick, rat, bovine, human, and mouse. Epitope localized to amino acids 316–333 present in CB-11 of the α1(II) chain.
Type IX collagen		
2C2	32	Recognizes the carboxyl-terminus (NC2 domain, *Figure 1*) of the HMW fragment of type IX collagen. Recognizes separately the α1(IX) (C2 fragment) and α2(IX) (C3 fragment) chains on immunoblotting (Brewton, R. G. and Mayne, R., unpublished observations).
4D6	32	Recognizes the amino terminus (NC4 domain) of the HMW fragment of type IX collagen. Requires fragments of the α1(IX) (C2 fragment) and α3(IX) (C5 fragment) chains to interact to form the epitope during immunoblotting (31). Does not recognize type IX collagen if the downstream promoter is used to synthesize the α1(IX) chain (31).
Type X collagen		
AC9	43	Recognizes an epitope located in a triple-helical sequence by rotary shadowing. Weakly reactive to α1(X) chains during immunoblotting.
1A6 6F6	44	Both antibodies recognize an epitope located in the amino terminal (non-triple helical) domain of type X collagen. Both antibodies are reactive during immunoblotting.
Type XII collagen		
75d7	34	Prepared against a synthetic peptide located in the NC1 domain of the α1(XII) chain. Reactive during immunoblotting and can be used to monitor the purification of type XII collagen from gel filtration columns.

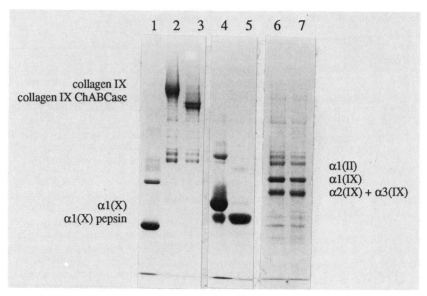

Figure 3. SDS-PAGE (4.5–15% gradient gel) of intact collagen molecules isolated from cultured chondrocytes. Lanes 1–5, without reduction; Lanes 6 and 7, with reduction using 2% β-mercaptoethanol. Lanes 2 and 6, type IX collagen; Lanes 3 and 7, type IX collagen after digestion with chondroitinase ABC; Lane 4, type X collagen; Lane 5, type X collagen after digestion with pepsin; Lane 1, chondroitinase ABC alone. Note:

(a) the reduction in molecular weight of type IX collagen after chondroitinase digestion (lanes 2 and 3) so that the α2(IX) chain now migrates with the α3(IX) chain (lanes 6 and 7);

(b) the type IX collagen preparation contains small amounts of type II and XI collagens due to the formation of cross-links not derived from allysine;

(c) type X is reduced in molecular weight after pepsin digestion (compare lanes 4 and 5).

Protocol 3. Isolation of type IX collagen from organ cultures

Equipment and reagents

- Culture medium: high glucose (4.8 g/l) DMEM supplemented with sodium ascorbate (50 μg/ml), BAPN (50–100 μg/ml), penicillin (100 units/ml), streptomycin (100 μg/ml)
- [^{14}C]proline (Sp. Act. >250 mCi/mmol) or [2,3-^{3}H]proline (Sp. Act. 25–55 Ci/mmol)
- Fetal bovine serum, 10%, dialysed (see step 4 below)
- Extraction buffer: 1 M NaCl, 100 mM Tris–HCl, pH 7.5, containing 20 mM EDTA, 10 mM NEM, 10 μM PMSF
- $(NH_4)_2SO_4$

- Storage buffer: 0.4 M NaCl, 100 mM Tris–HCl, pH 7.5
- Deionized 4 M urea
- DEAE–cellulose column (1.5 × 10 cm)
- Buffer A: 50 mM Tris–HCl, pH 7.4, 2.0 M urea
- Buffer B: Buffer A + 200 mM NaCl
- Buffer C: Buffer A + 250 mM NaCl
- Buffer D: Buffer A + 400 mM NaCl
- 1.3 % Potassium acetate in ethanol
- Polycarbonate (Nalgene) plasticware
- Polytron homogenizer

Method

1. Incubate embryonic chick sterna as whole organs, but cut larger mammalian cartilages into 0.5 mm thick slices before incubation.

2. Culture either preparation in 10 volumes of culture medium.

3. Replace the medium every 48 h.

4. Label the collagens by the addition of 1 μCi/ml [^{14}C]proline or 10 μCi/ml [^{3}H]proline. Labelling with [^{35}S]O$_4$ (10 μCi/ml) is also possible since collagen IX is a proteoglycan. Tissue viability can be improved by inclusion of fetal bovine serum (10%, dialysed) in the culture medium, but this will prevent purification of collagens released into the medium.

5. Use plasticware (polycarbonate, Nalgene) during the extraction and purification steps and perform all procedures at 4 °C.

6. Briefly homogenize the tissue (Polytron), and extract 10 g of tissue for 24 h with the extraction buffer.

7. Remove tissue debris by centrifugation (27 000 g 20 min) and add (NH$_4$)$_2$SO$_4$ (176 mg/ml) to the supernatant.

8. After at least 4 h, collect the precipitate by centrifugation (27 000 g, 30 min), and dissolve in 25 ml storage buffer.

9. Clarify the solution by centrifugation and dilute the supernatant with 25 ml of deionized 4 M urea.

10. Chromatograph the collagens on a column of DEAE–cellulose equilibrated in 200 ml buffer B.

11. Collect 5 ml fractions and monitor the eluant either at 220 nm or by scintillation counting of a radiolabelled sample.

12. Wash the column with 100 ml buffer B and then with 100 ml of buffer C.

13. Elute type IX collagen by application of 100 ml buffer D.

14. Dialyse the sample overnight against 500 ml storage buffer, and precipitate essentially pure type IX collagen by the addition of ammonium sulfate (176 mg/ml), or, for proteins from culture media, by the addition of 3 volumes of 1.3% potassium acetate in ethanol.

15. Using this procedure, organ cultures of about 10 g of cartilage usually yield 1–5 mg of type IX collagen and several hundred micrograms can be isolated from the culture medium.

i. Type IX collagen from chondrocytes grown in agarose gels

This procedure will select for type IX molecules with longer chondroitin sulfate chains, and the material eluting from the DEAE column with buffer B will also contain appreciable amounts of type IX collagen but with shorter glycosaminoglycan chains. It is possible to use this fraction to obtain highly enriched preparations of type IX collagen by eluting from DEAE–cellulose with a salt gradient and then monitoring the fractions by SDS-PAGE.

Protocol 4. Isolation of type IX collagen from chondrocytes grown in agarose gels

Equipment and reagents

- DMEM supplemented with 1 mM cysteine (for all procedures)
- PBS: 0.15 M NaCl, 10 mM sodium phosphate, pH 7.4
- Crude clostridial collagenase (Worthington, grade CLS, Boehringer–Mannheim)
- Low-melt preparative grade agarose (Bio-Rad 162–0017)
- Double strength DMEM

- Culture medium: DMEM, 10% fetal bovine serum, sodium ascorbate (50 μg/ml), BAPN (50–100 μg/ml)
- Extraction buffer: 1 M NaCl, 100 mM Tris-HCl, 20 mM EDTA, pH 7.5
- Solid NaCl
- Storage buffer: as *Protocol 3*
- DEAE–cellulose column: as *Protocol 3*
- 60 mm and 100 mm Petri dishes

Method

1. For all procedures DMEM is supplemented with 1 mM cysteine.

2. Rinse sterna from two dozen chick embryos several times in PBS, and then divide into two 60 mm Petri dishes with 3 ml DMEM containing 1–2 mg/ml of crude clostridial collagenase.

3. Digest sterna overnight at 37 °C in 5% CO_2/95% air atmosphere.

4. After collagenase digestion, combine cell suspensions, and collect the cells by centrifugation (600 *g*, 10 min, room temperature).

5. Remove the supernatant by aspiration, suspend the cells in 25 ml PBS, and wash the cells twice in PBS. Use of serum during washing is not recommended as the cells will tend to aggregate.

6. Suspend the cells in 2.5 ml DMEM, count and dilute to 4×10^6 cells/ml. Typically $6–8 \times 10^7$ cells are obtained.

7. Prepare the agarose by suspending 400 mg in 20 ml of H_2O, autoclave, and mix with 20 ml double-strength DMEM.

8. Prepare 100 mM Petri dishes by coating each dish with 5 ml of 1% agarose autoclaved in H_2O.

9. Mix the cell suspension (step 6) with an equal volume of agarose/DMEM mixture (step 7) to give a concentration of 2×10^6 cells/ml in 0.5% agarose/DMEM. Maintain at 37 °C.

10. Add 5 ml of this suspension to each of the precoated Petri dishes (step 8).

11. Solidify the agarose by brief exposure to 4 °C.

12. Incubate overnight at 37 °C in a 5% CO_2/95% air atmosphere.

13. Overlap the cultures with culture medium, and maintain cultures for two weeks replacing the medium three times per week.

14. After freezing at $-20\,^{\circ}C$ and thawing the cultures, remove the upper agarose layer, and suspend in 200 ml PBS.

15. Extract at room temperature in this solution for 20 min followed by centrifugation (27 000 g; 15 min; room temperature).

16. Extract the pellet with PBS and centrifuge as in steps 14, 15. The supernatants should not contain any appreciable amounts of collagen. Perform all further steps at $4\,^{\circ}C$.

17. Extract the pellet with 200 ml of extraction buffer for 24 h.

18. Remove the agarose by centrifugation at 27000 g for 30 min and precipitate the collagens by the slow addition of solid NaCl (35 g).

19. Collect the collagens by centrifugation at 27000 g for 30 min, dissolve in storage buffer, and purify by ion-exchange chromatography on DEAE–cellulose as described in *Protocol 3*. The yield of type IX collagen is 10–20 mg. Analysis of a preparation is shown in *Figure 3*.

3.2.3 Type X collagen

Intact type X collagen can be isolated either from the medium of chick chondrocyte cultures (26) or extracted from fetal bovine growth cartilage with 1 M NaCl, 10 mM DTT at neutral pH (27).

Protocol 5. Isolation of type X collagen from the medium of chick chondrocyte cultures

Reagents

- DMEM + 20% fetal bovine serum (FBS) + 0.01% hyaluronidase (type I–S; Sigma)
- DMEM + 20% FBS
- 0.05% Trypsin
- DMEM + 10% FBS + 50 µg/ml ascorbate, 100 µg/ml BAPN
- Protease inhibitors
- 5 mM EDTA, 1 mM NEM, 1 mM PMSF

- Ammonium sulfate (solid)
- Neutral buffer: 0.4 M NaCl, 50 mM Tris–HCl, pH 7.5
- Dialysis solutions:
  A: 0.7 M NaCl, 0.5 M HOAc
  B: 1.2 M NaCl, 0.5 M HOAc
  C: 2.0 M NaCl, 0.5 M HOAc

Method

1. Obtain cells from zones 2 and 3 of a dozen 12-day embryonic chick tibiotarsi.

2. Plate the cells at a concentration of 3×10^6 cells/100 mm plate in DMEM containing 20% fetal bovine serum (FBS) plus 0.01% hyaluronidase. (Hyaluronidase increases the plating efficiency of the cells and is removed the next day with a change of medium.)

Protocol 5. *Continued*

3. Grow cells in medium containing 20% FBS until confluency, release with 0.05% trypsin, and subculture into four new plates. To retain the differentiated phenotype of the cells (i.e. continued synthesis of type II and type X collagen) the cells must be grown at high density and the number of passages kept to a minimum (usually no more than two). Cells are cultured in 20% FBS until confluency and from a dozen tibiotarsi enough cells can be generated to obtain ten 150 mm plates after two weeks of culture.

4. At confluency, change the medium to 10% FBS, 50 μg/ml ascorbate, 100 μg/ml BAPN. The medium which contains type X collagen can then be harvested for at least two months.

5. Collect the conditioned medium at one- or two-day intervals depending on how quickly it becomes acidic. Type X collagen represents 25–50% of the total collagen synthesized and is mostly (> 90%) secreted into the medium. The cells will eventually stratify to form a layer 5–10 cells deep with the cells closest to the plastic substratum appearing hypertrophic.

6. Add protease inhibitors to the spent culture medium and add solid ammonium sulfate (176 mg/ml) to 30% saturation.

7. Centrifuge the suspension at 10 000 *g* for 30 min and dissolve the precipitate in neutral buffer.

8. Repeat step 7 to remove proteoglycans and serum proteins.

9. Dialyse the sample against solution A to precipitate type I and II collagens and subsequently against solution B to precipitate type XI collagen. Both type IX and type X collagens are precipitated with solution C, but most of the type IX collagen has already been lost since it remains with the cell layer.

i. Additional purification of type X collagen

Further purification of type X collagen is achieved by gel filtration on a column of Sephacryl S-500 (26) and subsequently by affinity chromatography on an anti-type X monoclonal antibody–Sepharose column. The native form of type X collagen (59 kDa) will interact with the Sephacryl S-500 gel matrix while the pepsin-resistant domain (45 kDa) does not interact, presumably due to the hydrophobic character of the large globular domain at the carboxyl terminus of the molecule. However, under acid conditions (0.5 M HOAc), the native form of type X also binds to the Sephacryl column. This interaction can, however, be minimized by inclusion of ethylene glycol in the chromatography buffer (26). Routinely, a yield of 50 mg of type X collagen is obtained by this procedure and analysis of a preparation is shown in *Figure 3*.

Type X collagen apparently binds to many surfaces in a non-specific manner, and for many assays to be performed successfully detergents must be included in the buffer. The minimal concentrations of detergents necessary to inhibit the binding of type X collagen to polystyrene ELISA plates are, separately, CHAPS (1–5 mg/ml), SDS (1–5 mg/ml), and Triton X-100 (0.01%). It is also important to include 1 mg/ml CHAPS during pepsin digestion to obtain quantitative recovery of small amounts of radiolabelled type X collagen after dialysis.

ii. Immunoaffinity purification of type X collagen
Type X collagen can also be purified by affinity chromatography using a column in which a monoclonal antibody (AC9, see *Table 1*) is coupled to CNBr-activated Sepharose CL-4B.

Protocol 6. Immunoaffinity chromatography of type X collagen

Reagents

- AC9 antibody
- 0.2 m Sodium bicarbonate buffer, pH 9.5
- CNBr-activated Sepharose
- Tris buffer
- 1 M NaCl, 50 mM Tris–HCl, pH 7.5, 1% Triton X-100, pH 7.5, **or**

- 1 M NaCl, 50 mM Tris–HCl, pH 7.5, 1–5 mg/ml CHAPS, pH 7.5
- Elution buffer: 2 M guanidine hydrochloride, 50 mM Tris–HCl, pH 7.5
- 2 M Tris–HCl, pH 9.0

Method

1. If not using commercially prepared CNBr-activated Sepharose, perform the activation of Sepharose reaction at 4 °C, for 1–2 min as detailed (51).

2. React 40 mg of the antibody at a concentration of 4 mg/ml in sodium bicarbonate buffer, with 10 ml packed volume of CNBr-activated Sepharose. Perform the coupling reaction at 4 °C overnight.

3. Make the slurry 1 M in Tris buffer using 2 M Tris–HCl, pH 9.0 and incubate for an additional 4 h at 25 °C to block remaining active sites on the gel. An antibody coupling efficiency of 80% gives 3.2 mg antibody/ml Sepharose.

4. Prepare protein solutions in 1 M NaCl, 50 mM Tris–HCl, 1% Triton X-100 **or** 1–5 mg/ml CHAPS, pH 7.5, and clarify by centrifugation at 10 000 g for 30 min at 4 °C before loading on the column.

5. Chromatograph samples on a 1.5 cm diameter column at a flow rate of 0.25 ml/min.

6. Add elution buffer to obtain collagen type X as a single peak. The antibody–antigen complex is also dissociated at pH values of < 2.0 or > 11.0, but is not dissociated by 3 M KSCN, pH 6.0.

3.2.4 Type II procollagen

The procedure described in detail elsewhere (28) is recommended for the isolation of type II procollagen from embryonic chicken sterna in sufficient amounts for biochemical analysis.

4. Isolation of collagen fibrils from cartilage

In cartilage, the fibrils are normally extensively cross-linked so that they have a very high tensile strength and resistance to denaturation by urea or guanidine hydrochloride, as well as to digestion by trypsin. High sheer stress will break collagen fibrils, and individual collagen molecules will protrude from the ends (29). The resulting fibril fragments are subsequently purified by a few simple washing steps but, in pure form, tend to aggregate laterally particularly after repeated pelleting at high centrifugal forces. The fibrils are composed largely of type II, IX, and XI collagens, but additional material is also present which cannot be removed after washing with chaotropic agents.

The vitreous humour is also a useful source of collagen fibrils which can be isolated from the rest of the vitreous by high-speed centrifugation. Structually, the fibrils are closely related to cartilage fibrils (30) and contain both type II and type IX collagen (23, 31).

Protocol 7. Preparation of cartilage fibril fragments

Equipment and reagents

- Homogenization buffer: 1 M NaCl, 100 mM Tris–HCl, pH 7.5, containing 20 mM EDTA, 10 mM NEM, 10 µM PMSF
- Polytron homogenizer
- Extraction buffer: 150 mM NaCl, 50 mM Tris–HCl, pH 7.5, containing **either** 8 M urea **or** 4 M guanidine–HCl
- Vortex mixer

Method

1. Finely mince 10 g of cartilage with a scalpel and suspend at 4°C in 10 ml of the homogenization buffer.

2. Homogenize the mixture by repeated and vigorous treatment with a Polytron homogenizer. Between bursts cool the mixture to 4°C.

3. Remove coarse tissue fragments by centrifugation at 27 000 g at 4°C for 20 min, and centrifuge the clear supernatant at 115 000 g at 4°C for 2 h.

4. Suspend the pellet in 10 ml extraction buffer (containing urea **or** guanidine–HCl. Vigorous shaking on a vortex mixer is required to disentangle the pelleted fibril fragments which are recovered by centrifugation at 115 000 g for 2 h at 4°C.

5. Repeat the extraction twice in buffer containing urea or guanidine hydrochloride. The fibril fragments are essentially free of soluble collagen molecules, with the yield being in the milligram range but depending on the type of cartilage used.

5. Rotary shadowing of collagens and fibrils

Electron microscopic analysis by rotary shadowing has proved to be a very useful technique in determining:

(a) the purity of a collagen preparation;

(b) the structure of type IX collagen including the location of the prominent kink at NC3 (32);

(c) the location of epitopes for monoclonal antibodies to different collagen types;

(d) the organization of fragments of type X collagen obtained after digestion with mammalian collagenase (33);

(e) the structural organization of collagen fibrils and the location of type IX collagen on the surface of the fibril (24, 29). A selection of electron micrographs of different cartilage collagens and fibrils after rotary shadowing is shown in *Figure 4*.

6. Type XII and type XIV collagens

Type XII and type XIV collagens are closely related homotrimeric molecules that are present in several non-mineralized connective tissues including skin and tendon (2, 3). In addition, type XII collagen is found in perichondrium and periosteum (34), and both type XII and type XIV are found in articular cartilage (5).

Figure 1 shows a diagram of the structure of type XII collagen in comparison to type IX collagen. Type XIV collagen has a very similar structure to type XII by rotary shadowing. The molecule of either type XII or XIV contains two triple-helical domains called COL1 and COL2. The COL1 domains of type IX, XII, and XIV are homologous in amino acid sequence (35), but the COL2 domains of type XII and XIV are unrelated to type IX collagen.

6.1 Pepsin digestion

After pepsin digestion, both type XII and XIV yield mainly two disulfide-bonded fragments, one of about 10 kDa which contains the COL1 domain, and a second of more variable size (13.5–19 kDa), which contains the COL2

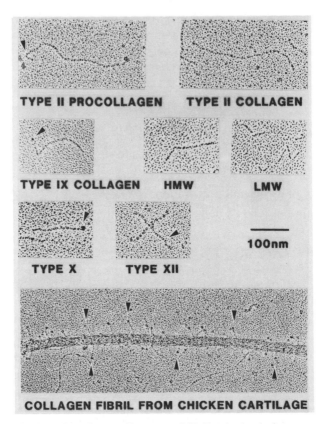

Figure 4. A selection of cartilage collagens and fibrils obtained after rotary shadowing. Note for type II procollagen the presence of a prominent kink at the amino terminus (arrowhead) and a prominent knob at the carboxyl terminus. For type IX collagen, the NC4 domain is observed as a knob (arrowhead) and a prominent kink is observed at NC3 (see *Figure 1*). For HMW the NC4 domain is lost after pepsin digestion and cleavage occurs at NC2 releasing the LMW fragment which is observed as a short rod. For type X a knob is observed at the carboxyl terminus (arrowhead) and for type XII there is a single collagen arm with a kink at NC2 (arrowhead) and three prominent non-collagenous arms (NC3 domains). For the collagen fibril from chicken cartilage, type IX molecules can be observed along the surface of the molecule with the COL3 and NC4 domains projecting from the fibril (arrowheads).

domain. Most experiments are performed with the COL1 fragment which is obtained during differential salt precipitation between 1.2 and 2.0 M NaCl, 0.5 M HOAc. It can easily be detected in a crude pepsin extract of tissues by two-dimensional gel electrophoresis when the first dimension is run in the unreduced state and the second after reduction with β-mercaptoethanol (*Figure 5*).

Protocol 8. Isolation and purification of type XII and type XIV collagen

Equipment and reagents

- Freezer mill (SPEX Ind., Metuchen, NJ)
- 10 mM EDTA, 5 mM NEM, 0.5 mM PMSF, 10 μM p-aminobenzamidine
- 0.5 M HOAc
- Pepsin (Sigma)
- 5 M NaOH
- Dialysis solutions:
  —0.7 M NaCl, 0.5 M HOAc
  —1.2 M NaCl, 0.5 M HOAc
  —2.0 M NaCl, 0.5 M HOAc

Method

1. Grind 30 g (wet weight) of tissue in liquid N_2 in a freezer mill.

2. Suspend the powder in 300 ml of distilled water containing 10 mM EDTA, 5 mM NEM, 0.5 mM PMSF, and 10 μM p-aminobenzamidine.

3. Wash the homogenate five times with the same solution and pellet between each wash by centrifugation at 12 000 g at 4°C for 30 min.

4. Suspend the residue in 450 ml of 0.5 M HOAc and treat with pepsin (0.5 mg/ml) for 24 h at 4°C.

5. Adjust the pH to 8.5 by adding 5 M NaOH.

6. Centrifuge (12 000 g, 30 min at 4°C) to eliminate insoluble material.

7. Fractionate the supernatant by differential salt precipitation by dialysing against increasing concentrations of NaCl (0.7, 1.2, and 2.0 M) in 0.5 M HOAc at 4°C.

8. Collect the precipitates at each step by centrifugation (12 000 g, 45 min at 4°C), dialyse against 0.5 M HOAc, and store frozen at −20°C.

6.1.1 Additional purification of types XII and XIV collagen

Further purification of the fragments is achieved by reverse-phase HPLC in which the fragments are first run without reduction and then separated from contaminating peptides after reduction with β-mercaptoethanol and S-pyridyl-ethylation as described below. The peptide fragments from the 2.0 M NaCl precipitate are separated by HPLC using 10 mM heptafluorobutyric acid as the ion-pairing agent and a gradient of acetonitrile (12.8–44.8%, 60 min). We use a C18 Vydac TP 201 (4.6 × 250 mm) column (The Separation Group) protected with a guard column filled with pellicular C18 matrix (Waters Associates). Other C18 columns (30 nm pore size) give comparable results, and the elution profile normally shows a very broad peak. One minute fractions are collected and analysed by gel electrophoresis under non-reducing and reducing conditions. The fractions containing reducible fragments (10–16 kDa after reduction) are dried and dissolved in a solution of 6 M guanidine–HCl buffered with 0.25 M Tris–HCl, pH 8.5, containing 1 mM

Richard Mayne et al.

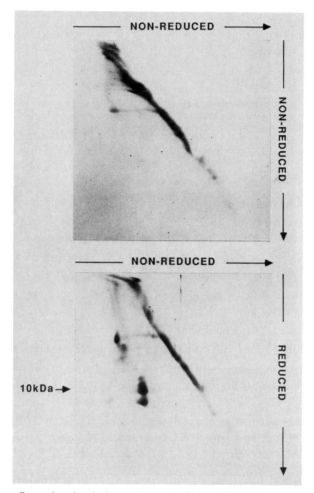

Figure 5. Two-dimensional gel electrophoresis of the 2 M NaCl precipitate of type XII collagen isolated from bovine periodontal ligament. First dimension, no reduction; second dimension, no reduction (top), with reduction (bottom). The 10 kDa fragment is indicated on the left of the figure. The molecular mass is based on CNBr-derived peptides from the α1(I) collagen chain (not shown).

EDTA. Reduction is performed in the presence of 0.5% β-mercaptoethanol for 2 h at room temperature in the dark under nitrogen. 4-Vinylpyridine (Sigma) is added (40 μl/ml) and the solution further incubated for 2 h. (The *S*-pyridylethylated cysteines are easier to identify in microsequencing than other cysteine derivatives.) The fractions are then loaded on the HPLC column and runs performed using the elution conditions as described above. Occasionally, we have noticed that peptides obtained in this manner are not pure and we have rerun the fragment on the same column with the same gradient, but in the presence of 9 mM trifluoroacetic acid as an ion-pairing agent. This is usually sufficient to remove any contaminant.

After this procedure, the amino-termini of the COL1 domains from either α1(XII) or α1(XIV) chains are blocked. It is, however, possible to 'trim' the N-terminus of native fragments with trypsin so that denatured chains obtained after purification are suitable for sequencing (B. Dublet and M. van der Rest, unpublished).

This purification method gives very reproducible fragments of COL1 for the α1(XII) and α1(XIV) chains, but the COL2 domain is less reproducible in size apparently because the cysteine residues are located outside the triple-helical domain and are variably cleaved from COL2 after pepsin digestion. The above procedure has been applied successfully to skin and tendon of embryonic chicks and fetal calves as well as the periodontal ligament of adult cattle.

6.2 Purification of intact type XII collagen and similar molecules

To purify intact type XII collagen (or from some tissues what is probably a mixture of type XII and type XIV collagens) we have used a two-step chromatography procedure involving affinity chromatography on concanavalin A–Sepharose followed by Sephacryl S-500 gel permeation chromatography.

Protocol 9. Purification of type XII collagen on Sephacryl S-500

Equipment and reagents

- Buffer 1: 50 mM Tris–HCl, pH 8.0, containing 1 mM NaCl and proteinase inhibitors (1 mM PMSF, 1 mM *p*-aminobenzamidine, 10 mM NEM, 10 mM EDTA)
- Buffer 2: 50 mM Tris–HCl, pH 7.4
- Buffer 3: 50 mM Tris–HCl, pH 7.4, 0.5 M NaCl

- Buffer 4: 50 mM Tris–HCl, pH 7.4, 0.5 M NaCl, 1 M methyl α-D-glucopyranoside
- Concanavalin A–Sepharose (Pharmacia) column (2.6 × 12 cm)
- Sephacryl S-500 column (2.6 × 90 cm)

Method

1. Extract approximately 30 g of tissue (leg tendons dissected from 17-day chick embryos, or hairless fetal calf skin) for 48 h at 4 °C with Buffer 1.

2. Centrifuge the suspension at 12 000 *g* for 20 min at 4 °C, and store the supernatant at − 20 °C.

3. Dilute the extract by adding an equal volume of Buffer 2, and chromatograph on the Con A–Sepharose column (4 ml/h) at room temperature. (We have observed that cleaner preparations can be obtained when new gel is used.)

4. Wash the column with Buffer 3 until the absorbance at 214 nm reaches zero and then elute with Buffer 4.

Protocol 9. *Continued*

5. Collect 2 ml fractions and analyse as described (see below).

6. Pool the fractions containing type XII collagen and chromatograph on a Sephacryl S-500 column (16 ml/h) at room temperature in Buffer 3.

7. Monitor the effluent at 214 nm, collect 8-ml fractions, and store desired fractions at −20 °C.

We have found that on occasion type XII will bind irreversibly to the Sephacryl S-500 beads and we are, therefore, developing an alternative procedure in which type XII-like collagens are eluted from CM-cellulose as described below (47) with additional modifications (48).

Protocol 10. Purification of type XII collagen by CM-cellulose chromatography

Equipment and reagents

- Isotonic sucrose
- 1 M NaCl
- 0.25 M NaCl
- Equilibration buffer: 25 mM NaCl, 10 mM Tris–HCl, pH 7.4, 1% isopropanol (v/v)

- 1 M methyl α-D-glucopyranoside
- CM-cellulose column
- Concanavalin A–Sepharose column (see *Protocol 9*)

Method

1. Suspend powdered fetal skin in 10 volumes of isotonic sucrose for 15 min and after centrifugation, extract the pellet with 1 M NaCl for 7 h and centrifuge at 20000 *g* for 30 min, all at 4 °C. The following steps are performed at room temperature.

2. Dilute the supernatant 12.5 times and apply to an equilibration-buffered CM-cellulose column.

3. Elute the bound material with 0.25 M NaCl.

4. Adjust the NaCl concentration of the desired fractions to 0.5 M, chromatograph on a Con A–Sepharose column.

5. Elute the retained proteins with 1 M methyl α-D-glucopyranoside dissolved in the buffer. This procedure has been attempted with embryonic chick tendons but with limited success.

Other methods were also recently described for the successful isolation of type XII and/or type XIV collagen which involve binding and elution from heparin-Sepharose (49, 50).

6.2.1 Characterization of type XII collagen

In order to characterize chicken type XII collagen, monoclonal antibody 75d7 is used extensively to follow column purification by immunoblotting (*Table*

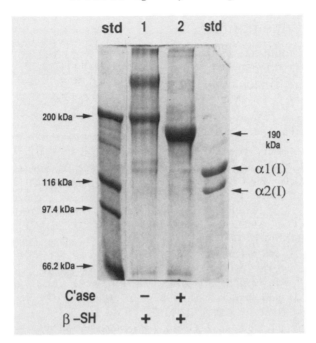

Figure 6. Gel electrophoresis of collagenase digestion of type XII collagen from chicken embryo tendon. A fraction after Sephacryl S-500 chromatography which contained type XII collagen was treated with bacterial collagenase (C'ase) and β-mercaptoethanol (β-SH) as indicated. Lane 1, undigested control; lane 2, digested sample. Molecular weight markers are shown at the extremities of the gel; right, pepsinized type I collagen; left, globular standards.

1). The tissue form of type XII collagen is characterized by bands migrating at 220 kDa and 270–290 kDa under reducing conditions which, after collagenase digestion, give rise to a single band of 190 kDa (*Figure 6*).

This method has been used to demonstrate that chick tendon contains mostly type XII collagen, but other tissues appear to contain a mixture of type XII and type XIV molecules. The yield from fetal calf skin was 30 μg/g of tissue. Preparations can be obtained of sufficient purity to be analysed by rotary shadowing (*Figure 4*) and to prepare tryptic peptides for amino acid sequencing.

7. Monoclonal antibodies against chicken cartilage collagens

Table 1 lists the monoclonal antibodies currently prepared against chicken cartilage collagens. All these antibodies are now characterized by more than one laboratory and have been shown to be specific for a single collagen type. The hybridomas are stable and can be used to prepare large amounts of antibody.

Acknowledgements

The authors would like to thank Pauline M. Mayne, Maxine H. Rudolph, Zhao Xia Ren, Randolph G. Brewton, and Bernard Dublet, for their assistance in the preparation of this manuscript.

References

1. Mayne, R. (1990). *Ann. N.Y. Acad. Sci.*, **599**, 39.
2. van der Rest, M. and Garrone, R. (1990). *Biochimie*, **72**, 473.
3. Gordon, M. K. and Olsen, B. R. (1990). *Curr. Opin. Cell Biol.*, **2**, 833.
4. Chen, Q., Gibney, E., Fitch, J. M., Linsenmayer, C., Schmid, T. M., and Linsenmayer, T. F. (1990). *Proc. Natl. Acad. Sci. USA*, **87**, 8046.
5. Lunstrum, G. P., McDonough, A. M., Keene, D. R., Morris, N. P., and Burgeson, R. E. (1990). *J. Cell Biol.*, **113**, 963.
6. Reese, C. A. and Mayne, R. (1981). *Biochemistry*, **20**, 5443.
7. Lee, S. L. and Piez, K. A. (1983). *Collagen Rel. Res.*, **3**, 89.
8. Chacko, S., Abbott, J., and Holtzer, H. (1969). *J. Exp. Med.*, **130**, 417.
9. Benya, P. D. and Shaffer, J. D. (1982). *Cell*, **30**, 215.
10. Watt, F. M. (1988). *J. Cell Sci.*, **89**, 373.
11. Watt, F. M. and Duddhia, J. (1988). *Differentiation*, **38**, 140.
12. Kucharska, A. M., Kuettner, K. E., and Kimura, J. H. (1990). *J. Orthopaed. Res.*, **8**, 781.
13. Grigoriadis, A. E., Aubin, J. E., and Heersche, J. N. M. (1989). *Endocrinology*, **125**, 2103.
14. Horton, W. E., Cleveland, J., Rapp, U., Nemuth, G., Bolander, M., Doege, K., Yamada, Y., and Hassell, J. R. (1989). *Exp. Cell Res.*, **178**, 457.
15. Aulthouse, A. L., Beck, M., Griffey, E., Sanford, J., Arden, K., Machado, M. A., and Horton, W. A. (1989). *In vitro*, **25**, 659.
16. Burgeson, R. E. and Hollister, D. W. (1979). *Biochem. Biophys. Res. Commun.*, **87**, 1124.
17. van der Rest, M. and Mayne, R. (1987). In *Structure and function of collagen types* (ed. R. Mayne, and R. E. Burgeson), p. 195. Academic Press, Orlando.
18. Ricard-Blum, S., Tollier, J., Garrone, R., and Herbage, D. (1985). *J. Cell Biochem.*, **27**, 347.
19. Furuto, D., Bhown, A. S., and Miller, E. J. (1989). *Matrix*, **9**, 353.
20. Miller, E. J. (1971). *Biochemistry*, **10**, 1652.
21. Noro, A., Kimata, K., Oike, Y., Shinomura, T., Maeda, N., Yano, S., Takahashi, N., and Suzuki, S. (1983). *J. Biol. Chem.*, **258**, 9323.
22. Bruckner, P., Vaughan, L., and Winterhalter, K. H. (1985). *Proc. Natl Acad. Sci. USA*, **82**, 2608.
23. Yada, T., Suzuki, S., Kobayashi, K., Kobayashi, M., Hoshino, T., Horie, K., and Kimata, K. (1990). *J. Biol. Chem.*, **265**, 6992.
24. Shimokomaki, M., Wright, D. W., Irwin, M. H., van der Rest, M., and Mayne, R. (1990). *Ann. N.Y. Acad. Sci.*, **580**, 1.
25. Bruckner, P., Hörler, I., Mendler, M., Houze, Y., Winterhalter, K. H., Eich-Bender, S., and Spycher, M. (1989). *J. Cell Biol.*, **109**, 2537.

26. Schmid, T. M. and Linsenmayer, T. F. (1983). *J. Biol. Chem.*, **258**, 9504.
27. Kirsch, T. and von der Mark, K. (1990). *Biochem. J.*, **265**, 453.
28. Pesciotta, D. M., Curran, S., and Olsen, B. R. (1982). In *Immunochemistry of the Extracellular Matrix*, Vol. 1 (ed. H. Furthmayr), p. 91. CRC Press, Boca Raton.
29. Vaughan, L., Mendler, M., Huber, S., Bruckner, P., Winterhalter, K. H., Irwin, M. H., and Mayne, R. (1988). *J. Cell Biol.*, **106**, 991.
30. Ren, Z. X., Brewton, R. G., and Mayne, R. (1991). *J. Struct. Biol.*, **106**, 57.
31. Brewton, R. G., Wright, D. W., and Mayne, R. (1991). *J. Biol. Chem.*, **266**, 4752.
32. Irwin, M. H., Silvers, S. H., and Mayne, R. (1985). *J. Cell Biol.*, **101**, 814.
33. Schmid, T. M., Mayne, R., Jeffrey, J. J., and Linsenmayer, T. F. (1986). *J. Biol. Chem.*, **261**, 4184.
34. Sugrue, S. P., Gordon, M. K., Seyer, J., Dublet, B., van der Rest, M., and Olsen, B. R. (1989). *J. Cell Biol.*, **109**, 939.
35. Dublet, B. and van der Rest, M. (1991). *J. Biol. Chem.*, **266**, 6853.
36. Bernard, M., Yoshioka, H., Rodriguez, E., van der Rest, M., Kimura, T., Ninomiya, Y., *et al.* (1988). *J. Biol. Chem.*, **263**, 17159.
37. Yoshioka, H. and Ramirez, F. (1990). *J. Biol. Chem.*, **265**, 6423.
38. Kimura, T., Cheah, K. S. E., Chan, S. D. H., Lui, V. C. H., Mattei, M.-G., van der Rest, M., Ono, K., Solomon, E., Ninomiya, Y., and Olsen, B. R. (1989). *J. Biol. Chem.*, **264**, 13 910.
39. Linsenmayer, T. F. and Hendrix, M. J. C. (1980). *Biochem. Biophys. Res. Commun.*, **92**, 440.
40. Fitch, J. M., Mentzer, A., Mayne, R., and Linsenmayer, T. F. (1988). *Develop. Biol.*, **128**, 396.
41. Holmdahl, R., Rubin, K., Klareskog, L., Larsson, E., and Wigzell, H. (1986). *Arthritis Rheum.*, **29**, 400.
42. Burkhardt, H., Holmdahl, R., Deutzmann, R., Wiedemann, H., von der Mark, H., Goodman, S., and von der Mark, K. (1991). *Eur. J. Immunol.*, **21**, 49.
43. Schmid, T. M. and Linsenmayer, T. F. (1985). *J. Cell Biol.*, **100**, 598.
44. Summers, T. A., Irwin, M. H., Mayne, R., and Balian, G. (1988). *J. Biol. Chem.*, **263**, 581.
45. Häuselmann, H. J., Aydelotte, M. B., Schumacher, B. L., Kuettner, K. E., Gitelis, S. H., and Thonar, E. J.-M. A. (1992). *Matrix*, **12**, 116.
46. Mayne, R., Mayne, P. M., Ren, Z-X., Accavitti, M. A., Gurusiddappa, S., and Scott, P. G. (1994). *Connect. Tiss. Res.*, **30**, 1.
47. Aubert-Foucher, E., Font, B., Eichenberger, D., Goldschmidt, D., Lethias, C., and van der Rest, M. (1992). *J. Biol. Chem.*, **267**, 15759.
48. Font, B., Aubert-Foucher, E., Goldschmidt, D., Eichenberger, D., and van der Rest, M. (1993). *J. Biol. Chem.*, **268**, 25015.
49. Koch, M., Bernasconi, C., and Chiquet, M. (1992). *Eur. J. Biochem.*, **207**, 847.
50. Brown, J. C., Mann, K., Wiedemann, H., and Timpl, R. (1993). *J. Cell Biol.*, **120**, 557.

Matrix components (types IV and VII collagen, entactin, and laminin) found in basement membranes

BILLY G. HUDSON, SRIPAD GUNWAR, ALBERT E. CHUNG, and ROBERT E. BURGESON

1. General introduction

Basement membranes are specialized extracellular matrices that play key roles in diverse biological processes. The processes include: orchestration of embryonic development, maintenance of tissue architecture during remodelling and development, and ultrafiltration of blood. They appear as an amorphous sheet-like structure that is positioned between a cell layer and a thick collagenous stroma, as exemplified by the dermal–epidermal junction of human skin (*Figure 1A*) or between two layers of cells, as exemplified by the renal glomerular basement membrane (*Figure 1B*). In general, basement membranes are composed of several macromolecular constituents, including type IV collagen, laminin, entactin, and heparan sulfate proteoglycan (perlecan) which interact to form a supramolecular structure (*Figure 1C*). In the case of skin, this suprastructure is connected to the underlying stroma by anchoring fibrils that are composed of type VII collagen.

This chapter focuses on strategies and techniques for the isolation, purification, and characterization of type IV collagen, entactin, and laminin of basement membranes and type VII collagen of anchoring fibrils.

4A. TYPE IV COLLAGEN

SRIPAD GUNWAR and BILLY G. HUDSON

2. Introduction

Type IV collagen is the major constituent of mammalian basement membranes such as the renal glomerular and lens basement membranes. Collagen IV

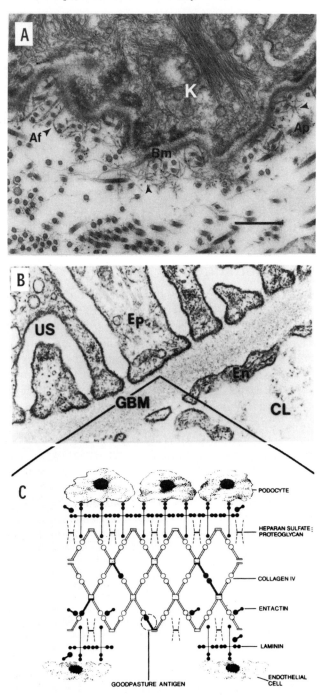

Figure 1. Basement membrane morphology and molecular structure. Panel A is an electron micrograph of the human dermal–epidermal junction basement membrane zone. A portion of a keratinocyte (K) is shown upon the basement membrane (Bm). The basement membrane is attached to the underlying stroma by anchoring fibrils (Af), depicted by arrowheads. A distinct anchoring plaque (Ap) is located at the right of the field. Bar equals 400 nm. Panel B is an electron micrograph of the capillary wall of renal glomerulus. The glomerular basement membrane (GBM) is juxtapositioned between epithelial cells (Ep) and endothelial cells (En), urinary sinus (US), and capillary lumen (CL). Panel C illustrates the structure of the major components and their preferential distribution in GBM. Protomers of type IV collagen containing the classical $\alpha1$ and $\alpha2$ chains are represented by the undarkened molecules and those containing the new chains ($\alpha3$, $\alpha4$, and $\alpha5$) are represented by the darkened molecules. Panels B and C are adapted from reference 3 with permission.

exists as a supramolecular structure and it is thought to serve as the scaffold for the binding and alignment of other basement membrane constituents (*Figure 1C*). The protomeric form of collagen IV comprises three α chains, and it is characterized by three distinct structural domains (*Figure 2*): the non-collagenous domain (NC1) at the carboxyl terminus; the triple-helical domain in the middle region; and the 7S domain at the amino terminus. In the suprastructure, protomers associate in a head-to-head fashion to form dimers and in a tail-to-tail fashion to form tetramers, resulting in an insoluble matrix.

Studies of the structure and function of collagen IV of authentic basement membranes have relied extensively upon the use of proteolytic cleavages and chemical extractions for the excision of domains and chains from the insoluble matrix (*Figure 2*). Pepsin excises fragments of the collagenous domain with retention of triple-helical structure, but destroys the NC1 domain. Bacterial collagenase, on the other hand, excises the NC1 domain with retention of hexamer structure and the 7S domain with retention of tetramer structure, but destroys the triple-helical domain. Extraction with 8 M urea in the presence of reducing agents solubilizes the intact chains, but without retention of triple-helical structure. These strategies, isolation procedures, and techniques for characterization of the collagen IV domains and chains are described in Part A of this chapter.

3. Isolation of 7S and NC1 domains from bovine glomerular basement membranes

This protocol is based on a procedure previously described for the isolation of 7S and NC1 domains from bovine glomerular basement membranes (GBM) (1), lens basement membranes (LBM) (1) and alveolar basement membranes (ABM) (2).

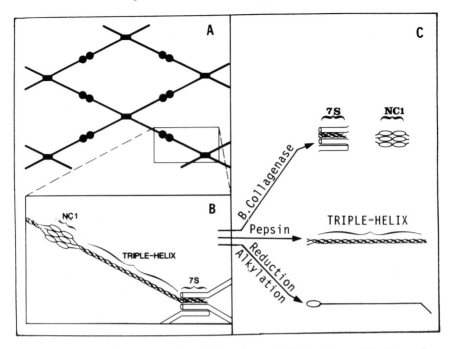

Figure 2. Strategies for the isolation of domains and chains of type IV collagen from basement membrane. Panel A illustrates the supramolecular structure of type IV collagen. Panel B is an enlargement of a region of the suprastructure showing the protomer structure and ways in which it is connected in the insoluble matrix. The protomer is a triple-helical molecule comprising three polypeptide chains and it has three distinct structural domains. These are: the 7S domain at the amino-terminus, the triple-helical domain in the middle region, and the NC1 domain at the carboxyl-terminus. Protomers associate in a head-to-head fashion to form dimers and in a tail-to-tail fashion to form tetramers. Panel C illustrates strategies for isolating the 7S domain (tetramer) and NC1 domain (hexamer), under non-denaturing conditions by enzymatic cleavage, and individual chains by chemical extraction.

Protocol 1. Excision of 7S and NC1 domains of type IV collagen from basement membrane by digestion with collagenase

Equipment and reagents

- Digestion buffer: 50 mM Hepes, pH 7.5, 0.01 mM CaCl$_2$, 4 mM NEM, 1 mM PMSF, 5 mM benzamidine–HCl, 25 mM 6-aminohexanoic acid
- 50 mM Tris–HCl, pH 7.5

- Bacterial collagenase (*Clostridium histolyticum*, CLSPA, Worthington Biochemical Corporation)
- Sorvall RC-5B fitted with GSA rotor
- Spectrapor dialysis membrane tubing (16 mm)

Method

1. Disperse 1 g of GBM in 100 ml of digestion buffer. To this, add and disperse 2 mg of bacterial collagenase. Alternatively, suspend 100 anterior lens capsules directly in 20 ml of digestion buffer and add 0.5 mg of collagenase to the suspension.

2. Perform the digestion at 37 °C for 20 h with slow stirring on a magnetic stirrer. Centrifuge the suspension at 6500 *g* in a Sorvall RC-5B using a GSA rotor, for 30 min at room temperature.

3. Collect the supernatant solution and dialyse against four changes of 50 mM Tris–HCl buffer (pH 7.5). Use for DEAE–cellulose chromatography (*Protocol 2*).

3.1 Purification of 7S and NC1 domains by DEAE–cellulose chromatography

Protocol 2. DEAE-52 cellulose anion-exchange chromatography of collagenase-solubilized 7S and NC1 domains of collagen IV

Equipment and reagents

- DEAE-52 cellulose (Whatman Ltd)
- 2.2 × 18 cm column
- 50 mM Tris–HCl, pH 7.5
- Collagenase digest (*Protocol 1*, step 3)
- Amicon Ym-10 (250 ml capacity) concentrator

Method

1. Prepare a DEAE-52 cellulose anion-exchange column as follows: Wash sufficient preswollen DEAE-52 cellulose for the column three times with 50 mM Tris–HCl, pH 7.5, to remove any fines from the cellulose.

2. Pack the column with DEAE-52 cellulose, and wash the column with 10 bed volumes of Tris buffer.

3. Load the dialysed collagenase digest on to the column and elute with Tris buffer. The 7S and NC1 domains elute in the unbound fraction. Collect this fraction, concentrate it to 10–12 ml using the concentrator, and dialyse it overnight against four changes of Tris buffer.

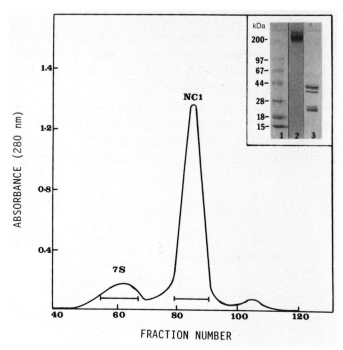

Figure 3. Purification of 7S tetramer and NC1 hexamer of type IV collagen by chromatography on a Sephacryl S-300 column. A collagenase digestion mixture from GBM was first chromatographed on a DEAE-52 column. The unbound fraction from this column was then fractionated on a Sephacryl S-300 column. The 7S tetramer and NC1 hexamer elute as noted on the chromatogram. The inset shows the electrophoretic pattern of these domains as analysed by SDS-PAGE.

3.2 Purification of 7S and NC1 domains by molecular-sieve chromatography

---

Protocol 3. Final purification of 7S tetramer and NC1 hexamer by chromatography on Sephacryl S-300

Equipment and reagents

- Sephacryl S-300 (Sigma) column (2.2 × 100 cm)
- Tris-buffer; 50 mM Tris–HCl, pH 7.5, 0.15 M NaCl, 0.05% sodium azide
- DEAE-52 fraction (*Protocol 2*, step 3)
- Amicon YM-10 concentrator

Method

1. Pack a Sephacryl S-300 column and equilibrate it with the buffer.

2. Apply the unbound DEAE-52 fraction (*Protocol 2*, step 3) to the column

---

and elute with Tris buffer at a flow rate of 20 ml/h. Collect 3.0–3.5 ml fractions and measure absorbance at 280 nm. The column profile is shown in *Figure 3*. The 7S tetramer elutes in peak I and NC1 hexamer in peak II. The inset in *Figure 3* shows the SDS-PAGE pattern of 7S and NC1 domains.

3. Concentrate peak I and II fractions using an Amicon YM-10 filter.

4. Isolation of type IV collagen by limited pepsin digestion

Protocol 4. Isolation of the triple-helical domain of type IV collagen by limited pepsin digestion of basement membrane

Equipment and reagents

• 0.5 M HOAc
• Polytron homogenizer

• Pepsin (Worthington Biochemical Corp.)

Method

1. Perform all steps at 4 °C.

2. Disperse GBM or LBM in 0.5 M HOAc at a concentration of 5 mg/ml using a Polytron homogenizer.

3. Prepare a solution of pepsin (10 mg/ml) in water and add to the basement membrane suspension at a ratio of 1:10 (pepsin:substrate).

4. After 36 h of digestion at 4 °C, centrifuge the suspension at 10 000 *g* for 20 min. Precipitate the collagenous peptides with NaCl, up to 20% concentration, and centrifuge at 25 000 *g* for 1 h.

5. Dissolve the pellet in 0.5 M HOAc and precipitate again with NaCl. Redissolve the pellet in 0.5 M HOAc, dialyse against the same solvent, lyophilize, and store at −20 °C.

The SDS-PAGE patterns for the final product from GBM are shown in *Figure 4*. The triple-helical domain is composed of several fragments varying in size from 78 kDa to > 340 kDa that are held together by disulfide cross-links in the native state.

105

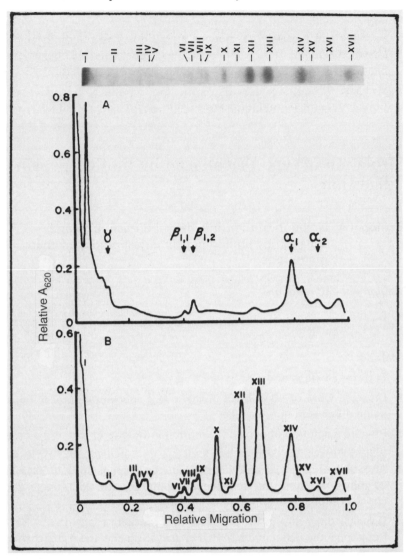

Figure 4. Analysis of the triple-helical domain of type IV collagen by SDS-PAGE. The triple-helical domain of type IV collagen was excised from GBM by pepsin digestion for 24 h at 4°C. The supernatant solution was analysed by SDS-PAGE under non-reducing and reducing conditions. The gel scans are shown in panel A (unreduced) and panel B (reduced). After reduction, the triple-helical domain dissociates into 17 peptic fragments, as noted in panel B. This figure was adapted from reference 4 with permission.

4.1 Isolation of type IV collagen by denaturation and reduction

---

Protocol 5. Isolation of intact chains of type IV collagen from GBM by denaturation and reduction

Equipment and reagents

- Buffer 1: 8 M urea, 10 mM Tris–HCl, pH 8.5
- Buffer 2: 8 M urea, 300 mM Tris–HCl, pH 8.5
- Buffer 3: 8 M urea, 100 mM Tris–HCl, pH 7.5
- β-Mercaptoethanol
- Iodoacetamide (Sigma)
- Sepharose CL.4B column (5.0 × 90 cm)
- DEAE–cellulose column (2.5 × 10 cm)

Method

1. Suspend GBM in a freshly prepared solution of Buffer 1. Stir for 6 h at 37 °C and centrifuge at 10 000 g for 15 min at room temperature. Resuspend the pellet in the same buffer and repeat the procedure three times.

2. Disperse the insoluble pellet in Buffer 2 at room temperature. Add β-mercaptoethanol (1% v/v) to the suspension, flush with nitrogen, and stir for 24 h at 37 °C.

3. Alkylate sulfhydryls with a 3-fold molar excess of iodoacetamide for 2 h at room temperature.

4. Centrifuge the suspension at 10000 g for 15 min at room temperature and dialyse the supernatant solution against Buffer 3.

5. Apply 250 mg of the alkylated sample on a column of Sepharose CL-4B equilibrated in Buffer 3. Elute the column at a flow rate of 60 ml/h, collect 12 ml fractions, and measure the absorbance at 280 nm.

6. Pool the fractions containing α chains (*Figure 5A*) dialyse against buffer 1 and apply to a DEAE–cellulose column, equilibrated in Buffer 1, at room temperature. Under these conditions, the chains do not bind to the column (*Figure 5B*).

7. Collect the unbound fractions, dialyse against water, and lyophilize. This sample contains the α chains (M_r 185 kDa and 175 kDa) and minor amounts of chains with M_r = 164 kDa and 152 kDa (*Figure 5C*). There is a variation in the amounts of 164 000 and 152 000 polypeptides.

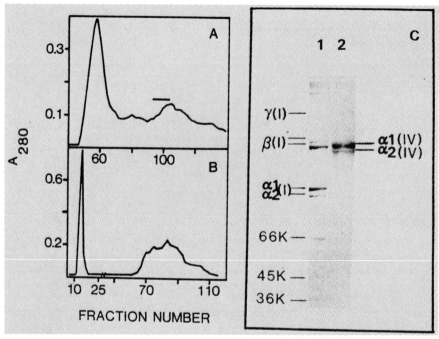

Figure 5. Isolation of intact α1 and α2 chains of type IV collagen. Panel A, Sepharose C1-4B gel filtration chromatography of GBM solubilized after reduction and alkylation of disulfide bonds. The column buffer was 8 M urea, 0.1 M Tris–HCl at pH 7.5. Fractions containing the chains (indicated by bar) were pooled and concentrated. Panel B, DEAE–cellulose chromatography of pooled fractions from A. The concentrated pooled fractions from A were dialysed into 8 M urea, 0.01 M Tris–HCl, pH 8.5, and chromatographed on a column equilibrated in the same buffer at room temperature. Panel C, SDS-PAGE of collagen chains eluted in the unbound fraction in panel B. Lane 1, type I collagen standard and lane 2, reduced and alkylated collagen IV alpha chains. The gels were stained with Coomassie Blue. This figure was adapted from reference 5 with permission.

4B. LAMININ

ALBERT E. CHUNG

5. Introduction

Laminin is the major non-collagenous component of the basal lamina. The typical molecule is made up of an A chain (400 kDa), a B1 and a B2 chain (each approximately 220 kDa) which associate to form a cross-shaped structure. The three chains are held together by disulfide and non-covalent bonds. The carboxyl ends of the B1 and B2 chains together with a structurally analogous region of the A chain form a triple-helical rod which constitutes the long arm of the molecule. The long arm terminates in a globular knob contributed by the carboxyl end of the A chain. The short arms are derived from the amino terminal regions of the three chains each of which is organized in alternating cysteine-rich and cysteine-poor domains. Although most of the current information on structure–function relationships have been obtained with this typical laminin there are variants which may be tissue specific and functionally distinct.

Laminin interacts with extracellular matrix molecules such as type IV collagen, entactin, and perlecan heparan sulfate proteoglycan. It also binds to cell membranes through integrin receptors and other plasma membrane-associated molecules. Laminin and supramolecular complexes containing laminin stimulate neurite outgrowth, promote cell attachment, chemotaxis, cell differentiation, and neuronal survival (see Chapter 12). Biological functions have been assigned to specific amino acid sequences or fragments of laminin. The E8 fragment derived by proteolytic cleavage of laminin or the laminin–entactin complex is particularly active.

The major source of laminin is the murine EHS (see Chapter 12) tumour which is rich in extracellular matrix molecules. Other sources include placenta and parietal endoderm-like cells that are maintained in culture. Laminin is readily isolated from the EHS tumour or parietal endoderm cells as a stoichiometric complex with entactin. Although the two molecules are not covalently linked, the complex can only be dissociated with chaotropic agents that may adversely affect the properties of laminin. Furthermore, it is not a routine exercise to isolate laminin entirely free of entactin and its proteolytic fragments. Therefore, caution is advised in the interpretation of data obtained with laminin samples that have not been rigorously characterized.

6. Preparation of laminin–entactin complex from EHS tumour

This protocol is as described by Paulsson *et al.* (6). All manipulations are to be performed at 4°C unless otherwise indicated.

Protocol 6. Preparation of the laminin–entactin (nidogen) complex from EHS tumour

Equipment and reagents
For A

- TBS: 0.05 M Tris–HCl, pH 7.4, 0.15 M NaCl, 2 mM NEM, 2 mM PMSF
- Extraction buffer (TBS–EDTA): TBS, 10 mM EDTA
- Sample buffer: 5% β-mercaptoethanol, 2% SDS, 10% glycerol, 0.05% Bromophenol blue in 0.0625 M Tris–HCl, pH 6.8

- Polytron homogenizer (Brinkman Instruments) with PTA 2TS head or equivalent
- Sorvall RC centrifuge with Sorvall GSA rotor or equivalent
- Sepharose CL-6B (Pharmacia) or Bio-Gel A5m (Bio-Rad) column (3 × 135 cm)

Protocol 6. *Continued*

- Amicon ultrafiltration cell with PM 30 membrane
- 5–15% gradient polyacrylamide–SDS gel
- Coomassie Blue R250

For B

- DEAE–Sephacel (Pharmacia) column (12.3 × 5 cm)
- Equilibration buffer: 0.05 M Tris–HCl, pH 7.4, 2.5 mM EDTA, 0.5 mM PMSF, 0.5 mM NEM
- NaCl gradient (0–0.8 M) in Equilibration buffer
- Amicon ultrafiltration cell (see *Protocol 6A*)

A. *Initial preparation*

1. Homogenize 75 g of frozen tumour tissue (see Chapter 13) for 1 min in 1.5 l of TBS NEM and 2mM PMSF in Polytron homogenizer.

2. Place the homogenate in 250 ml bottles and centrifuge at 8300 *g* for 20 min at 4°C. Discard the supernatant solution.

3. Homogenize the residue in 800 ml of TBS for 1 min. Repeat step 2.

4. Collect the residue and homogenize for 1 min in 375 ml extraction buffer.

5. Stir the homogenate for 1 h at 4°C. Collect and save the supernatant solution by centrifugation as in step 2.

6. Repeat steps 4 and 5 to obtain additional material.

7. Pack a 3 × 135 cm column with Sepharose CL-6B or Bio-Gel A5m and equilibrate it at room temperature with 2–3 l of TBS-EDTA.

8. Apply 100 ml of the extract to the column, elute with the extraction buffer and collect the protein peak in the void volume. Monitor the absorbance of the effluent at 280 nm.

9. Concentrate the protein solution to 1 mg/ml by ultrafiltration with an Amicon ultrafiltration cell. The yield should be 1–3 mg/g wet weight of tumour tissue.

10. Boil an aliquot of the preparation for 2 min with 2 volumes of sample buffer, and examine by electrophoresis on a 5–15% gradient polyacrylamide–SDS gel. Staining with Coomassie Blue R250 should reveal major bands at 400 kDa and 220 kDa for laminin and 150 kDa for entactin.

B. *Further purification may be obtained by DEAE–Sephacel chromatography*

1. Perform all steps at 4°C.

2. Pack the column and equilibrate it with the Equilibration buffer.

3. Dilute the concentrated extract with one-half volume of the Equilibration buffer to lower the NaCl concentration to 0.1 M, and apply it to the DEAE–Sephacel column.

4. Wash the column with Equilibration buffer and collect the flow-through protein peak. This will contain most of the laminin–entactin complex. Monitor the effluent at 280 nm.

5. Elute the column with a linear sodium chloride gradient from 0 to 0.8 M (1 litre) in the Equilibration buffer. Collect the early eluent which contains additional amounts of the complex.

6. Pool the samples containing the complex, concentrate to 1–2 mg/ml and store at $-20\,°C$.

EHS tumour tissue may be obtained from Dr Hynda Kleinman, NIDR, NIH, Bldg. 30, Rm 414, Bethesda, MD 20892. The protocol can be modified to accommodate smaller quantities of tissue.

6.1 Isolation of laminin from the laminin–entactin complex

This procedure is as described by Paulsson *et al.* (6) in which the laminin–entactin complex is dissociated by dialysis against 2 M guanidine–HCl and the dissociated molecules separated by molecular-sieve chromatography in the presence of a chaotropic agent.

Protocol 7. Isolation of laminin from the laminin–entactin complex

Equipment and reagents

- Laminin–entactin complex (*Protocol 6*)
- Dissociating buffer: 2 M guanidine–HCl in 0.05 M Tris–HCl, pH 7.4, containing 2 mM EDTA, 0.05 mM PMSF, 0.5 mM NEM
- Sepharose CL-4B 3 × 130 cm column

Method

1. Dialyse 50 ml of the concentrated laminin–entactin complex (1 mg/ml) overnight at $4\,°C$ against 1 l of dissociating buffer.

2. Prepare a 3 × 130 cm Sepharose CL-4B column and equilibrate in the dissociating buffer.

3. Apply the dialysed sample to the column and develop with the dissociating buffer. Collect 10 ml fractions and monitor absorbance of the protein effluent at 280 nm. The enriched laminin peak elutes just after the void volume of about 300 ml and is collected in about 15 fractions.

4. Analyse the fractions by SDS–polyacrylamide gel electrophoresis and by Western blotting.

The laminin obtained in this way may still contain small quantities of entactin and concentration and rechromatography may be necessary.

7. Isolation of laminin from M1536-B3 cells

This is based on a procedure described by Chung *et al.* (7). The procedure is simple and can be carried out on a small scale from cell cultures with minimum effort. The product is satisfactory for biological assays. M1536-B3 cells are grown in suspension where they form spherical aggregates consisting of a monolayer of cells attached to a core of extracellular matrix made up predominantly of laminin and entactin. The cells are removed by treatment of the aggregates with cytochalasin B, and the matrix isolated by repeated centrifugation at low speed.

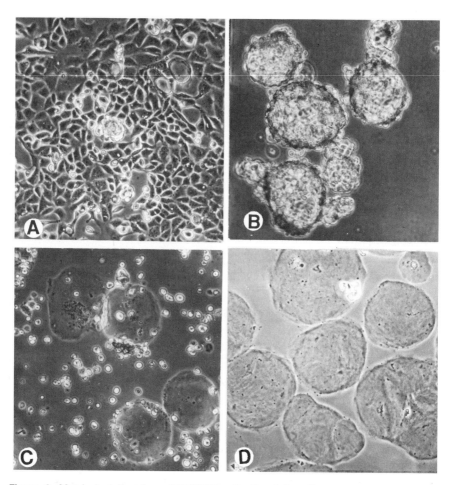

Figure 6. Matrix isolation from M1536-B3 cells. Panel A, cells grown as a monolayer; panel B, cell aggregates grown in suspension culture; panel C, cells treated with cytochalasin B; panel D, purified matrix.

Protocol 8. Isolation of the laminin–entactin complex from M1536-B3 cells

Equipment and reagents

- M1536-B3 cells
- DMEM high-glucose medium + 10% FBS
- PBS + Ca^{2+} + Mg^{2+}, pH
- Cytochalasin B (Aldrich) stock solution: 25 mg/ml in DMSO
- Beckman J6B centrifuge
- Conical 125 ml centrifuge tubes

- Protease inhibitors (Sigma):
  —Leupeptin: 1000 × = 1 mg/ml water
  —Pepstatin A: 1000 × = 0.7 mg/ml methanol
  —E64: 1000 × = 1 mg/ml 50% ethanol
- Extraction buffer: 50 mM Tris–HCl, pH 7.5, 5 mM EDTA, 2 M urea, 0.5 M NaCl

Method

1. Seed M1536-B3 cells[a] at a cell density of $1-2 \times 10^5$ cells/ml in DMEM high-glucose medium supplemented with 10% fetal bovine serum. Cells can be grown in roller bottles, flasks, or tissue culture dishes at 37°C in humidified 5% CO_2/95% air atmosphere. For 100 mm dish use 10 ml of medium (*Figure 6A*).

2. After 2 days harvest cells by trypsinization. Count the cells and seed at the same density as in step 1 and plate in bacterial type Petri dishes. Twenty 100 mm dishes will yield enough material for routine studies.

3. The cells will form spherical aggregates that increase in size (*Figure 6B*). After 10 days collect the medium with the cells and place in conical 125 ml glass centrifuge tubes. Rinse each dish with 4–5 ml of PBS containing calcium and magnesium, add to the tubes and collect the spheres by centrifugation at 600 *g* for 2 min at 4°C in a Beckman J6B centrifuge.

4. Gently resuspend the spheres in PBS containing magnesium and calcium to which cytochalasin B at 20 µg/ml has been added. Use 10 ml of PBS/cytochalasin B solution for the cells from 10 dishes.

5. Incubate the suspension at 37°C for 3.5–4 h. Periodically swirl the tubes to resuspend the aggregates.

6. After the incubation, vortex the tube vigorously for 20 sec and allow the tube to stand for 5 min. The suspension will form two layers, a turbid upper layer and an opaque sediment with a whitish surface. If this is not observed, incubate the suspension for another hour then proceed as before.

7. Vortex the tube vigorously for 20 sec and centrifuge at 600 *g* for 1 min at 4°C. Carefully remove the upper turbid layer by aspiration and discard (*Figure 6C*). Take care not remove the lower layer which contains the matrix. Suspend the pellet in PBS containing calcium and magnesium but without the cytochalasin, using the same volume as used originally, *i.e.* 10 ml/10 dishes.

Protocol 8. *Continued*

8. Vortex the suspension as in step 7 and repeat the washing 7–8 times. Periodically examine a droplet of the suspension by phase-contrast microscopy. The extracellular matrix sacs will be obvious and there should be very few cells or debris visible (*Figure 6D*).

9. Collect the purified matrix by centrifugation at 1400 r.p.m. Resuspend in PBS/Ca^{2+}/Mg^{2+} containing protease inhibitors and store at $-$ 20°C.

10. The matrix may be solubilized with the extraction buffer.

11. Analysis of the matrix on SDS-PAGE reveals $>$ 90% laminin and entactin.

Larger quantities of material may be obtained by culturing cells in 2-litre spinner bottles containing 1 litre of medium. The best results are obtained with Bellco bottles equipped with Teflon paddles. The paddles are rotated at 55–70 r.p.m. by using a Bellco magnetic stirring table. The procedures for matrix extraction are as described above.

[a] These cells can be obtained from: Albert E. Chung, Department of Biological Sciences, University of Pittsburgh, Pittsburgh, PA 15260, USA.

7.1 Small-scale isolation of laminin

An alternative protocol for the isolation of laminin on a small scale can be achieved by molecular-sieve chromatography by FPLC or HPLC on a DuPont Zorbax GF450 column.

Protocol 9. Isolation of laminin on a small scale

Equipment and reagents

- Extraction buffer: 4 M urea in PBS, pH 7.0
- Zorbax G450 100 × 250 mm column (DuPont)
- Eppendorf centrifuge
- HPLC or FPLC apparatus
- Speed-Vac concentrator (Savant Instruments)
- Anti-entactin antiserum (UBI)

Method

1. Extract an aliquot of matrix obtained in *Protocol 8* with 5 volumes of 4 M urea in PBS at 4°C overnight.

2. Equilibrate the Zorbax column with 5 column volumes of the extraction buffer at a flow rate of 0.8 ml/min.

3. Centrifuge the extract in an Eppendorf centrifuge at maximum speed (1000 *g*) for 5 min at 4°C to remove undissolved material. Carefully remove and save the supernatant fraction.

4. Inject 0.5 ml of the extract on to the column and elute at 0.8 ml/min. Collect 1 min fractions. The laminin will appear as the first large peak. This procedure takes approximately 20 min.

5. Analyse each fraction by SDS-PAGE. The laminin should be essentially pure. If the samples are contaminated with entactin, pool and dialyse overnight against water. Concentrate to a small volume in a Speed-Vac and add 4 M urea/PBS.

6. Repeat the chromatography (step 4).

7. Test for contaminating entactin by Western blotting with an anti-entactin antiserum.

7.2 Metabolic labelling of laminin

Laminin is readily labelled in cell cultures with ^{35}S for preparative electrophoresis of the laminin subunits.

The procedures are essentially as described by Chung *et al.* (8).

Protocol 10. Metabolic labelling of laminin with ^{35}S and the isolation of laminin A, B1, and B2 chains by preparative gel electrophoresis

Equipment and reagents

- M1536-B3 cells (*Protocol 8*)
- Tran ^{35}S-label (> 1000 Ci/mmole, containing methionine and cysteine; ICN Biochemicals)
- Solubilizing buffer: 0.0625 m Tris–HCl, pH 6.8, containing 10% glycerol (v/v), 2% SDS (w/v), 5% 2-mercaptoethanol (v/v)
- Bio-Rad or Hoeffer apparatus for 14 × 20 cm gel
- 15% polyacrylamide gel
- 30% sucrose solution
- 7.5% polyacrylamide gel
- 5% polyacrylamide stacking gel
- Peristaltic pump
- Fraction collector
- Eluting buffer, pH 8.3: 0.025 M Tris base, 0.192 M glycine, 0.1% SDS, 0.5% 2-mercaptoethanol
- Bromophenol blue
- Liquid scintillation fluid (e.g. EcoLite; ICN Biochemicals)
- Gel running buffer: eluting buffer without 2-mercaptoethanol

Method

1. Seed M1536-B3 cells in 100 mm tissue culture dishes as described in *Protocol 8*. Add Tran ^{35}S-label at 1 μCi/ml of medium.

2. Allow the cells to grow for 10 days and isolate the matrix as described in *Protocol 8*. Care should be taken for the proper handling of radioisotopes.

3. Dissolve the matrix in solubilizing buffer by heating for 5 min in a boiling water bath. This can be stored at − 20 °C until needed.

4. Prepare the separating polyacrylamide gel plates in a Bio-Rad or Hoeffer apparatus capable of accommodating a 14 × 20 cm gel. Assemble the glass plates with 3 mm preparative spacers. Each of the two spacers has a 1 mm hole drilled through it approximately 2.5 cm from the bottom edge to allow continuous flow of buffer through a channel in the gel.

Protocol 10. *Continued*

5. Pour a layer of 15% polyacrylamide solution into the glass sandwich so that it is about 1 mm below the holes in the spacers. Allow the gel to polymerize.

6. Layer a 30% sucrose solution (dissolved in the gel running buffer) on top of the polymerized 15% gel to a height of 3–4 mm.

7. Gently layer on top of this sucrose 7.5% polyacrylamide to give a separating gel 3–4 cm in height. After the gel has polymerized pour a 5% polyacrylamide stacking gel on top to 1 cm in height.

8. The above arrangement creates a channel in the slab gel that allows continuous elution during the electrophoresis. After the apparatus is assembled connect the outlet in one of the spacers to a peristaltic pump and the other outlet to a fraction collector. Pump eluting buffer through the channel at a rate of 2 ml/min to remove the sucrose layer. Care should be taken to check that the elution channel is not blocked.

9. Add Bromophenol blue (0.001%) to the solubilized matrix and apply 4–5 ml on top of the gel.

10. Run the gel at 200 V for the first 7 h and at 300 V for the final 3 h.

11. Begin collecting the effluent after the dye front has reached the open channel. Maintain a flow rate of 2 ml/min and collect 4 ml fractions.

12. Locate the protein peaks by removing a 0.2 ml aliquot of each fraction and determine its radioactivity by liquid scintillation counting after mixing with liquid scintillation fluid (such as EcoLite).

13. Pool the radioactive peaks and concentrate by ultrafiltration as described in *Protocol 6*. Four radioactive peaks should be obtained, the first appears just after the dye front, the second is entactin, the third contains the laminin B chains, and the fourth the A chain.

14. Examine an aliquot of the concentrated fractions by SDS-PAGE (*Figure 7*). The laminin A and B chains should be essentially pure. The yield from the run should be > 50%.

8. Isolation of E8 fragment of laminin

This procedure is as described by Paulsson *et al.* (6). The E8 fragment is obtained by elastase digestion of the laminin–entactin complex followed by molecular–sieve chromatography. The E8 fragment contains the carboxyl terminal regions of the three chains of laminin and consists of two subunits, E8-A derived from the A chain and E8-B derived from the B1 and B2 chains. The latter two fragments are disulfide linked.

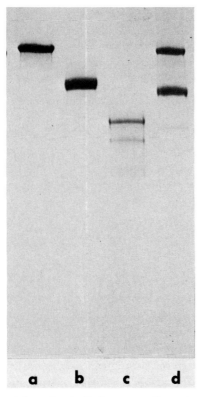

Figure 7. Purification of laminin and entactin by preparative gel electrophoresis. Lane d, matrix prepared from M1536-B3 cells; lane a, laminin A chain; lane b, laminin B1 and B2; lane c, entactin A and B.

Protocol 11. Isolation of E8 fragment of laminin from laminin–entactin complex

Equipment and reagents

- Dialysis buffer: 0.15 M NaCl in 0.05 M Tris–HCl, pH 7.4, containing 2 mM EDTA, 0.5 mM NEM
- Elastase (Serva)
- PMSF dissolved in ethanol: 0.1 g in 5 ml
- Sepharose CL-6B 3.2 × 110 cm column
- Amicon ultrafiltration concentrator (*Protocol 6*)

Method

1. Dialyse 200 ml of the laminin–entactin complex (1 mg/ml), isolated from the EHS tumour as described in *Protocol 6*, against the dialysis buffer.

117

Protocol 11. *Continued*

2. Add 1 mg of elastase and incubate for 24 h at 25 °C.

3. Stop the reaction by adding PMSF ethanol.

4. Prepare a 3.2 × 110 cm column of Sepharose CL-6B and equilibrate it with the dialysis buffer.

5. Load a 70 ml aliquot of the elastase-treated sample on the column and elute with the dialysis buffer. Collect 10 ml fractions and monitor the effluent at 280 nm for protein. The third peak from the column contains the E8 fragment.

6. Pool the fractions, concentrate by ultrafiltration as described previously (*Protocol 6*) and rechromatograph on the same Sepharose CL-6B column.

Analysis of the protein on a non-reducing SDS gel should yield two protein bands E8-A, 140 kDa, and E8-B, 80 kDa. After reduction three bands—140, 45, and 32 kDa—should be obtained.

9. Cell attachment to laminin

The approach described may be used for purified extracellular matrix components or their fragments (see Chapter 12). A convenient way to compare fractions or different preparations under the same conditions is to apply droplets of each sample on different marked areas of a single dish and proceeding as detailed in *Protocol 12*. This is a particularly good technique for observing neurite outgrowth if neonatal mouse cerebellar cells are used (9).

Protocol 12. Cell attachment to laminin–entacin complex

Equipment and reagents

- Nitrocellulose strip 1 × 5 cm (Schleicher and Schuell)
- PBS, pH 7.4
- 3% BSA in PBS (use BSA of the highest purity available)
- [^{3}H]thymidine, Sp. Act. = 2 Ci/mmol
- Serum-free DMEM + 0.05% BSA
- Trypsin

Method

1. Dissolve the strip of nitrocellulose in 6 ml of methanol.

2. Add 0.1 ml of the nitrocellulose solution to a 35 mm Falcon Petri dish and spread the solution evenly over the surface. Allow the methanol to evaporate in the hood. This leaves a uniform transparent layer of nitrocellulose on the dish.

3. Dilute the urea-extracted matrix (*Protocol 9*) to a concentration of 10–100 µg of matrix proteins/ml of PBS. A concentration of urea 0.1–0.5 M

should not interfere. However, the matrix can be dialysed for a few hours against PBS to remove the urea. Add 1 ml of the diluted solution to each dish and allow to stand at 4 °C overnight. Block non-specific binding sites by incubating in 3% BSA in PBS for an additional 24 h. It is important to use BSA of the highest purity.

4. Quantitative data can be obtained for cell attachment if the cells are prelabelled for 12–24 h with [^{3}H]thymidine at a concentration of 2.5 μCi/ml of medium. Grow cells to semi-confluency before replacing the medium with fresh medium to which the label has been added. Harvest labelled cells by trypsinization and resuspend in serum-free DMEM containing 0.05% BSA. Take a cell count and count an aliquot in a liquid scintillation counter to determine the average radioactivity/cell.

5. Add 1×10^5 cells/dish. Dishes preincubated with 3% BSA in PBS alone serve as controls. Incubate the dishes for 2–4 h at 37 °C in a humidified 5% CO_2/95% air atmosphere.

6. Remove the unattached cells, after the incubation, by rinsing the dish twice with 1 ml aliquots of PBS. Remove the attached cells by incubating with 1 ml trypsin solution. Monitor detachment by phase-contrast microscopy. Mix the cells with liquid scintillation fluid and determine the radioactivity. Perform the attachment assay in triplicate.

7. Calculate the fraction of input cells attached from the difference between the average counts/min obtained for the experimental and the control dishes divided by the counts/min in the input cells.

4C. TYPE VII COLLAGEN

ROBERT E. BURGESON

10. Introduction

Type VII collagen is a major constituent of anchoring fibrils. Anchoring fibrils are short (about 800 nm long), centrosymmetric structures that underlie certain external epithelia, such as the dermal–epidermal junction of skin, the amniotic epithelium, the oral and vaginal mucosa, and the cornea and sclera. The anchoring fibrils form a network between the basement membrane and anchoring plaques that physically entraps fibrous elements of the underlying matrix and secures the basement membrane to the stroma.

Type VII procollagen is synthesized and secreted by keratinocytes and epithelial cell lines. The biosynthetic product contains a relatively small

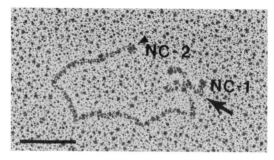

Figure 8. Rotary shadowed image of a type VII procollagen molecule. The small NC2 domain (arrowhead) and the large, complex NC1 domain (large arrow) are indicated.

C-terminal non-triple-helical domain (NC2; M_r 95 000); a 424 nm long triple-helical domain (M_r 510 000) containing several non-helical interruptions; and a large non-triple-helical domain (NC1; M_r 450 000) at the amino-terminus (*Figure 8*). Denaturation of the procollagen molecule produces three identical alpha-chains, M_r 352 000. The intact procollagen molecule has been isolated as an intact, native molecule only from the medium of cultured cells (10). The triple-helical domain, or its major pepsin fragments (P1, M_r 280 000, 93 000 per alpha-chain; P2, M_r 230 000, 77 000 per alpha-chain), can be isolated from tissues following pepsin digestion (11, 12). The NC1 domain has been isolated from tissue following solubilization using clostridial collagenase (13).

The preparation of the triple-helical domain can be accomplished by solubilization of amnion using pepsin. The solubilized collagens are partially fractionated by differential salt precipitation, where type VII co-precipitates with type V collagen. Separation of type VII from type V can be accomplished by CM-cellulose chromatography, following removal of contaminating proteoglycans by DEA–cellulose chromatography. This product is sufficiently pure for routine use. If further purification is necessary, the pure native structure can be obtained by preparative velocity sedimentation. Pure denatured alpha-chains are obtained following HPLC reversed-phase chromatography.

10.1 Initial isolation of pepsin-derived type VII collagen

Protocol 13. Preparation of type VII triple-helical domain from amniotic membranes: pepsin digestion and differential salt precipitation

Note:

(a) All membrane preps should be performed using between 180–200 g wet weight of membrane, which amounts to the membranes from 30–40 placentas;

(b) All volumes should be determined using a graduated cylinder;

(c) Precool all solutions to 4 °C.

Equipment and reagents

- Sorvall RC-3B (low speed) with H600A rotor
- Sorvall RC-5B (high speed) with G5A rotor
- Polytron homogenizer (Brinkman Instruments) with PTA 2TS probe
- Solid NaCl
- 1 m NaCl
- 0.5 M HOAc
- Solid pepsin (Sigma P6887)
- 1 M NaCl, 0.1 M Tris–HCl, pH 8.1
- 2 × DEAE (50 ml): 25 ml 4 M urea, 1.46 g NaCl, 1.21 g Tris base, adjust pH to 8.3 with HCl, and add water to 50 ml final volume

Method

1. Strip membranes from placentas. Store membranes in ice water. Freeze membranes in H_2O until enough are accumulated to begin the preparation. Pool membranes in a 4-l beaker. Homogenize membranes with a Polytron homogenizer 500 ml each time in H_2O and ice to a fine slurry. The final volume should be 3.5 litres.

2. Using 4 × 1 l bottles, centrifuge at 4500 r.p.m. (5240 *g*) in the Sorvall RC-3B using the H-4000 rotor for approximately 20 min at 4 °C. Decant and discard the supernatant solution and suspend the pellet in 3 l of cold 1 M NaCl. Stir the pellet at 4 °C until it completely separates. Repeat step 2 using 3 l of cold H_2O to wash out any remaining NaCl from the pellet. Record the wet weight of the final pellet.

3. Resuspend the pellet in 1 l of cold 0.5 M HOAc and add solid pepsin to 100 mg/100 g pellet wet weight. Stir at 4 °C for 16 h. The yield will depend upon the pepsin activity at the time of use. Freshly purchased pepsin may be too active at the concentration indicated above. During the digestion, much of the solubilized type VII will be converted to P1 and P2 and lost. Far out-of-date pepsin may have too little activity. The optimal amount of pepsin can only be determined empirically using the recommended concentration as a guide.

4. Clarify the resulting solution using the Sorvall RC-5B at 13 000 r.p.m. (2750 *g*) for 20 min at 4 °C. Retain the supernatant solution and determine its volume. *Slowly* add solid NaCl to the clarified solution to 10% (w/v), and stir for approximately 20 min. Harvest the precipitate at 13000 r.p.m. as above for 30 min. Discard the supernatant salt solution and combine the pellets, redissolving them in 1 l of 1 M NaCl, 0.1 M Tris–HCl, pH 8.1. Be sure to check the pH of the redissolved pellets. Stir at 4 °C overnight to inactivate any residual pepsin.

5. To the dissolved pellet, *slowly* add solid NaCl to 2.7 M. After stirring for 60 min, clarify the solution by centrifugation as above, and filter the supernatant solution through a large funnel loosely packed with glass wool. Discard the pellet.

Protocol 13. *Continued*

6. To the supernatant solution, *slowly* add solid NaCl to 4 M. Stir for 60 min. Harvest the pellet as in step 4 and dissolve it in 250 ml of 0.5 M HOAc. Stir at 4 °C until dissolved.

7. Add solid NaCl to 4% (w/v), and stir for 60 min. Clarify the supernatant by centrifugation as in step 4. Discard the pellet.

8. Add solid NaCl to 8% (w/v) to the supernatant, and stir 60 min. Harvest the pellet as in step 4, and redissolve in 50 ml cold 2 × DEAE buffer.

10.2 Purification of pepsin-derived type VII collagen

Protocol 14. DEAE–cellulose chromatography of salt precipitated type VII collagen triple-helix

Equipment and reagents

- Column: DE-52 (Whatman), 2.5 × 12 cm, room temperature
- Sorvall RC-5B, GSA and SS-34 rotors
- *Buffers* (degas buffers under vacuum with constant stirring):
- 8 M Urea stock solution. Hint: deionize urea stock solution immediately before use by passing through a mixed bed ion-exchange resin (IONAC NM-60H$^+$/OH$^-$ form, type 1, J. T. Baker Inc.)
- 500 ml 1 × DEAE buffer: 125 ml urea stock solution, 7.31 g NaCl, 6.06 g Tris base, HCl to pH 8.3, water to 500 ml
- 300 ml DEAE-acid wash: 6.25 ml HCl, 35.1 g NaCl, water to 300 ml
- 50 ml 2 × CMC buffer: 25 ml urea stock solution, 4 ml of LiOAc solution, 3 ml of 1 M LiCl solution (see *Protocol 15* for stock solutions)

Method

1. Wash column with 300 ml of DEAE-acid wash and equilibrate the resin with 200 ml of 1 × DEAE buffer. Check the effluent pH.

2. Dilute the sample dissolved in 2 × DEAE buffer 1:1 with H$_2$O.

3. Apply the sample and collect the column effluent in a beaker. Immediately apply 100 ml of 1 × DEAE buffer and collect in the same container. Measure the volume of the combined effluent and acidify with acetic acid to 0.5 M by the formula: 0.0286 × *volume collected* = *volume of glacial acetic acid to add*.

4. Add solid NaCl to 14% (w/v) to the acidified effluent. Collect precipitate at 13 000 r.p.m. for 40 min at 4 °C in a Sorvall RC-5B, GSA rotor (27 500 g). Dissolve the pellet in 50 ml cold 0.5 M HOAc.

5. Add 200 ml ice-cold absolute ethanol and collect the pellet at 16 000 r.p.m. in the SS-34 rotor (30 679 g) for 20 min at 4 °C. Redissolve the pellet in 50 ml 2× CMC buffer.

10.3 CM-cellulose chromatography

Protocol 15. Carboxymethyl-cellulose chromatography of partially purified type VII triple-helical domain

Equipment and reagents

Column:
- CM-52 (Whatman), 2.5 × 12 cm, room temperature. Monitor column effluent by UV absorbance at 214 nm
- Sorvall RC-5B, SS-34 rotor

Stock solutions:
- 1 M lithium acetate, pH 4.8
- 1 M lithium chloride
- 8 M urea (freshly deionized, see *Protocol 14*)

Buffers:
- 1 l 1 × CMC buffer: 250 ml urea solution, 40 ml LiOAc solution, 30 ml LiCl solution, 680 ml water. Degas under vacuum with constant stirring.
- 100 ml 2 × CMC buffer: 50 ml urea solution, 8 ml of 1 M LiOAc solution, 6 ml LiCl solution, 36 ml water

- 125 ml CMC salt wash: add 5.14 g LiCl to 125 ml 1 × CMC buffer, water to 125 ml
- CMC gradient:
  —250 ml 1 × CMC buffer
  —250 ml 1 × CMC buffer plus 5.34 g LiCl
  Use a gradient maker to generate a linear increase in LiCl concentration.
- 125 ml CMC base wash: 31.25 ml urea solution, 15.3 g LiCl, 0.60 g LiOH, water to 125 ml. Hint: base wash column every seventh run. Check effluent of column to be certain pH becomes alkaline. Re-equilibrate immediately with 2 × CMC buffer until pH returns to 8.3, and 1 × buffer to conductivity of 1 × buffer.

Method

1. Wash the column with 125 ml of the salt wash buffer.

2. Equilibrate the column with about 300 ml of 1 × CMC buffer. Check the conductivity of the column effluent to be certain the column has returned to the conductivity of the 1 × CMC buffer before applying the sample.

3. Dilute the sample dissolved in 2 × CMC buffer 1:1 with H_2O. Clarify by centrifugation (16 000 r.p.m. (30 680 g), 10 min, in a Sorvall RC-5B, SS-34 rotor) at room temperature and carefully decant the supernatant solution into a clean tube.

4. Apply the clarified sample to the column and wash with 200 ml of 1 × CMC buffer (flow rate = 1 ml/min).

5. Apply the LiCl gradient and collect approximately 80 × 6 ml fractions. The pattern should resemble that shown in *Figure 9*.

6. Evaluate the elution pattern by polyacrylamide gel electrophoresis under reducing conditions. The UV pattern may not show good resolution. However, a portion of the type VII containing peak should be free of contaminating type V (*Figure 10*). Pool these fractions and dialyse against three changes of 0.5 M HOAc, 4 l per change. Lyophilize to concentrate.

7. Store the type VII collagen as a lyophylate, or better, dissolve in 0.5 M HOAc and store at −120 °C. If stored in solution at 4 °C, the sample should be treated with DFP to prevent proteolysis.

Robert E. Burgeson

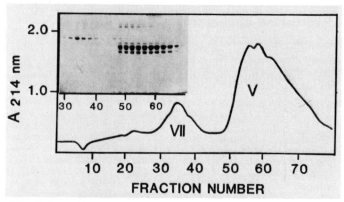

Figure 9. CM-cellulose chromatographic resolution of type V and type VII collagens partially purified by differential salt precipitation. The electrophoretic profiles of alternative individual fractions on a 5% polyacrylamide gel following disulfide bond reduction are shown (insert). From reference 12, with permission.

10.4 Additional considerations in the purification of type VII collagen helical molecules

The product of CM-cellulose separation (*Figure 10*) of type VII collagen helix is about 70–80% pure on a total protein basis, but is only 30–50% pure by mass. The non-protein contaminant has not been identified. The yield from CM-cellulose chromatography is 1–3 mg of type VII collagen per 200 g (wet weight) membrane. Further purification of the native triple-helix requires preparative velocity sedimentation. The CM-cellulose product is dissolved in 2 M deionized urea, 0.8 M NaCl, 50 mM Tris–HCl, pH 7.5, and 200 µl are carefully layered atop each of six 5–20% (w/v) sucrose gradients made in the same buffer. Each gradient has a total volume of 3.5 ml and is made in polyallomer tubes designed for use in a SW60 Ti rotor (Beckman Instruments). Sedimentation is performed in an L8-80 ultracentrifuge (Beckman Instruments), or equivalent, that can accommodate this rotor for 16 hours at 54 000 r.p.m. (20 709 g) at 12 °C. At the end of the run, the rotor should be allowed to coast to a complete stop, and the gradients immediately dripped from the bottom to yield 34 to 35 fractions. Under these conditions, the triple helix should peak at approximately fraction 10 from the bottom of the gradient. The fractions can be readily monitored by SDS-gel electrophoresis.

The denatured alpha-chain can be isolated from the CM-cellulose purified materials by standard high-pressure liquid chromatography (HPLC). The CM-cellulose chromatography product is dissolved in 10 mM trifluoroacetic acid (TFA), 11% (v/v) acetonitrile, 1 mM 2-mercaptoethanol, and denatured at 60 °C for 15 min. The sample is then filtered and applied to a 5 × 250 mm, C-18 reverse-phase column (Vydac) and equilibrated with 10 mM TFA, 11% acetonitrile at room temperature. The column is eluted with a linear 11–56% acetonitrile gradient in 10 mM TFA. The elution profile is monitored by UV

124

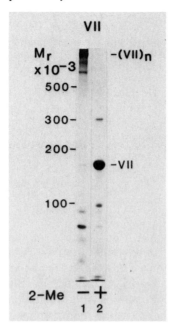

Figure 10. SDS–polyacrylamide gel pattern of the pooled CM-cellulose purified type VII collagen. The unreduced (lane 1) and reduced (lane 2) patterns are shown. Contaminating proteins are readily apparent, but the majority of the Coomassie Blue stained materials are type VII collagen triple-helix. From reference 12, with permission.

absorption at 214 nm, and fractions are evaluated by SDS–polyacrylamide electrophoresis.

The NC1 domain can also be isolated from amniotic membranes according to *Protocol 16*. The procedure involves solubilization of the amnion with collagenase and fractionation of the solubilized proteins by antibody-affinity chromatography and molecular sieve chromatography. Conventional chromatographic procedures could potentially be devised to circumvent the need for large amounts of antibody, but no such protocol has yet been published.

11. Purification of NC1 domain of type VII collagen

Protocol 16. Purification of the type VII collagen NC1 domain from human placenta

Equipment and reagents

- Digestion buffer: 5 mm $CaCl_2$, 10 ml inhibitor solution (see below), 50 mM Tris–HCl, pH 7.8
- Extraction buffer: 2 M deionized urea, 5 mM EDTA, 10 ml inhibitor solution (see below), 25 mM Tris–HCl, pH 7.8

125

Protocol 16. *Continued*

- Inhibitor solution: 625 mg NEM and 150 mg PMSF first dissolved in 10 ml isopropanol
- DEAE buffer: 5 mM EDTA, 0.2 M NaCl, 2 M deionized urea, 50 mM Tris–HCl, pH 7.8
- Clostridial collagenase (obtained from Worthington Biochemicals), reconstitute in 5 ml extraction buffer
- PBS
- NP-185 antibody affinity column: prepare the affinity column from Protein-G purified ascites fluid containing the NP-185 antibody (Telios, La Jolla). Conjugate purified antibody to 10 ml of packed CNBr-activated Sepharose 4B according to the manufacturer's instructions (Pharmacia).
- FPLC Mono Q column (Pharmacia): 1 ml, 1 ml/min flow rate, room temperature
- FPLC Superose 6 preparation grade column (Pharmacia): 50 × 1.6 cm, equilibrated and eluted with 100 mM NaCl, 50 mM Tris–HCl, pH 8.3, eluted at 0.5 ml/min into 1 ml fractions, monitored by UV absorbance at 220 nm.
- Sorvall RC-5B centrifuge with GSA rotor
- Polytron homogenizer
- Whatman DE-52 resin
- L8-80 ultracentrifuge, No. 19 rotor (Beckman Instruments)

Method

1. *DAY 1.* Prepare 400 g membranes (from 60–80 placentas) as described in steps 1 and 2, *Protocol 13*. Dissolve the membrane pellet in 400 ml of digestion buffer per 200 g wet weight. Stir in the cold until the pellet is disrupted into a uniform slurry. Add 4000 units clostridial collagenase per 200 g wet weight and stir at room temperature for 24 h.

2. *DAY 2.* Add an additional 4000 units clostridial collagenase per 200 g wet weight and stir at room temperature for 24 h.

3. *DAY 3.* Clarify the resulting solution by centrifuging at 13000 r.p.m. (27500 g) for 1 h at room temperature in the GSA rotor of a Sorvall RC-5B centrifuge. Decant the supernatant solution into a 2 l graduated cylinder and measure the volume. Discard the pellet.

4. Pour the measured supernatant solution into a 2 liter vacuum flask. Slowly (over 30 min) add 300 g $(NH_4)_2SO_4$ per litre with constant stirring. Degas briefly under vacuum and incubate the mixture on ice overnight.

5. *DAY 4.* Harvest the pellet at 4 °C as in step 3. Decant and discard the supernatant solution. Determine and record the approximate volumes of the pellets, and add extraction buffer to each centrifuge bottle in an amount equal to one-half the volume of the decanted supernatant (presumably about 200 ml = half of the original 400 ml bottle). Resuspend the pellets using a Polytron homogenizer in the centrifuge bottles and then transfer the suspensions into a total of four centrifuge bottles. Allow the suspension to stir for 30 min at 4 °C, and then clarify the suspension by centrifugation as in step 3. Filter the solution through two layers of Whatman No. 1 filter paper.

6. Adjust the conductivity of the solution to the conductivity of the DEAE buffer by adding NaCl or DEAE buffer *lacking* NaCl as needed. Wash 100 g Whatman DE-52 resin on a Buchner funnel over Whatman No. 1

paper with 500 ml of DEAE buffer, three times. Add the washed resin to the conductivity adjusted solution, and stir or gently shake the suspension at 4 °C for 10 min. Remove the resin by centrifugation as in step 3. Decant the supernatant solution into a 2 l beaker on ice. Resuspend the DEAE resin in 300 ml DEAE buffer per bottle and collect the resin by centrifugation as above. Add this wash to the original, and filter through two layers of Whatman No. 1 filter paper. Concentrate this solution by precipitation with 50% $(NH_4)_2SO_4$ as in step 4.

7. *DAY 5.* Harvest the precipitate as in step 3 and dissolve in 600 ml PBS and dialyse extensively versus PBS. Clarify the dialysate by centrifugation (18 500 r.p.m., (27 500 *g*) No. 19 rotor, 2 h, 4 °C, L8-80 ultracentrifuge). Apply the clarified sample to the NP-185 antibody affinity column. The antibody is conjugated to the column at a concentration of 1 ml of antibody solution (1–2 OD_{280} units per ml) per ml of swollen beads. Wash the column with two or more column volumes of PBS, and elute bound NC1 with 0.5 M HOAc, collecting the eluent in 5 ml fractions. Immediately neutralize the fractions by the addition of solid Tris-base to pH 8.0. Treat protein containing fractions with diisopropyl fluorophosphate (5 μg/ml) and dialyse extensively against 100 mM NaCl, 50 mM Tris–HCl, pH 8.3. Immediately wash the affinity column with PBS to retain antibody activity. Concentrate the dialysed sample by application to FPLC Mono Q, eluted with the same buffer containing 0.3 M NaCl.

8. *DAY 6.* Final purification is achieved by molecular sieve chromatography on FPLC Superose 6 (*Figure 11*). The final product (*Figure 12*) is essentially pure as judged by SDS–polyacrylamide gel electrophoresis. The yield is about 1.5 mg/400 g of membranes.

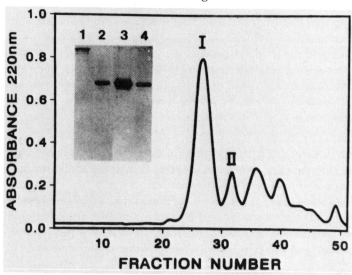

Figure 11. Fast protein liquid molecular-sieve chromatography of affinity-purified NC1 on Superose 6. Immunoreactive NC1 was detected in both peaks I and II. Inset: electrophoretic analysis of peak I (lanes 1 and 3) and of peak II (lanes 2 and 4) before (lanes 1 and 2) and after (lanes 3 and 4) reduction with 2-mercaptoethanol. 7.5% Polyacrylamide gel stained with Coomassie Blue. From reference 13, with permission.

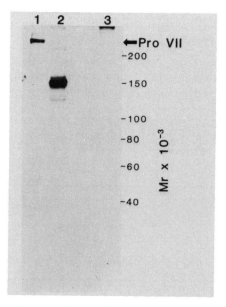

Figure 12. SDS–7.5%-polyacrylamide gel electrophoretic profile of type VII procollagen (lane 1), of isolated NC1, reduced (lane 2), and unreduced (lane 3). From reference 10, with permission.

128

References

1. Langeveld, J. P. M., Wieslander, J., Timoneda, J., McKinney, P., Butkowski, R. J., Wisdom, B. J., Jr, and Hudson, B. G. (1988). *J. Biol. Chem.*, **263**, 10481–8.
2. Gunwar, S., Bejarano, P., Kalluri, R., Langeveld, J. P. M., Wisdom, B. J., Jr, Noelken, M. E., and Hudson, B. G. (1991). *Am. J. Respir. Cell Mol. Biol.*, **5**, 107–12.
3. Hudson, B. G., Gunwar, S., Wisdom, B. J., Jr, Hudson, M. D., and Noelken, M. E. (1991). In *Nephrology* (ed. M. Hatano), pp. 852–6. Springer Verlag, Tokyo.
4. West, T. W., Fox, J. W., Jodlowski, M., Freytag, J. W., and Hudson, B. G. (1980). *J. Biol. Chem.*, **255**, 10451–9.
5. Dean, D. C., Barr, J. F., Freytag, J. W., and Hudson, B. G. (1983). *J. Biol. Chem.*, **258**, 590–6.
6. Paulsson, M., Aumailley, M., Deutzmann, R., Timpl, R., Beck, K., and Engel, J. (1987). *Eur. J. Biochem.*, **166**, 11–19.
7. Chung, A. E., Freeman, I. L., and Braginski, J. E. (1977). *Biochem. Biophys. Res. Commun.*, **37**, 859–67.
8. Chung, A. E., Jaffe, R., Freeman, I. L., Vergnes, J.-P., Braginski, J. E., and Carlin, B. E. (1979). *Cell*, **16**, 277–87.
9. Chakravarti, S., Tam, M. F., and Chung, A. E. (1990). *J. Biol. Chem.*, **265**, 10597–603.
10. Lunstrum, G. P., Sakai, L. Y., Keene, D. R., Morris, N. P., and Burgeson, R. E. (1986). *J. Biol. Chem.*, **261**, 9042–8.
11. Bentz, H., Morris, N. P., Murray, L. W., Sakai, L. Y., Hollister, D. W., and Burgeson, R. E. (1983). *Proc. Natl. Acad. Sci. (USA)*, **80**, 3168–72.
12. Morris, N. P., Keene, D. R., Glanville, R. W., Bentz, H., and Burgeson, R. E. (1986). *J. Biol. Chem.*, **261**, 5638–44.
13. Bächinger, H. P., Morris, N. P., Lunstrum, G. R., Keene, D. R., Rosenbaum, L. M., Compton, L. A., and Burgeson, R. E. (1990). *J. Biol. Chem.*, **265**, 10095–101.

5

Matrix components produced by endothelial cells: type VIII collagen, SPARC, and thrombospondin

E. HELENE SAGE and PAUL BORNSTEIN

1. Introduction

The development, growth, and response to injury of the blood vasculature depend in part on interactions of vessel wall cells with their extracellular matrix (ECM). In addition to providing cues for other cells, the endothelial ECM directs many of the functions and responses of the endothelium itself. Since changes in endothelial cell behaviour (for example attachment, proliferation, biosynthetic activity, and/or migration) have been causally linked to one or more specific ECM macromolecules, identification of matrix-associated proteins secreted by endothelial cells has become critical to our understanding of vascular cell biology. Few proteins, with the exception of von Willebrand protein, have been shown to be specific to endothelial cells. Although several of them were either initially described or studied extensively as products of endothelial cells *in vitro*, extracellular proteins such as fibronectin, laminin, plasminogen activators and their inhibitors, various collagen types, thrombospondin (TSP), and SPARC, have been ascribed functions within a broad range of cells and tissues. Because of their potential role in the regulation of endothelial cell growth, three of these proteins have been chosen as subjects for this chapter: type VIII collagen, TSP, and SPARC.

The relevance of these proteins to endothelial growth was suggested by their appearance in the culture medium as major biosynthetic products of subconfluent bovine aortic endothelial cells (BAEC). SPARC, a Ca^{2+}-binding glycoprotein associated with cells undergoing morphogenesis and remodelling, has been described as an inhibitor of cell spreading that binds to ECM (1, 2). TSP, a large, trimeric glycoprotein with several functional domains, acts as an attachment factor as well as a growth inhibitor for endothelial cells (3, 4). In contrast to SPARC and TSP, type VIII collagen has a restricted distribution *in vivo* that includes embryonic blood vessels and heart (5, 6). As shown in the concluding section of this chapter, levels of mRNA and protein corresponding to these macromolecules are modulated during the assembly of endothelial tubes *in vitro*.

It is important to mention that, despite their apparent relevance to the endothelium, the best sources of these proteins are not vascular endothelial cells. For example, type VIII collagen is most prevalent in the corneal Descemet's membrane (DM), which is produced, in part, by the corneal endothelium; TSP is generally purified from platelets. In several cases it has been instructive to isolate different forms of these molecules from tissues and from cultured cells, including BAEC. This chapter does not deal with the culture of endothelial cells *per se*. Several different strains of these cells are readily available and can be cultured according to standard techniques. Where appropriate, we have referred the reader to published research articles for background and additional information.

2. Type VIII collagen: purification and localization *in vitro* and in tissues

This section deals with the purification of native (triple-helical) and denatured (single chain) type VIII collagen. Intact, secreted molecules (generally referred to as procollagens) can be recovered from tissue culture medium, while significantly larger amounts of truncated molecules are extracted from certain tissues. For each protocol we have discussed its relative advantages, and, where possible, have evaluated other published procedures. A general approach for the purification of procollagens and collagens from cells and tissue can be found elsewhere (7).

Sources of specialized reagents have been provided. Other reagents are available from standard chemical suppliers (for example Sigma, Aldrich) and must be of Reagent grade. Type VIII collagen, especially as recovered *in vitro*, is a highly labile protein that adheres readily and irreversibly to many surfaces. Yields of native protein are maximized by the use of proteinase inhibitors, maintenance of early purification steps at 0–4°C, and treatment of surfaces (for example tubes, flasks) with a surfactant (ProSil-28, PCR Inc.). Individual chains of type VIII collagen are purified from denatured, triple-helical molecules. Although two M_r classes have been reported for intact type VIII collagen chains (8, 9), a protocol has been included for the purification of the larger chains, designated VIII-1 (M_r 180 000) and VIII-2 (M_r 125 000) (these have been referred to, respectively, as EC-1 and its derivative, EC-2 (8)). An intact chain of M_r 61 000 (9) has been shown by amino acid sequence analysis to correspond to 50K-B (VIII), a fragment of type VIII collagen extracted from corneal Descemet's membrane (DM) (10). The relationship of a second fragment from DM, 50K-A (VIII), to the type VIII collagen chains secreted by endothelial cells *in vitro* has not yet been established. We also do not know whether the different chains of type VIII collagen exist as homo- or heterotrimers and whether their relative levels of expression are specific to certain tissues.

2.1 Sources of type VIII collagen

Table 1 lists several cell types from which type VIII collagen has been prepared in reasonable quantities (8). Yields of VIII-1 and VIII-2 are given as radiochemical amounts on the basis of cell number, as absolute values for recovered, purified protein are difficult to determine. Rodent type VIII collagen can be obtained from murine brain capillary endothelium (13). The human osteosarcoma line MG-63 (ATCC CRL 1427) can be used as a source of human type VIII collagen. DM has so far proven to be the best source of type VIII collagen, although smaller amounts have been extracted from perichondrium (15). Digestion with pepsin appears to be necessary to release type VIII collagen from higher molecular weight complexes, since intact molecules could not be extracted from DM with the denaturant guanidinium–HCl (16).

Table 1. Best sources of type VIII collagen

Cell[a] or tissue	Comments	Reference
1. Bovine aortic endothelial cell	1–2% of total medium proteins; $1–2 \times 10^4$ d.p.m./10^6 cells; cells are readily available and easy to grow	11
2. Bovine corneal endothelial cell	1–7% of total medium proteins; 0.7×10^4 d.p.m./10^6 cells; cells are difficult to obtain in large quantity	12
3. Human Ewing's sarcoma cell (EW-A2)	>80% of total medium collagens; 4×10^6 c.p.m./70 ml medium[b]	13
4. Human astrocytoma cell (U-251 MG)	5% of total medium protein; $2–6 \times 10^6$ c.p.m./300 ml medium[c]; cells grow vigorously	14
5. Human hepatocellular carcinoma cell (Hep 3B21-7)[d]	>90% of total medium collagens; 1.25×10^4 c.p.m./10 ml medium	13
6. Bovine Descemet's membrane	approx. 10 mg native (pepsin-treated) 50K (VIII) from 100 DM; tissue may be difficult to obtain	15

[a] Recoveries are based on [^{3}H]proline c.p.m. incorporated into culture medium protein by ascorbate-supplemented cells over a period of 16–24 h, and are corrected for the proline content of collagen. The numbers are approximations only and depend largely on the secretory activity of the cultures, the amount and specific activity of the isotope (see reference), and the precautions taken during purification. Purification has been described in *Protocol 1*.
[b] 17–35 cm^2 flasks.
[c] 30–150 mm dishes, each containing 10 ml labelling medium.
[d] ATCC HB 8064.

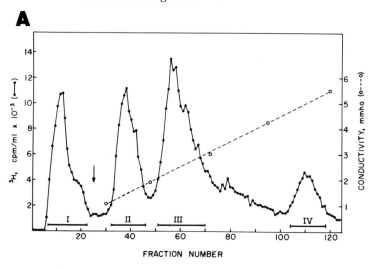

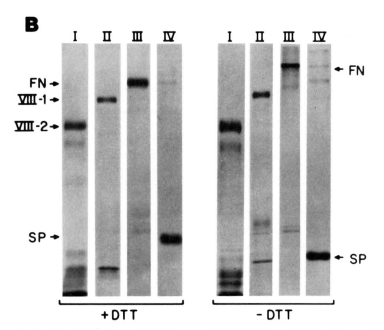

Figure 1. Fractionation of type VIII collagen and SPARC from BAEC culture medium. A, [³H]Proline-labelled proteins that were precipitated from culture medium at a concentration of ammonium sulfate between 20% and 50% (w/v) were chromatographed on DEAE-cellulose at 4°C with an NaCl gradient (arrow) of 0–200 mM, as described in *Protocol 1*. Roman numerals indicate pooled fractions. B, Fractions I–IV (shown in A) were resolved on a 6%/10% slab gel with and without reduction (±DTT), and proteins visualized by fluorescent autoradiography. Two chains of type VIII collagen (VIII-1 and VIII-2), SPARC (SP), and fibronectin (FN) are identified. (Reproduced with permission, from reference 11.)

134

2.2 Purification of type VIII collagen from cell culture medium

Both BAEC and U-251 MG cells have proved to be excellent sources of type VIII collagen, in part because they grow readily under standard conditions of tissue culture. In *Protocol 1*, steps A1–B3 describe the isolation of the native molecule from BAEC; the same procedure is followed with medium from U-251 MG cells, except that 30% $(NH_4)_2SO_4$ is used to precipitate type VIII collagen (step A6, with the elimination of step A7) (14). The elution pattern of BAEC culture medium proteins from DEAE–cellulose is shown in *Figure 1A*, and analysis of peak fractions by standard SDS-PAGE (with and without prior reduction of the samples) is seen in *Figure 1B*. Note that VIII-2 does not bind to the resin, while VIII-1 elutes at the inception of the NaCl gradient (these proteins appear as non-disulfide-bonded single chains due to the denaturing conditions of SDS-PAGE).

Alternatively, native, intact type VIII collagen can be purified directly from culture medium by chromatography on anti-type VIII collagen IgG linked to Sepharose (5). This procedure, while producing radiochemically pure type VIII collagen, requires large amounts of antibody for relatively low yields of purified protein.

Part C in *Protocol 1* describes the purification of VIII-1 and/or VIII-2 chains on CM-cellulose. The corresponding elution profile and SDS-PAGE analysis are shown in *Figure 2*. Chromatography under similar conditions can also be used to purify type VIII collagen that has been treated with pepsin (see Section 2.3 and reference 8).

Protocol 1. Purification of type VIII collagen from cell culture medium

Equipment and reagents

- BAEC cells
- Plastic tissue culture dishes (150 mm) (Falcon)
- DMEM (Gibco) + 10% FCS (Hyclone) containing penicillin G (100 units/ml) and streptomycin SO_4 (100 µg/ml) (Gibco)
- DMEM/antibiotics containing 50 µg/ml Na ascorbate, 64 µg/ml β-APN, 50 µCi/ml L-[2, 3, 4, 5-³H]proline (~ 100 Ci/mmol) (Amersham)
- PBS
- Clinical centrifuge
- Protease inhibitor cocktail (PIC): stock solution of 2 mM PMSF in absolute EtOH, NEM added in crystalline form directly to the medium; giving a final concentration of 0.2 mM PMSF, 10 mM NEM, 2.5 mM EDTA (pH 7.5) (PIC-1)

- Solid $(NH_4)_2SO_4$
- DE-1 buffer: 6 M urea (ultrapure, BRL enzyme grade), 50 mM Tris–HCl, pH 8.0, 0.2 mM PMSF, 2.5 mM EDTA
- 0.1 M acetic acid
- DE-52 cellulose (Whatman) column (2 × 20 cm)
- NaCl gradients: 0–200 mM in DE-1 buffer and 0–80 mM in CMC buffer
- 0.5 M NaCl in DE-1 buffer
- Pepstatin A (Peninsula labs): stock solution of 10 mg/ml in 100% DMSO
- CMC buffer: 6 M urea. 40 mM Na acetate, pH 4.8
- CM-cellulose (Whatman) water-jacketed column (2 × 8 cm)

Protocol 1. *Continued*

A. *Radiolabelling of BAEC and initial processing of culture medium*

1. Grow 20–50 150 mm plastic tissue culture dishes of BAEC in DMEM/FCS/antibiotics, until approximately 80% subconfluent.

2. Change medium to DMEM/antibiotics for 30 min. Incubate cells (10 ml/dish) in fresh DMEM/antibiotics containing 50 μg/ml Na ascorbate, 64 μg/ml β-APN, and 50 μCi/ml L-[2,3,4,5-^{3}H]proline for 18–24 h.

3. Remove medium from cells, wash cell layers once with 2 ml PBS and add to medium; discard cells.

4. Clarify medium by pelleting cells and debris in a clinical centrifuge (5 min).

5. To supernate add protease inhibitor cocktail (PIC) to produce a final concentration of 0.2 mM PMSF, 10 mM NEM, and 2.5 mM EDTA (pH 7.5) (PIC-1), and chill on ice to 0–4 °C. *All subsequent procedures are performed at 4 °C.*

6. Add solid $(NH_4)_2SO_4$ slowly (over several hours) to a final concentration of 20% (w/v) and stir overnight. Do not let the solution foam.

7. Centrifuge solution (48 000 *g*, 30 min) and bring supernate (containing type VIII collagen and SPARC) to 50% $(NH_4)_2SO_4$ (w/v) as described in step 6. (Note: pellet contains type III collagen and fibronectin and may be saved by solubilization in approximately 10 ml DE-1 buffer, followed by dialysis vs. 0.1 M acetic acid and lyophilization.)

8. Centrifuge supernate as in step 7. Solubilize in 20 ml (for 200 ml labelled medium, or approximately 10% of the initial starting volume) of DE-1 buffer and dialyse vs. DE-1 buffer (3 changes at a 1:10 ratio of sample to dialysis buffer, over 24 h).

9. Remove the sample from the dialysis bag, centrifuge to remove insoluble material (48 000 *g*, 20 min), and determine total c.p.m. (aliquot 10–25 μl/3 ml liquid scintillant) by scintillation spectrophotometry.

B. *Chromatography on DEAE-cellulose*

1. Pump sample on to a column of DE-52 cellulose, equilibrated at 4 °C in DE-1 buffer, at a rate of 30 ml/h. Follow with 3 column volumes of DE-1 buffer and collect fractions of 3 ml (this unbound fraction contains VIII-2 collagen).

2. Elute bound protein with a gradient of NaCl (0–200 mM) in a total volume of 400 ml DE-1 buffer. Collect 3 ml fractions. (Note: at the end

of a run, the column should be stripped with 0.5 M NaCl in DE-1 buffer and re-equilibrated in DE-1 buffer.)

3. Monitor column effluent by scintillation counting (approximately 20 µl of every other fraction mixed thoroughly with 3 ml liquid scintillant). After peak-containing fractions have been pooled, read conductivity values on approximately 10 remaining fractions (see *Figure 1A*).

4. Add pepstatin A (0.5 µg/ml) to pooled fractions, dialyse vs. 0.1 M acetic acid (4 changes at 1:200, for 2 days), and lyophilize.

5. Analyse 2.5–5×10^5 c.p.m. of each sample by SDS-PAGE as shown in *Figure 1B*. Type VIII collagen (VIII-2, M_r 125 000 according to collagenous protein standards) is in the unbound fraction, and VIII-1 (M_r 180 000) elutes slightly after inception of the gradient. The mobilities of both are essentially unchanged in the presence of DTT.

C. *Chromatography on CM-cellulose*

1. At room temperature, dissolve the lyophilized fraction (from step B4) containing VIII-1 or VIII-2 (a minimum of 10^6 c.p.m.) in 5–10 ml CMC buffer containing pepstatin A (0.5 µg/ml) and dialyse vs. CMC buffer (2 changes, 1:10 for 1 day). Centrifuge at room temperature (48 000 *g*, 20 min) to remove particulates, and count 10 µl.

2. Denature for 10 min at 39 °C.

3. Apply sample (maintained at 39 °C) to a H_2O-jacketed column of CM-cellulose, equilibrated in CMC buffer at 42 °C. Follow with 3 column volumes of CMC buffer.

4. Elute bound proteins with a gradient of NaCl (0–80 mM) in 200 ml CMC buffer. Monitor effluent by scintillation counting (for example see *Figure 2*) and collect 2 ml fractions.

5. Pool peak fractions, add pepstatin A to each pool (0.5 µg/ml), dialyse vs. 0.1 M acetic acid at 4 °C (as in step B4 above), and lyophilize. Read conductivities on approximately 10 remaining fractions.

6. Analyse 2–4×10^5 c.p.m. of each pooled sample on SDS-PAGE (±DTT). A single band is obtained at approximately 2 mmho (for example see *Figure 2*, inset).

2.3 Purification of type VIII collagen from bovine Descemet's membrane (DM)

A scheme used successfully for the purification of type VIII collagen from bovine (15, 16) and ovine DM (17) is shown in *Protocol 2*. In our experience, fresh eyes from calves or yearling cattle are preferable to either frozen or adult tissue. If fresh material (obtained usually from a local abbatoir) is difficult to procure, Pel-Freez is a reliable source of frozen tissue. Analysis of

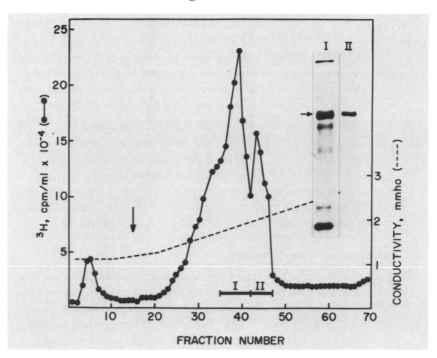

Figure 2. Purification of type VIII collagen by chromatography on CM-cellulose. [³H]Proline-labelled proteins from the culture medium of U-25I MG (astrocytoma) cells were precipitated in 30% ammonium sulfate and chromatographed under denaturing conditions on CM-cellulose, as described in *Protocol 1*. Gradient elution (arrow) was from 0–80 mM NaCl. *Inset*: Lanes correspond to pooled fractions, resolved by SDS-PAGE on a 6%/10% gel (− DTT); arrow indicates type VIII collagen chain (VIII–2). Reproduced with permission, from reference 14.

native collagens obtained at step A17 always shows a contamination of type VIII by type V collagen (however, no type I or III collagen should be present). It is important to keep this level of purity in mind if the preparation is to be used for the production or subsequent purification of anti-type VIII collagen antibodies.

Denaturation, followed by molecular-sieve chromatography, is necessary to remove type V collagen (*Protocol 2*, Part B, steps 1–4). *Figures 3A* and *3B* show the typical elution pattern and SDS-PAGE characteristics, respectively, of type VIII collagen chains, denoted as 50K (due to their M_r according to collagenous protein standards). Resolution of the apparent doublet in *Figure 3B (lane 4)* is achieved by HPLC (*Protocol 2C*). The two chains shown in *Figure 3C* (50K-A and 50K-B) are of sufficient purity for unambiguous sequence analysis (15).

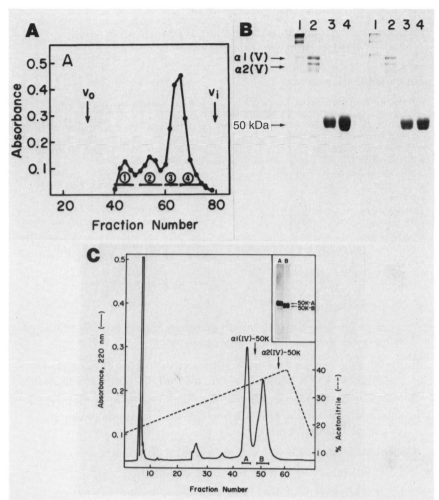

Figure 3. Purification of type VIII collagen from bovine Descemet's membrane. A, Proteins were solubilized by pepsin treatment of DM and precipitated in 1.5 M NaCl, as described in *Protocol 2*. Three to four milligrammes were denatured and chromatographed on Agarose A-1.5m. Pooled fractions are indicated. B, Fractions shown in A were resolved on an 8% polyacrylamide gel (– DTT, left and + DDT, right). The α chains of type V collagen and the 50 kDa chain of pepsin-treated type VIII collagen are indicated. C, A 200 μg sample as shown in B (lane 4) was chromatographed on a Vydac C18 column with a linear gradient of acetonitrile (20–40%) in 0.1% TFA. The elution positions of M_r 50 000 standard type IV collagen chains are indicated. *Inset*: Pooled fractions (A and B) were analysed on an 8% SDS-gel. 50 K-A and 50 K-B are identified. Reproduced with permission from reference 15.

139

Protocol 2. Purification of type VIII collagen from bovine Descemet's membrane

Equipment and reagents

- Fresh or frozen bovine eyes
- PIC-2 (final concentrations: 0.1 M caproic acid, 0.1 M EDTA, 5 mM NEM, 5 mM benzamidine–HCl, 1 mM PMSF)
- 10 mM Tris–HCl, pH 7.4
- 0.1% NaN$_3$
- 3% Triton X-100
- Solid NaCl
- 1 M NaCl
- DNase II (Sigma)
- 4% Na deoxycholate
- 0.5 M HOAc
- 0.5 M HOAc/0.7 M NaCl
- 0.1 M HOAc
- Pepsin (Sigma, 3 × crystallized)

- 1 M NaCl/50 mM Tris–HCl, pH 7.5
- Pepstatin A
- 1 M CaCl$_2$/50 mM Tris–HCl, pH 7.5 (2 M CaCl$_2$ stock solution, mixed with activated charcoal and filtered through Whatman 3M paper on a Buchner funnel)
- Microcentrifuge
- Agarose A-1.5m (Bio-Rad) column (1 × 90 cm)
- 0.1% Trifluoroacetic acid (TFA)
- Vydac C-18 reversed–phase HPLC column (Separation Group)
- 20–40% Acetonitrile gradient in 0.1% TFA
- SDS-PAGE analysis equipment

A. *Tissue preparation*

1. Soak or partially thaw 100 fresh or frozen bovine eyes in 3 litres of autoclaved, distilled (d) H$_2$O containing PIC-2.

2. Remove corneas with a razor blade, wash in H$_2$O/PIC-2, and store overnight at 4°C in 10 mM Tris–HCl, pH 7.4/PIC-2 (corneas will swell).

3. On ice, pull DM from corneal stroma with two pairs of toothed forceps, wash in running tap H$_2$O, and remove excess liquid with the aid of a large Buchner funnel.

4. Incubate sequentially at room temperature in 100 ml:

   (a) 0.1% NaN$_3$, 1–2 h, rinse in funnel with tap H$_2$O;

   (b) 3% Triton X-100, 1 h, rinse thoroughly in tap H$_2$O followed by dH$_2$O;

   (c) 1 M NaCl containing 0.5 mg/ml DNase II, 1 h; remove liquid;

   (d) 4% Na deoxycholate in 0.1% NaN$_3$, 4 h; remove liquid slowly to avoid foaming. Transfer DM to beaker on ice.

5. Suspend briefly in 300 ml ice-cold 0.5 M acetic acid, filter, and resuspend in 300 ml 0.5 M acetic acid.

6. Add pepsin (0.5 mg/ml) and stir 8–12 h on ice.

7. Clarify the extract by centrifugation at 10 000 *g*, 30 min, 4°C.

8. Freeze supernatant and lyophilize (do not allow to thaw).

9. Solubilize lyophilized residue at 1 mg/ml in 1 M NaCl/50 mM Tris–HCl (pH 7.5) by stirring for 1–4 h (*perform this and subsequent steps at*

4 °C). Check pH several times. Remove insoluble material by centrifugation as in step 7.

10. Add solid NaCl with constant stirring to a final concentration of 4 M; stir slowly overnight (do not let solution foam).

11. Collect white, flocculent precipitate by centrifugation as in step 7.

12. Solubilize precipitate (approximately 12 mg) at 10 mg (wet weight)/ml in 0.5 M acetic acid by stirring for 1–4 h. Add pepstatin A (10 μg/ml).

13. Dialyse vs. 0.5 M acetic acid, 12–24 h (3 changes at 1:100 (v/v)).

14. Dialyse vs. 0.5 M acetic acid containing 0.7 M NaCl, 12 h (or overnight). Remove contents from dialysis bag and centrifuge as in step 7. The pellet contains largely native type I collagen (approximately 60 mg).

15. Add solid NaCl slowly to the supernate to a final concentration of 1.5 M and stir for 12 h (or overnight).

16. Centrifuge as in step 7 and discard supernate.

17. Solubilize pellet at 1 mg wet weight/ml of 0.5 M acetic acid, dialyse vs. 0.5 M acetic acid (3 changes at 1:100 (v/v)), and lyophilize. Yield: approximately 10 mg each of types V and VIII-50K collagens.

B. *Molecular-sieve chromatography*

1. Dissolve 2–3 mg of protein from step A17 in 1 ml 1 M $CaCl_2$/50 mM Tris–HCl (pH 7.5).

2. Denature at 55 °C, 20 min.

3. Clarify solution by microcentrifugation at room temperature, 2 min.

4. Apply supernate to the agarose column equilibrated with step B1 buffer. Collect fractions of 1 ml, at a flow rate of approx. 12 ml/h at room temperature, and monitor the effluent by absorption at 230 nm.

5. Pool peak-containing fractions (see *Figure 3A*), dialyse vs. 0.1 M acetic acid, and lyophilize.

6. Analyse by SDS-PAGE (*Figure 3B*). Major, final peak (fraction 4) should contain only collagenous components of 50 kDa (these migrate with an apparent M_r of 65–70 kDa according to globular protein standards), the mobility of which is unchanged in the presence of DTT.

C. *HPLC*

1. Dissolve 200–500 μg of sample from step B6 in 1 ml 0.1% TFA and heat at 55 °C for 2–3 min.

2. Inject on to a Vydac C-18 reversed-phase HPLC column (Separations Group), and elute proteins in 0.6 ml fractions with a linear gradient of 20–40% acetonitrile in 0.1% TFA.

E. Helene Sage and Paul Bornstein

Protocol 2. *Continued*

3. Monitor fractions at 220 nm, pool peak fractions, dialyse vs. 0.1 M acetic acid at 4 °C, and lyophilize.

4. Analyse by SDS-PAGE (*Figure 3C*). The first major peak corresponds to 50K-A, and the second to 50K-B. On 5% or 8% gels, 50K-B exhibits a slightly greater mobility than 50K-A.

2.4 Immunolocalization of type VIII collagen

Procedures for the localization of type VIII collagen in cultured cells and in tissues are outlined in *Protocols 3* and *4*. Performance of these experiments relies on the use of affinity-purified anti-type VIII collagen IgG. Both monoclonal and polyclonal anti-type VIII collagen antisera have been generated successfully by several investigators (see refs. 5, 16, and 17 for further information). An example of staining in endothelial cells for type VIII collagen is shown in *Figure 4A* and *4B*. In subconfluent cells, most of the reaction

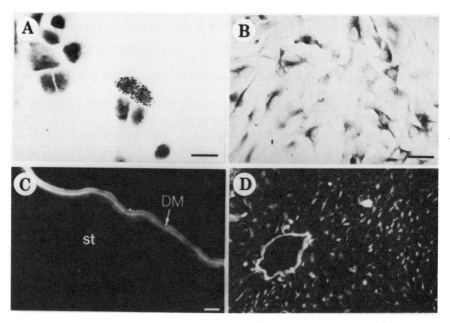

Figure 4. Localization of type VIII collagen in cultured endothelial cells, Descemet's membrane, and embryonic vessels. A, affinity-purified anti-type VIII collagen IgG was used to identify type VIII collagen in cultures of rat heart endothelium (18); B, porcine coronary artery endothelium (Dr C. Johnson, Mayo Clinic, Rochester, MN); C, bovine DM; D, blood vessel in 16 d embryonic mouse head mesenchyme. Immune complexes were detected by immunoperoxidase (A and B) or by immunofluorescence (C and D) (see *Protocol 4*). st, stroma. Bars = 20 μm.

product is seen associated with endoplasmic reticulum and secretory granules. Confluent cultures deposit type VIII collagen in the ECM (revealed by removing cells with a 0.1% solution of Triton X-100 in PBS; see ref. 5 for further details).

Examples of immunohistochemistry are seen in *Figure 4C* and *4D*. The extensive staining of DM is uniquely characteristic for active preparations of anti-type VIII collagen IgG (DM is often used as a positive control tissue for this reason). Type VIII collagen is also synthesized by mesenchymal cells of the developing mouse and chicken heart (6). We have observed greater histological detail when staining was performed with the avidin-biotin-peroxidase technique (*Protocol 4,* part B, step 9), followed by counterstaining of the tissue with Toluidine blue, compared to an immunofluorescent technique. Frozen sections have also been used successfully for the localization of type VIII collagen (5).

Experiments involving immunolocalization should include the following controls (preferably on serial sections):

(a) substitution of normal (preimmune) rabbit IgG (Sigma) at the same concentration used for the primary antibody;

(b) use of secondary antibody alone;

(c) incubation with excess purified antigen (for example type VIII-50K collagen);

(d) use of a series of concentrations of more than one preparation of primary antibody (preferably from more than one animal); and

(e) use of antibodies directed against other proteins that are expected to have similar or clearly different distributions.

Protocol 3. Immunolocalization of type VIII collagen in cultured cells

Equipment and reagents

Note: Millipore-filter all solutions before use

- Glass coverslips or LabTek (Nunc) chamber slides
- Serum-free DMEM
- 3% Paraformaldehyde in PBS containing 1% sucrose, 1 mM MgCl$_2$, 0.1 mM CaCl$_2$, pH 7.3 (add 1.2 ml 1 M NaOH/100 ml, heat at 50°C, adjust pH with 1 M HCl)
- PBS
- 0.05 M Glycine in PBS
- 90% Glycerol/10% PBS containing 2% *n*-propylgallate, pH 7.5
- Clear nail polish or rubber cement

- Antibody 1: Polyclonal anti-bovine type VIII (50K) collagen IgG (15) was precipitated from whole antisera by addition of ammonium sulfate to final concentration of 20% (w/v) and adsorbed against type V collagen coupled to Sepharose CL-4B (Pharmacia). Affinity-purified anti-type VIII collagen IgGs, eluted from type VIII-50K collagen coupled to Sepharose CL-4B at typical concentrations of 5–10 mg/ml.
- Antibody 2: FITC-conjugated goat anti-rabbit IgG (Cappel)

Protocol 3. *Continued*

Method

1. Grow cells on glass coverslips or LabTek chamber slides under normal conditions until degree of confluence reaches approximately 80%.

2. Remove medium; incubate in serum-free DMEM, 20 min at room temperature.

3. Fix for 30 min in paraformaldehyde solution detailed above.

4. Rinse once in PBS for 10 min.

5. Rinse twice in 0.05 M glycine in PBS, 10 min each.

6. At this point, if cells are to be rendered permeable, put methanol or acetone (stored at $-20\,°C$) in a glass Petri dish seated on a slab of dry ice. Immerse coverslip or slide (cell side up—use fine-point, stainless steel forceps) for 15 sec; return coverslip to plastic dish.

7. Rinse three times in PBS, 5 min each.

8. Expose slide or coverslip to antibody 1, 30 min. Use approximately 300 µl per sample (or enough to cover the cells), and incubate in a plastic box containing wet paper towels (to prevent desiccation).

9. Rinse twice in PBS, 10 min each.

10. Expose to antibody 2, as in step 8, at a dilution of 1:1000 (Note: see manufacturer's instructions for optimal concentrations).

11. Rinse three times with PBS, 10 min each. After the last rinse, swab non-cell side with H_2O-soaked Q-tip (to remove salts).

12. Invert coverslip on a drop (approx. 50 µl) of 90% glycerol/10% PBS containing 2% *n*-propylgallate (pH 7.5), or remove gasket and place a long coverslip over slide.

13. Seal around edges of coverslip with clear nail polish or rubber cement. Slides can be kept covered with foil at $4\,°C$ for approximately 1 week.

Protocol 4. Immunohistochemistry of type VIII collagen

Note: For detection of antigen by immunofluorescence, rather than by immunoperoxidase, omit step 5 and substitute step 9 with step 9a.

Equipment and reagents

- Karnovsky's solution: 2.5% glutaraldehyde, 3% paraformaldehyde in 0.1 M Na cacodylate
- Xylene
- Paraplast (Sherwood Medical)
- 0.1 M Cacodylate buffer, pH 7.6
- Series of ethanol in water: 50%, 70%, 80%, 95%, 100%
- Avidin–biotin–peroxidase complex (ABC reagent, Vector Labs)

- 0.1% Poly-L-lysine precoated slides
- PBS
- 70% Methanol containing 3% H_2O_2
- 1% Goat serum/PBS
- Affinity-purified anti-type VIII collagen IgG
- Biotinylated goat anti-rabbit IgG (Vector Labs)

- 3,3'-Diaminobenzidine–4-HCl: 3 mg/ml in 0.05 M Tris–HCl, pH 7.6, containing H_2O_2
- 90% Glycerol/10% PBS containing 2% *n*-propylgallate
- 0.5% Toluidine blue
- Permount
- Colour-film for photomicroscopy

Method

1. Fix tissues in Karnovsky's solution for 3 h at 4 °C, with constant stirring.

2. Wash tissues in cacodylate buffer, dehydrate in a graded series of ethanol and H_2O (50%, 70%, 80%, 95%, 100%) (4 h total), and rinse twice in xylene (1 h total).

3. Embed tissues in Paraplast, cut 5–6 μm sections, place on precoated slides, dry overnight at 37 °C, and store at 4 °C (maximum of 1 year).

4. Remove paraffin from sections by two washes in xylene (5 min each), rehydrate in a graded series of ethanol solutions (2 × 100%, 2 × 95%, 1 × 80%, and 2 × 70%), and wash four times in PBS (3 min each).

5. Incubate sections for 30 min in methanol/H_2O_2 (to inactivate endogenous peroxidases), followed by PBS.

6. Incubate sections in 1% goat serum/PBS for 2 h at 4 °C (to reduce non-specific binding).

7. Add affinity-purified anti-type VIII collagen IgG (20–50 μg/ml) to sections. Place in a humidified chamber for 1–2 h at 4 °C.

8. Rinse slides extensively in PBS.

9. Expose sections sequentially (at 4 °C) to:

    (a) Biotinylated goat anti-rabbit IgG for 1 h;

    (b) 3–4 rinses in PBS (3 min each);

    (c) avidin–biotin–peroxidase complex for 30 min;

    (d) 3–4 rinses in PBS;

    (e) 3, 3'-diaminobenzidine-4-HCl/H_2O_2 for 8–10 min;

    (f) 3–4 rinses in PBS.

OR

9a. Incubate the sections with goat anti-rabbit IgG conjugated to FITC in a humidified chamber for 1–2 h at 4 °C. Rinse three times with PBS (5 min each) and mount with 90% glycerol/10% PBS containing 2% *n*-propylgallate.

10. Rinse sections in tap H_2O and counterstain with 0.5% Toluidine blue.

11. Dehydrate sections in a graded series of ethanol solutions (2 × 70%, 1 × 80%, 2 × 95%, and 2 × 100%, 5 min each), clarify twice in xylene, 10 min each, and mount in Permount.

Protocol 4. *Continued*

12. Perform photomicroscopy with Kodak ASA 400 Ektachrome (for immunofluorescence) or ASA 160 (for immunoperoxidase) colour film.

Note. These procedures can be used for immunolocalization of SPARC. Anti-mouse SPARC IgG is used at 10–50 μg/ml for cells and tissues (1) (see also footnote a, Table 6). Fixation of tissue in Bouin's reagent (0.9% picric acid and 9% formaldehyde in 5% acetic acid) produced optimal results. Polyclonal anti-bovine type VIII (50K) collagen IgG (15) was precipitated from whole antisera by addition of ammonium sulphate to a final concentration of 20% (w/v) and was adsorbed against type V collagen coupled to Sepharose CL-4B (Pharmacia). Affinity-purified ant-type VIII collagen IgGs were eluted from type VIII-50K collagen coupled to Sepharose CL-4B at typical concentrations of 5–10 mg/ml.

3. SPARC: purification and localization *in vitro* and *in vivo*

SPARC (termed 43K protein) was initially purified from BAEC culture medium (19). It was subsequently found that almost all cultured endothelial cells secreted SPARC, which comprised 0.8–1.1% of the total radiolabelled protein in the medium after 18–24 h. Although several laboratories have purified SPARC (osteonectin, BM-40) from tissues, we have developed a protocol which yields up to 100 μg (per 10^7 cells) of native, undegraded protein (1, 2; see also references therein for osteonectin and BM-40). This procedure, outlined in *Protocol 5* utilizes a rapidly growing, readily available cell line derived from a murine parietal yolk sac carcinoma, PYS-2. Serum-free conditions are necessary to minimize BSA and a 70 kDa protein, both of which bind to SPARC. SPARC can be radiolabelled efficiently with [^{35}S]methionine; however, for most purposes unlabelled SPARC can be produced.

3.1 Purification of SPARC from cell culture medium

Protocol 5. Purification of SPARC from PYS cell culture medium

Equipment and reagents

- Falcon plastic TC plates (150 mm) or flasks
- DMEM ± 10% FCS
- [^{35}S]Methionine (Amersham)
- PMSF
- NEM
- Solid $(NH_4)_2SO_4$
- DE-2 buffer: 4 M urea, 50 mM Tris–HCl, pH 8.0
- DE-52 cellulose column (2 × 20 cm)
- NaCl gradient (50–200 mM)
- 1.5 mM monobasic Na phosphate buffer, pH 5.5
- TBS (50 mM Tris–HCl, 150 mM NaCl, pH 7.5), containing 0.2 mM PMSF
- Sephadex G-200 (Pharmacia) column (1 × 100 cm)
- 0.05 M acetic acid
- SDS-PAGE ± DTT analysis equipment
- Coomassie Blue

A. *Radiolabelling of PYS cells and initial processing of culture medium*

1. Grow 20–150 mm plates or flasks of PYS cells (approximately 7–10 × 10^6 cells per plate) to 50–70% confluence. (Note: it is critical that the cells do not reach confluence, as production of SPARC essentially ceases at this stage of growth.) Cells grow rapidly (> 1 population doubling/24 h) in DMEM/10% FCS.

2. Incubate cells for 24 h in serum-free DMEM (10 ml per dish). To monitor purification, add [^{35}S]methionine (500 μCi/culture dish) to one dish and process in combination with non-labelled-medium from the other dishes.

3. Collect medium, remove cellular debris in a clinical centrifuge, add PMSF and NEM (final concentrations of 0.2 mM and 10 mM, respectively), and stir on ice until medium reaches 4 °C.

4. Add solid $(NH_4)_2SO_4$ over a period of several hours to a final concentration of 50% (w/v). Stir at 4 °C for a minimum of 12 h.

5. Centrifuge medium (48 000 g, 30 min) and discard supernate. *Perform this and subsequent steps at 4 °C.*

B. *DEAE-cellulose*

1. Dissolve pellets in approximately 20 ml DE-2 buffer and dialyse vs. DE-2 buffer (2–3 changes at 1:10, 4–6 h each).

2. Chromatograph on a DE-52 cellulose column equilibrated at 4 °C in DE-2 buffer. Discard initial eluate (representing unbound material). Elute bound proteins with a linear gradient of 50–200 mM NaCl (400 ml total volume). Alternatively, two sequential step gradients of 75 mM NaCl (100 ml) and 175 mM NaCl (100 ml) can be used. Collect 3 ml fractions.

3. Monitor column by scintillation counting (50 μl from every other fraction per 3 ml scintillant). Pool fractions which eluted as a peak at 150–175 mM NaCl (see *Figure 1A*, equivalent to 4.5–5 mmho).

4. Dialyse pooled fractions (generally 20 ml) vs. 2 × 6 l of Na phosphate buffer (EDTA must not be present at any time during the procedure). Then dialyse vs. distilled (d) H_2O (2 × 6 l). After 2–3 d, a precipitate enriched in SPARC will appear in the bag. Decant entire contents into centrifuge tube. If no precipitate occurs in step B4, proceed with lyophilization (Step B5). The lyophilized protein can be redissolved in 25% of the original volume and step B4 repeated.

5. Centrifuge (48 000 g, 30 min), resuspend pellet in 2–4 ml d H_2O, and lyophilize. (Note: supernate can also be lyophilized to monitor efficiency of precipitation.)

C. *Chromatography on Sephadex G-200*

1. Dissolve 0.25–0.5 mg lyophilized protein from step B5 (H_2O precipitate) in 1 ml TBS/PMSF (this amount should correspond to 1–10 × 10^6 total

Protocol 5. *Continued*

c.p.m.). Stir for 4–6 h at 4 °C and clarify by microcentrifugation for 1–2 min.

2. Apply supernate to a column of Sephadex G-200, equilibrated at 4 °C in TBS. Column effluent is pulled by a peristaltic pump at approximately 10 ml/h. Collect 80 fractions of 1 ml each and monitor effluent by absorbance at 280 nm and/or scintillation counting.

3. Pool peak fractions as shown in Figure 5B, dialyse vs. 0.05 M acetic acid, and lyophilize. Alternatively, samples can be frozen in TBS at −70 °C, or dialysed directly vs. PBS (for cell culture studies) and then stored at −70 °C.

4. Analyse lyophilized samples (1–5 µg) by SDS-PAGE ± DTT (see *Figure 5B*). Stain with Coomassie Blue. In addition, fluorescent autoradiography can be performed; use approximately 10^4 c.p.m./sample. A single broad band, or occasionally a doublet, should be obtained with an apparent M_r of 39 000 (−DTT) and 43 000 (+DTT, co-migration with ovalbumin protein standard).

Yield: 250–500 µg purified SPARC from 5×10^7 subconfluent PYS cells (1).

As an alternative to PYS cells, SPARC can be purified from BAEC culture medium (see *Protocol 1*, steps A1–B4 with the following modifications: ascorbate and β-APN can be omitted, and DE-2 buffer (see *Protocol 5*, Part B, step 1) can be substituted for DE-1 buffer). As seen in *Figure 1A*, SPARC is eluted from DEAE–cellulose at 150–175 mM NaCl (4.5–5 mmho). On SDS-PAGE, it exhibits a characteristic shift in mobility in the presence of DTT (from M_r 30 000 to 43 000 after reduction) (*Figure 1B*, lane IV). Although SPARC isolated from PYS cells appears radiochemically pure (*Figure 5A*, lane 5), BSA, laminin, and other serum proteins are also present (lane 1). A large proportion of these contaminants can be removed by precipitation of SPARC at low ionic strength and acidic pH (*Protocol 5*, Part B, steps 4–5) (*Figure 5A*, lane 2). Further purification is achieved by molecular sieve chromatography (*Protocol 5*, Part C, steps 1–4 and *Figure 5B*). The purified protein (shown in *Figure 5A*, lanes 3 and 7) is suitable for sequence analysis. The degree of native structure can be assessed by circular dichroism (2) and by the ability of the protein to inhibit spreading of BAEC (2). SPARC is relatively stable and can be stored at −70 °C in TBS containing 4 mM Ca^{2+}, or lyophilized at 4 °C for up to 6 months.

3.2 Immunochemical procedures for the detection of SPARC

Western (immuno) blotting of SPARC is accomplished by a standard pro-

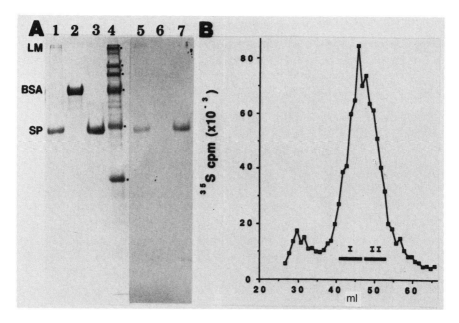

Figure 5. Purification of SPARC from PYS cell culture medium. [^{35}S]Methionine-labelled proteins from PYS cell culture medium were precipitated in 50% ammonium sulfate and initially chromatographed on DEAE-cellulose as described in *Protocol 5*. A fraction enriched in SPARC (eluted at 150 mM NaCl) was dialysed against H$_2$O (pH 5.5) at 4°C. A, Analysis by SDS-PAGE on a 10% gel (+ DTT) of: lane 1, starting material prior to dialysis; lane 2, supernate after dialysis; lane 3, precipitate after dialysis and subsequent chromatography on Sephadex G-200 (see B, pool I); lane 4, protein molecular mass standards (dots indicate 205, 116, 97, 66, 45, and 29 kDa); lanes 5–7 represent fluorescent autoradiograms of lanes 1–3. SPARC (SP), BSA, and laminin (LM) are identified after staining with Coomassie Blue. B, Protein precipitate after dialysis was chromatographed on Sephadex G-200 in TBS; pooled fractions are indicated by Roman numerals. Peak II is shown in A, lanes 3 and 7. (Reproduced with permission from references 1 and 2.)

tocol as outlined in *Protocol 6*. Anti-SPARC antisera have been produced in several laboratories (1, and references therein for further information). Although the protein is highly antigenic, a high degree of purity of the immunogen is critical to the production of specific antibodies. In *Figure 6* are shown representative examples of the distribution of SPARC in PYS cells (A) and in the endothelium lining the maternal sinuses of murine placenta (B). In culture, the protein does not appear in the ECM, and in tissues its location is principally intracellular (1).

Protocol 7 describes a procedure for the extraction of total RNA from BAEC and detection of SPARC mRNA by Northern blotting. Since BAEC transcribe appreciably high levels of SPARC, it is not necessary to enrich for SPARC poly A$^+$ RNA by further purification on columns of oligo dT.

E. Helene Sage and Paul Bornstein

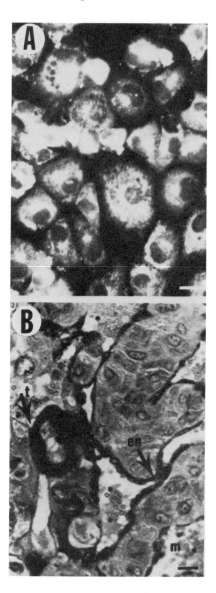

Figure 6. Immunolocalization of SPARC in cultured cells and tissues. A, PYS cells were rendered permeable and reacted sequentially with anti-SPARC IgG and FITC-goat anti-rabbit IgG as described in *Protocol 3*. Note cytoplasmic, vesicular staining and lack of reactivity in the ECM. Bar = 20 μm. B, Paraffin sections of 10 d mouse embryo and placenta were exposed to anti-SPARC IgG, and immune complexes were visualized by an avidin–biotin–peroxidase method, as described in *Protocol 4*. Specific staining was seen in endothelial cells (en) lining the maternal sinuses (m) as well as in trophoblastic giant cells (t). Bar = 50 μm. (Reproduced with permission, from reference 1.)

Protocol 6. Immunoblotting of SPARC protein

Equipment and reagents

- Prestained M_r standards (BRL)
- SDS gel ± DTT
- Nitrocellulose sheet (Schleicher and Schuell)
- 1% Amido Black dissolved in 10% HOAc/ 20% MeOH
- Coomassie Blue

- MT buffer: 0.1% non-fat dry milk, 0.05% Tween-20, 0.01% NaN_3 in PBS
- Anti-SPARC IgG [a]
- Seal-A-Meal bags (DAZEY)
- [^{125}I]Protein A
- 0.05% Tween–20/PBS
- Cassette with X-ray film

Method

1. Run samples including a positive control (5–50 ng SPARC) and pre-stained M_r standards on an SDS-gel (± DTT).

2. Transfer proteins electrophoretically at 4°C from the gel to a nitrocellulose sheet (approximately 0.5 amp for 1–2 h).

3. Stain nitrocellulose in 1% Amido black, and destain in 10% acetic acid/ 20% methanol. Photograph if desired (the transferred gel can be stained with Coomassie Blue and destained to check efficiency of transfer).

4. Block non-specific binding sites on nitrocellulose sheet by incubation (1–2 h at room temperature or overnight at 4°C) in MT buffer. Change buffer once.

5. Seal sheet with 10 ml MT buffer containing anti-SPARC IgG in a Seal-A-Meal bag, and place on rocking shaker for 1–2 h at room temperature.

6. Remove sheet from bag, and wash 2–3 times in MT buffer (10 min each).

7. Seal sheet in a new bag with 10 ml MT buffer containing [^{125}I]Protein A (approximately 10 μCi or 10^7 c.p.m.) and shake for 1–2 h at room temperature.

8. Remove sheet, wash twice in MT buffer, twice in 0.05% Tween–20/ PBS, and air dry.

9. Place in a cassette with X-ray film at −70°C. Film can be developed after 1–2 d, and the sheet should be re-exposed several times for accurate densitometry readings.

Note: this same procedure can be used for identification of type VIII collagen from tissues or cells. Use anti-type VIII collagen IgG at a concentration of 20–50 μg/ml. Load ~10 ng purified 50 kDa (VIII) as a positive control; the amount of tissue or protein containing unknown quantities of type VIII collagen that can be loaded depends on the size of the gel (usually the upper limit is ~50 μg). More efficient transfer of collagens to nitrocellulose is achieved by blotting at 4°C for 16 h.

[a] Anti-SPARC antiserum is used at a minimum dilution of 1:100 (v/v). Anti-SPARC IgG was precipitated from whole antisera by addition of ammonium sulfate to a final concentration of 20% (w/v). Antibody concentrations are determined by absorbance of IgG fractions at 280 nm and extrapolation to a standard curve for normal rabbit IgG. IgG specific for SPARC, purified by affinity chromatography on Sepharose CL-4B coupled to purified SPARC, ranged in concentration from 0.1–0.225 mg/ml (1) and is used at dilutions from 1:250 to 1:500.

Protocol 7. Detection of SPARC mRNA by Northern blotting

Note: Use gloves when working with RNA. All aqueous solutions should be made with $DepH_2O$ and autoclaved. The use of molecular biology grade, RNase-free reagents (e.g. IBI) is strongly recommended.

Equipment and reagents

- Solution A: 293 ml DEPC-treated H_2O ($DepH_2O$), 250 g guanidinium thiocyanate, 17.6 ml 0.75 M Na citrate, 26.4 ml 10% sarkosyl. Filter sterilize. Solution is stable for 3 months
- Solution B: 50 ml solution A + 360 µl 100% β-mercaptoethanol
- Serum-free DMEM
- Polypropylene tubes (Falcon)
- 2 M Na acetate, pH 4.0
- Phenol: saturated with $DepH_2O$
- Chloroform:isoamylalcohol (25:1, v/v)
- Isopropanol
- 1.2% Agarose gel (10 × 14 cm) (Agarose Molecular Biology grade, IBI)
- 10 × MOPS buffer stock solution (1 litre): 4.18% MOPS, 16.6 ml 3 M Na acetate, 20 ml 0.5 M EDTA, pH 8.0, and 3 ml 37% formaldehyde, pH ≥ 4.0
- Solution N (1 ml): 225 µl $DepH_2O$, 500 µl deionized formamide, 100 µl 10 × MOPS, 175 µl 37% formaldehyde
- RNA gel buffer: 50% glycerol, 1 ml EDTA, 0.04% bromophenol blue, 0.04% xylene cyanol in 1 × MOPS solution

- Ethidium bromide
- Nitrocellulose sheet
- SSC (1 ×): 0.15 M NaCl, 0.015 M Na citrate, pH 7.0
- Denaturing buffer: 50 mM NaOH, 10 mM NaCl
- Neutralizing buffer: 0.1 M Tris–HCl, pH 7.4
- Vacuum (Vacugene LKB)
- RNA M_r markers (Boehringer–Mannheim or BRL)
- UV irradiation source (Stratalinker, Stratagene)
- Seal-A-Meal plastic bags
- Solution H: 50% deionized formamide, 30% 20 × SSC, 50 mM sodium phosphate, 10 µg yeast total RNA, 4% 50 × Denhardt's solution
- Denhardt's solution (50 ×): 1% Ficoll, 1% polyvinylpyrrolidone, 1.1% BSA
- ^{32}P-labelled probe (cDNA probes are nick-translated with a Multi-Prime Kit (Amersham) and are purified with Gene-Clean (Bio 101)). Specific activities should be 10^8–10^9 c.p.m./µg DNA).

A. *Preparation of total RNA from BAEC (modified from reference 20. This same protocol can be used for Northern blotting of TSP)*
   Perform steps 1–7 at room temperature.

1. Rinse cells 3 times with serum-free DMEM (5 min each).

2. Add solution B (1 ml/60 mm dish) to promote cell lysis.

3. After 1 min, pipette contents of dish up and down (2–3 times): avoid bubbles. Transfer to a 15 ml sterile polypropylene tube.

4. Shear DNA by aspirating 5 times with a 22 G needle and a 3–5 ml syringe.

5. Add 0.1 ml 2 M Na acetate (pH 4.0) (final concentration is 0.2 M Na acetate); mix thoroughly by inverting the tube.

6. Add 1 ml phenol (saturated with DepH$_2$O) and mix as in step 5.

7. Add 0.2 ml of chloroform:isoamylalcohol. Mix vigorously by shaking for 10 sec. Put on ice for 15 min.

8. Centrifuge (2500 g), 20 min at 4 °C.

9. Take the aqueous (top) layer and transfer it with a sterile, plastic pipette to a clean 15 ml polypropylene tube.

10. Add an equal volume of isopropanol and put at − 20 °C for a minimum of 3 h (or overnight).

11. Centrifuge as in step 8.

12. Dissolve pellet in 300 μl solution B. Transfer to an Eppendorf tube (Note: baking glassware, tubes, and Eppendorf tips at 180 °C for at least 2 h is strongly recommended for labware used for isolating RNA).

13. Add an equal volume of isopropanol and leave overnight at − 20 °C.

14. Centrifuge (10 000 g in a microcentrifuge) at 4 °C for 15 min.

15. Rinse pellet gently with 70% ethanol. Aspirate drops on the side of tube with a pulled-point Pasteur pipette.

16. Let pellet dry for 5 min on the bench top.

17. Resuspend pellet in DepH$_2$O (usually 50 μl is optimal for a 60 mm dish of cells).

18. Measure concentration of total RNA in a spectrophotometer at 260 nm. An absorbance of 1.0 at 260 nm corresponds to 40 μg RNA/ml. For an initial reading, dilute 1 μl RNA solution to 400 μl DepH$_2$O. Use a 1 ml quartz cuvette reserved for this purpose: residual protein in the preparation can be estimated by reading the solution at 280 nm.

B. *Analysis of SPARC mRNA by Northern blot*

1. Prepare a horizontal 1.2% agarose gel by combining 1.2 g agarose with 87 ml DepH$_2$O. Heat until agarose is completely dissolved (1–2 min in a microwave oven), cool to 65 °C, and add 10 ml of 10 × MOPS buffer.

2. Resuspend 10 μg of BAEC total RNA into 20 μl of solution N.

3. Seal the tube and heat to 55 °C for 15 min.

4. Chill immediately on ice and add 2 μl of RNA gel buffer.

5. Place the gel in the electrophoresis apparatus, cover with 1 × MOPS, and load the RNA samples.

6. Perform electrophoresis at 80 V for 4 h. At this point the first dye should be approximately 5 cm from the bottom of gel.

Protocol 7. *Continued*

7. Remove the gel from the elecrophoresis apparatus and place in a glass container. Add DepH$_2$O containing 0.5 μg/ml ethidium bromide (EtBr) to cover the gel and stain for 5 min. Remove staining solution and treat with bleach to inactivate EtBr. Wash RNA gel twice with DepH$_2$O (15 min each) and leave the gel 8–16 h in DepH$_2$O to remove formaldehyde and EtBr.

8. Examine gel under a UV transilluminator to verify the integrity of the RNA. At this point the gel can be photographed.

9. Cut a piece of nitrocellulose slightly smaller than the size of the gel. Hydrate the sheet in DepH$_2$O, followed by 10 × SSC for 15 min.

10. Soak the gel (5 min for vacuum transfer or 1 h for capillary transfer, see step 11) first in denaturing buffer, followed by a neutralizing buffer (5 min or 1 h, as for denaturing buffer), and finally in 10 × SSC (5 min or 15 min), prior to transfer. Position the nitrocellulose carefully over the gel and remove all bubbles.

11. Perform the transfer by capillary action (for at least 12 h) or vacuum (2 h). Verify efficiency of the transfer by examining both nitrocellulose and gel under UV. The gel should contain no RNA. Mark, with a lead pencil, the positions of the 28S and 18S rRNAs (these can be used as indicators of mRNA size in kb, but RNA M_r markers are preferable for this purpose).

12. Cross-link RNAs to the nitrocellulose by baking the sheet for 2 h at 80 °C or by UV-irradiation. The sheet can be stored between filter papers in a desiccated environment at 4 °C, prior to the next step.

13. Prehybridize nitrocellulose at 42 °C for 8–16 h in a Seal-A-Meal plastic bag with 10 ml of Solution H.

14. Replace solution with 10 ml Solution H and add ^{32}P-labelled probe (10^6 c.p.m./ml). Hybridize overnight at 42 °C.

15. Remove nitrocellulose from the bag and proceed with the following series of washes (20 min each):

    (a) 2 × SSC + 0.1% SDS at room temperature;

    (b) 1 × SSC + 0.1% SDS at room temperature;

    (c) 0.2 × SSC + 0.1% SDS at 55 °C;

    (d) 0.1 × SSC + 0.1% SDS at 65 °C (performed twice).

16. Place nitrocellulose on a filter paper (Whatman 3M) and cover it with Saran Wrap. Expose overnight to X-ray film. A single band of 2.2 kb that migrates slightly behind the 18S rRNA marker should be apparent.

4. Thrombospondin (TSP)

TSP was first identified as a component of α-granules in platelets. The protein can be released, together with other α-granule constituents, upon activation of platelets with a variety of compounds that include thrombin, the calcium ionophore A23187, and collagen. Although a fraction of TSP is bound to the platelet surface (it is known that TSP binds specifically to fibrinogen), the majority of the released protein can be recovered from the supernate of activated platelets. Subsequently, TSP was found in significant amounts in the culture medium of a variety of cells, including fibroblasts, smooth muscle cells, and endothelial cells. There is good evidence that TSP plays an important role in the second, irreversible phase of platelet aggregation, but its extravascular function is not well understood. TSP supports the attachment and spreading of keratinocytes, squamous carcinoma cells, and melanoma cells. However, in fibroblasts, smooth muscle cells, and endothelial cells, TSP also participates in the regulation of cellular proliferation, and in these cells interaction of the protein with the cell surface may be more complex. A number of potential cell surface receptors have been described for TSP; such ligand–receptor complexes might interfere with focal adhesions and promote changes in cell shape that precede cell division.

4.1 Purification from platelets and endothelial cells

TSP is a large modular glycoprotein consisting of three chains, each with a molecular weight of about 145 kDa, linked by disulfide bonds. As far as is known, the three chains are identical. The basic purification scheme takes advantage of the large size of TSP (435 kDa) and its ability to bind to heparin. Clezardin et al. (21) have provided evidence, based on peptide mapping studies, for differences between platelet and cell-derived TSP. Although these differences could result from post-translational changes, the question of whether more than one TSP exists should be considered an open one. However, it should be pointed out that current methods of purification would not recover related proteins with different chemical properties.

Protocol 8 summarizes a protocol for the purification of TSP from the supernate of activated platelets and Protocol 9 for purification from the culture medium of endothelial cells. All solutions contain CaCl$_2$, which is required for the native conformation of the protein. The TSP obtained by this procedure migrates largely as a single band on SDS-PAGE with an apparent molecular weight of about 180 000, but this molecular weight is known to be anomalously high. Although highly purified, this TSP is not pure and often contains small amounts of fibrinogen and fibronectin. In addition, it is now known that many preparations of TSP obtained from platelets contain small amounts of active and inactive TGF-β. Additional purification has been achieved on FPLC and gelatin–Sepharose, as in Protocol 8.

Protocol 8. Purification of TSP from platelets

Equipment and reagents

- Buffer A: 10% ACD (130 mM Na$_3$ citrate, 110 mM glucose), 20 mM Tris–HCl, pH 7.5, 145 mM NaCl, 5 mM KCl, 5 mM glucose
- Buffer B: 20 mM Tris–HCl, pH 7.5, 145 mM NaCl, 5 mM KCl, 5 mM glucose, 1 mM CaCl$_2$
- Human thrombin (Sigma, 1000 units/mg)
- 0.2 M PMSF (in absolute ethanol)
- NEM
- Sepharose CL–4B column (2.5 × 100 cm)
- Column buffer 1: 20 mM Tris–HCl, pH 7.5, 150 mM NaCl, 1 mM CaCl$_2$
- Heparin–Sepharose column (1 × 10 cm)
- Column buffer 2: 50 mM Tris–HCl, pH 7.5, 150 mM NaCl, 1 mM CaCl$_2$

Method

1. Centrifuge 4 units of fresh human platelets (less than 72 h old) in 40 ml polycarbonate tubes (2400 g, 25 °C, 10 min).

2. Suspend platelets gently in 10 ml buffer A using a 10 ml plastic pipette.

3. Centrifuge at low speed (120 g) for 5 min at 25 °C to pellet blood cells (platelets remain in the supernate). Supernates can be respun to remove additional blood cells and the pellets re-extracted with buffer A to recover platelets that sedimented with cells.

4. Centrifuge suspended platelets at 2400 g for 5 min at 25 °C.

5. Suspend platelets gently in 10 ml buffer B.

6. Add 70 units of human thrombin to platelet suspension and stir gently with a small magnet at 20 °C for 2 min.

7. Add 200 μl 0.2 M PMSF; stir for 2 min and place on ice.

8. Centrifuge at 27 000 g, 4 °C, 20 min. Collect the supernate (which contains the released TSP) and add NEM to a final concentration of 10 mM. The platelet 'releasate' can be stored at −20 °C at this stage.

9. Chromatograph on a Sepharose CL-4B column equilibrated with Column buffer 1 at 4 °C and monitor the effluent by absorbance at 280 nm. TSP elutes as the first broad peak within the included volume of the column, directly after a sharper peak at the void volume. When the procedure is first used, the position of elution of TSP should be established by analysing individual column fractions by SDS-PAGE.

10. Pool the fractions containing TSP (usually in a volume of 50–80 ml) and load on to a heparin–Sepharose column equilibrated with Column buffer 2 at 4 °C. Elute with 20 ml of the same buffer followed by 20 ml of the same buffer containing 0.25 M NaCl. These effluents contain little or no TSP. TSP is eluted with 30 ml of the same buffer containing 0.6 M NaCl.

11. Fractions containing TSP can be dialysed against lower ionic strength

buffers and stored at $-20\,°C$. Solutions of TSP can be concentrated by pressure filtration but some loss of protein due to binding to filters will be experienced. It is not advisable to lyophilize TSP.

12. If desired, further purification can be achieved by:

   (a) chromatography on gelatin–Sepharose to remove fibronectin;

   (b) FPLC with a Superose 12 column;

   (c) FPLC with a Mono-Q anion-exchange column;

   (d) affinity chromatography with monoclonal anti-TSP IgG linked to agarose.

   Reference 21 and references therein should be consulted for details.

Protocol 9. Purification of TSP from endothelial cells

1. BAEC are grown to 90% confluence in 10 150 mm (diameter) plastic dishes (see *Protocol 2* for conditions). Collect culture medium during the final 3-day period.

2. Clarify medium by centrifugation and add PIC as described in *Protocol 1*, Part A, step 5.

3. Concentrate TSP on a heparin–Sepharose column as described in *Protocol 8*, step 10. Note: The production of TSP by endothelial cells in culture is markedly reduced when the cells reach confluence.

4. Further purification is achieved by following *Protocol 8*, steps 9–12.

---

5. Relevance of type VIII collagen, SPARC, and thrombospondin to endothelial cell behaviour: tube formation *in vitro*

Strains of BAEC, as well as certain other endothelial cells, will occasionally undergo a morphological change during subculture and modulate their synthesis of ECM proteins (22, and references therein). For example, BAEC adopt a bipolar phenotype termed 'sprouting' (after the analogous process *in vivo*) and simultaneously initiate transcription of the abundant interstitial collagen, type I (22).

Sprouting cultures of BAEC form networks of cords (*Figure 7A*) or tubes, both of which grow under the monolayer of polygonal, slowly dividing cells (22). Antibodies specific for type VIII collagen (*Figure 7B*) or SPARC (*Figure 7C*) stained predominantly those cells that were involved in tubular or cord-like networks. TSP was localized to fibrillar arrays surrounding the

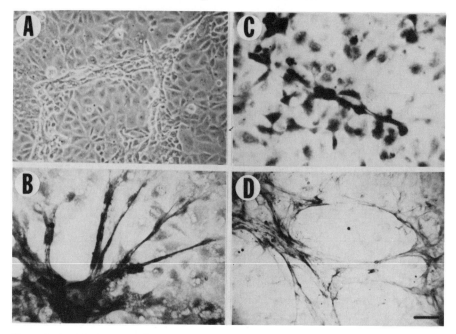

Figure 7. Localization of type VIII collagen, SPARC, and TSP in BAEC cultures containing endothelial cords and tubes. BAEC were grown until cords and tubes were apparent by phase-contrast microscopy (A). Cultures were fixed, rendered permeable, and exposed to anti-type VIII collagen IgG (B), anti-SPARC IgG (C), or anti-TSP IgG (D). Immune complexes were visualized by an avidin–biotin–peroxidase technique (see *Protocol 4*). Bar = 40 μm.

endothelial cords (*Figure 7D*). In *Figure 8C* are shown slot blots of mRNA from subconfluent BAEC (*Figure 8A*), confluent BAEC, and isolated tubes (*Figure 8B*), which were probed with TSP and SPARC cDNAs. mRNA levels for both these secreted glycoproteins were highest in subconfluent cultures; for comparison, the same RNA probed with type III collagen cDNA (the major procollagen product of BAEC) shows lesser differences among the three growth states (*Figure 8C*). It is clear, however, that BAEC in tubes transcribe both SPARC and TSP mRNA. Normalization for equal loadings of RNA to a 28S rRNA hybridization signal showed an increase in TSP mRNA of 8-fold in isolated tubes vs the confluent monolayers from which the tubes arose.

Since it has been proposed that TSP is an inhibitor of BAEC growth (4), this protein might limit the assembly and/or progression of angiogenesis *in vitro*. This presumption would be consistent with higher levels of TSP mRNA in mature endothelial tubes, as seen in *Figure 8C*. In contrast, further inspection of tube-forming cultures revealed selectively high mRNA and protein levels of SPARC in cells that were actively contributing to tube assembly and elongation (22). In view of the anti-spreading effect of SPARC on BAEC, as

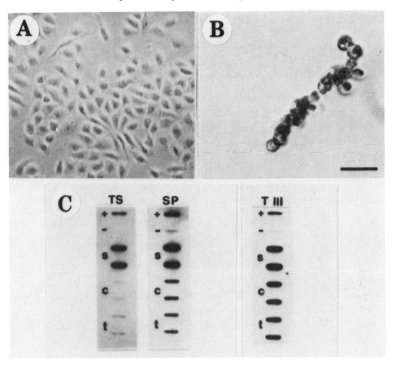

Figure 8. SPARC and TSP mRNAs are modulated during formation of endothelial tubes. Phase-contrast microscopy of subconfluent cultures of BAEC (A) and an endothelial tube isolated by treatment of a culture as shown in *Figure 8A* with 0.02% EDTA (B). (C), 10 μg total RNA from subconfluent cultures (s), confluent cultures (c), and isolated tubes (t) was slot-blotted on to a nitrocellulose sheet, and probed with ^{32}P-labelled SPARC cDNA (SP), TSP cDNA (TS), and type III collagen cDNA (TIII) (see *Protocol 7*). Levels of mRNA were quantitated by scanning densitometry and normalized to a signal for 28S rRNA to correct for equal amounts of RNA. 10 μg of total RNA from bovine fibroblasts (+) and murine F9 teratocarcinoma cells (−) were used as positive and negative controls. Bar = 40 μm.

well as its ability to interact with components of the ECM secreted by BAEC (1), it has been proposed that SPARC facilitates migration and changes in cell shape by reducing contacts between cells and their ECM (22).

A matrix protein which is probably integral to the process of vascular tube formation is type I collagen. Transcription of the α1(I) and α2(I) genes is initiated when BAEC begin to sprout; both mRNA and protein levels of type I collagen are high in BAEC engaged in the formation and stabilization of tubes (22). It is likely that type I collagen would predominate over the less abundant type VIII collagen in directing BAEC interactions. Clearly we can only speculate at this time about the roles of these selected ECM components on endothelial cell behaviour. Perturbation of the process of endothelial tube formation *in vitro* by deletion and/or modification of one or more of these proteins will allow us to delineate the complex processes by which endothelial cells make new vessels.

Acknowledgements

We are grateful to members of our laboratories for critical comments on the manuscript and for their help in the establishment of techniques described in this chapter. Special appreciation is due to Drs Luisa Iruela-Arispe and Jeffrey Yost for their suggestions and provision of data. We also thank Brenda Wood for assistance with the manuscript.

References

1. Sage, H., Vernon, R., Decker, J., Funk, S., and Iruela-Arispe, M.-L. (1989). *J. Histochem. Cytochem.*, **37**, 819.
2. Sage, H., Vernon, R., Funk, S., Everitt, E., and Angello, J. (1989). *J. Cell. Biol.*, **109**, 341.
3. Frazier, W. A. (1987). *J. Cell. Biol.*, **105**, 625.
4. Bagavandoss, P. and Wilks, J. W. (1990). *Biochem. Biophys. Res. Commun.*, **170**, 867.
5. Kapoor, R., Sakai, L. Y., Funk, S., Roux, E., Bornstein, P., and Sage, H. (1988). *J. Cell. Biol.*, **107**, 721.
6. Sage, H. and Iruela-Arispe, M.-L. (1990). *Ann. NY Acad. Sci.*, **580**, 17.
7. Sage, H. and Bornstein, P. (1982). In *Methods in enzymology* (ed. L. W. Cunningham and W. D. Frederickson), Vol. 82, p. 96. Academic Press, New York.
8. Sage, H. and Bornstein, P. (1987). *Biology of the extracellular matrix: structure and function of collagen types* (ed. R. Mayne and R. Burgeson), p. 173. Academic Press, Orlando.
9. Benya, P. D. and Padilla, S. R. (1986). *J. Biol. Chem.*, **261**, 4160.
10. Yamaguchi, N., Benya, P. D., van der Rest, M., and Ninomiya, Y. (1989). *J. Biol. Chem.*, **264**, 16022.
11. Sage, H., Pritzl, P., and Bornstein, P. (1980). *Biochemistry*, **19**, 5747.
12. Sage, H., Pritzl, P., and Bornstein, P. (1981). *Arteriosclerosis*, **1**, 427.
13. Sage H., Balian, G., Vogel, A., and Bornstein, P. (1984). *Lab. Invest.*, **50**, 219.
14. Alitalo, K., Bornstein, P., Vaheri, A., and Sage, H. (1983). *J. Biol. Chem.*, **258**, 2653.
15. Kapoor, R., Bornstein, P., and Sage, H. (1986). *Biochemistry*, **25**, 3930.
16. Jander, R., Korsching, E., and Rauterberg, J. (1990). *Eur. J. Biochem.*, **189**, 601.
17. Kittelberger, R., Davis, P. F., Flynn, D. W., and Greenhill, N. S. (1990). *Connect. Tiss. Res.*, **24**, 303.
18. Diglio, C. A., Grammas, P., Giacomelli, F., and Wiener, J. (1988). *Tiss. Cell.*, **20**, 477.
19. Sage, H., Johnson, C., and Bornstein, P. (1984). *J. Biol. Chem.*, **259**, 3993.
20. Chomczynski, P. and Sacchi, N. (1987). *Analyt. Biochem.*, **162**, 156.
21. Clezardin, P., Hunter, N. R., Lawler, J. W., Pratt, D. A., McGregor, J. L., Pepper, D. S., and Dawes, J. (1986). *Eur. J. Biochem.*, **159**, 569.
22. Iruela-Arispe, M.-L., Hasselaar, P., and Sage, H. (1990). *Lab. Invest.*, **64**, 174.

6

Enzymes involved in the post-translational processing of collagen

RICHARD A. BERG

1. Purification of vertebrate prolyl hydroxylase

1.1 Introduction

An affinity column technique for isolating enzymes or proteins bound to macromolecular substrates such as collagen was designed by covalently linking the macromolecule to a chromatography resin. This method has been developed for the purification of prolyl hydroxylase to homogeneity (1). The purified enzyme from a number of sources including chick embryos, human fibroblasts, and human liver was found to be a tetramer composed of two pairs of non-identical subunits α and β (2). Recently it has been found that the pair of β subunits of prolyl hydroxylase is identical to protein disulfide isomerase (3) which is non-covalently associated with the 2 α subunits of prolyl hydroxylase to produce an $\alpha_2\beta_2$ tetrameric structure.

1.2 Choice of affinity column

The original affinity column method for prolyl hydroxylase involved linking an underhydroxylated collagen substrate of the enzyme (collagen purified from *Ascaris* cuticles) to agarose using the technique of cyanogen bromide activation (4). Prolyl hydroxylase bound efficiently to such a matrix when a crude tissue homogenate fractionated by ammonium sulfate was passed through a column filled with the substituted agarose. The enzyme was eluted specifically from the column with a solution of a competing substrate (Pro–Pro–Gly)$_n$. Due to the difficulty of obtaining *Ascaris* cuticle collagen and the expense of obtaining (Pro–Pro–Gly)$_n$, an improvement on this method was developed in 1975 utilizing a competitive inhibitor, poly-L-proline, covalently bound to agarose (5). The affinity column was eluted by using a solution of lower molecular weight competing polypeptide inhibitor, poly-L-proline. This approach, however, contained several difficulties, including:

(a) The use of a poly-L-proline affinity column results in the co-purification of a profilactin (6, 7) along with the enzyme.

(b) The use of poly-L-proline in eluting prolyl hydroxylase from the affinity column creates an additional difficulty in that poly-L-proline cannot be readily separated from the purified enzyme using gel filtration as was the original competitive substrate (1).

(c) The poly-L-proline adsorbs to the agarose gel filtration column which was used to separate it from purified enzyme; hence the number of times that a given column can be reused is limited.

(d) Since poly-L-proline is not visualized by staining with Coomassie Brilliant Blue, it is impossible to determine whether or not it has been completely removed from prolyl hydroxylase by SDS-PAGE. Even trace amounts of poly-L-proline reduce the specific activity of prolyl hydroxylase since it is a competitive inhibitor.

The present method is an improvement on this technique which circumvents most of these problems, and has the further advantage of giving much larger amounts of purified enzyme than has previously been possible (8).

1.3 Preparation of affinity column

Protocol 1. Poly-L-proline coupling to agarose

Equipment and reagents

- Poly-L-proline of molecular weight greater than 30 kDa (Sigma)
- Agarose A-5m (100–200 mesh) (Bio-Rad Labs)

Method

1. Activate the agarose gel using cyanogen bromide by placing 100 ml packed agarose gel into a 200 ml flask.

    (a) Adjust pH to 11 followed by the immediate addition of 25 g CNBr.

    (b) Check the pH regularly and maintain above pH 10 with 1 M NaOH.

    (c) Maintain the temperature of the reaction below 37°C by adding ice to the reaction mixture.

    (d) Alternatively, obtain CNBr-activated Sepharose 4B from Sigma Chemical Co. Wash this product with 0.001 M HCl and then combine with the dissolved ligand in coupling buffer, see step 2.

2. Couple poly-L-proline to the activated agarose by adding 1.0 g of poly-L-proline (MW > 30 000) dissolved in 50 ml 0.1 M sodium bicarbonate containing 0.1 M NaCl. Stir the mixture gently at 4°C for 12 h. Remove any unbound polyproline by washing the gel with 10 columns of 0.1 M NaCl.

3. Estimate the amount of poly-L-proline bound to the agarose by amino

---

acid analysis of 1.0 ml gel hydrolysed in 6 M HCl. There should be approximately 800 mg proline bound per 100 ml of packed gel.

4. Pour 2 ml of the substituted agarose into a chromatography column having dimensions of 30 cm by 1.5 cm.

1.4 Affinity chromatography

The application of enzyme to the affinity column takes up to 24 h, after which the affinity column is washed with at least 1 litre of the dialysis buffer described in *Protocol 2*, or until the optical density of the effluent measures less than 0.30 at 230 nm. The affinity column is then eluted with 20 ml of dialysis buffer containing 3 mg per ml of poly-L-proline. The elution of enzyme activity is followed by monitoring the optical density at 280 nm. The entire peak containing material absorbing at 280 nm including the poly-L-proline and the eluted enzyme is concentrated with an Amicon ultrafiltration apparatus using a PM-30 membrane. The sample containing enzyme–inhibitor complex is concentrated to a volume of approximately 5 ml and stored frozen until further chromatography. The enzyme–inhibitor complex is stable for many months if stored frozen.

Protocol 2. Crude enzyme preparation and affinity chromatography

Equipment and reagents

- Poly-L-proline of molecular weight < 10 kDa (Sigma)
- Thirteen-day-old chick embryos
- Amicon ultrafiltrator
- Amicon membranes, PM-30
- Blender
- 0.2 M NaCl, 0.1 M Tris–HCl, pH 7.8
- Dialysis buffer: 100 mM NaCl, 0.2 M glycine, 10 mM DTT, and 0.1 M Tris–HCl, pH 7.8

Method

1. Homogenize thirteen-day-old chick embryos in 1 ml/g wet weight volume of 0.2 M NaCl, 0.1 M Tris–HCl, pH 7.8 at 4°C by blending at high speed for 60 seconds.

2. Centrifuge the homogenate at 10 000 *g* for 40 min at 4°C, and separate the supernatant from the pellet by decantation. Maintain the supernatant at 4°C for all remaining steps.

3. To the supernatant, add 176 mg/ml $(NH_4)_2SO_4$ to a final saturation of 30% by slowly adding the solid ammonium sulfate with constant stirring.

4. Centrifuge the $(NH_4)_2SO_4$ precipitated proteins at 12 000 *g* for 30 min. Separate the supernatant from the pellet.

Protocol 2. *Continued*

5. To the supernatant, add an additional 346 mg/ml ammonium sulfate slowly with constant stirring to a final saturation of 60%.

6. Centrifuge at 12 000 g for 30 min to obtain a pellet of protein precipitating between 30 and 60% saturation of ammonium sulfate. The pellet is resuspended in and dialysed against the dialysis buffer at 4 °C.

7. Store the dialysed supernatant at -20 °C until affinity chromatography is performed.

8. For affinity chromatography, thaw the frozen crude enzyme, centrifuge at 12 000 g for 30 min to remove protein precipitated by freezing, and dilute to 7.5 mg per ml with the dialysis buffer. Apply approximately 7.5 g of protein in one litre to the affinity column described above over a period of 12–24 h.

9. Wash the affinity column with the same buffer until the OD at 280 nm is less than 0.1; the affinity column is then eluted with a solution of 20 ml of 3 mg/ml poly-L-proline (molecular weight < 10 kDa) in the dialysis buffer followed by buffer alone to wash the polyproline out of the column.

10. Concentrate the fractions having absorbance of at least 0.15 OD units at 280 nm on an Amicon ultrafiltrator, using a PM-30 membrane, to approximately 5 ml and freeze.

1.5 Ion-exchange chromatography of affinity-purified prolyl hydroxylase

Protocol 3. Ion-exchange chromatography

Equipment and reagents

- DEAE-52 cellulose (Whatman)
- Amicon Ultrafiltrator
- Amicon membranes (PM-10)

- Buffer: 50 mM Tris–HCl, 50 mM NaCl, 10 mM DTT, 50 mM glycine, pH 7.4

Method

1. Concentrate up to six pooled eluates from the affinity column, each containing approximately 1.5 mg of prolyl hydroxylase, on an Amicon PM-10 filter under 20–30 lb (80–130 N) pressure of nitrogen at 4 °C.

2. Dialyse the sample overnight against the buffer at 4 °C, the same buffer is used to equilibrate the DEAE–cellulose column.

3. Equilibrate a DEAE–cellulose column (1 cm × 20 cm) with the buffer.

4. Apply the sample from step 1 to the DEAE column at a flow rate of

40 ml/h. After application, wash the column with approximately 40 ml of the buffer.

5. Develop the column with a 300 ml linear gradient of NaCl, from 0.05 M to 0.35 M, in the buffer.

6. Collect 4 ml fractions and remove specific fractions for analysis by SDS-PAGE.

7. Separate the fractions containing prolyl hydroxylase (peak II, see *Figure 1*) from profilactin (peak I, see *Figure 1*), and freeze. Enzyme at this stage must be stored frozen and is sensitive to freezing and thawing.

1.5.1 Profile of DEAE–cellulose chromatogram of prolyl hydroxylase

The elution profile from the DEAE-cellulose column (*Figure 1*) reveals a large breakthrough volume containing proteins which were bound to the affinity column and eluted with poly-L-proline as well as the polyproline (*Figure 2*, lane 2). Two peaks of protein are obtained after the polyproline. The first peak (*Figure 2*, lanes 3 and 4) eluted has been referred to as a poly-L-proline binding protein (7), which is now known to be profilactin (6). The second peak, peak II, contains 100% of the prolyl hydroxylase activity eluted from the column (*Figure 2*, lanes 5 and 6). Poly-L-proline binding protein

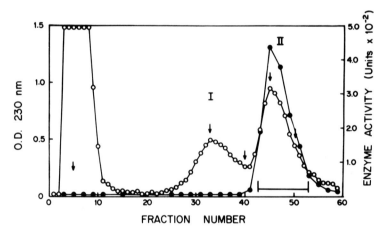

Figure 1. Elution profile of DEAE-cellulose chromatography of pooled affinity column eluates. The sample, having a volume of about 15 ml, was applied to the column (1 × 20 cm) with a flow rate of about 40 ml/h. The column was washed with about 40 ml of buffer and then eluted with a linear gradient of NaCl, 0.05–0.35 M. Open circles indicate optical density at 230 nm, while closed circles indicate enzyme activity. Arrows indicate fractions from which aliquots were taken for electrophoresis (see *Figure 2*). Bar indicates fractions which were pooled, concentrated on Amicon PM-30, and further purified.

(PBP) is not related structurally to prolyl hydroxylase, yet it can be precipitated from fibroblast cell extracts by antibodies specific for prolyl hydroxylase (9). PBP has been shown to be identical to a complex of actin and profilin which form a complex that binds polyproline. PBP is completely separated from prolyl hydroxylase by gel filtration. The peak containing prolyl hydroxylase (*Figure 1*, peak II) is concentrated and subjected to gel filtration.

1.6 Gel filtration of ion-exchange chromatographically purified prolyl hydroxylase

---

Protocol 4. Gel filtration

Equipment and reagents

- Amicon ultrafiltrator with PM-10 membranes
- Agarose (A-1.5m) 200–400 mesh (Bio-Rad Laboratories) column (1.5 × 85 cm)
- 0.1 M Tris–HCl, pH 7.8, 0.1 M NaCl, 0.2 M glycine, 10 mM DTT

Method

1. Pool the peak from the DEAE–cellulose column containing prolyl hydroxylase activity and concentrate on an Amicon PM-10 filter under 10–20 lb (40–80 N) of nitrogen at 4 °C. It is important to concentrate the sample to no more than 3–4 ml so the gel filtration step will have a high resolution and not be overloaded with sample.
2. Apply the concentrated sample to the agarose column equilibrated with the buffer specified above (*Figure 3*).
3. Collect 3 ml fractions and assay for prolyl hydroxylase activity and test for purity by subjecting aliquots to SDS-PAGE (*Figure 4*).
4. Assay prolyl hydroxylase activity using a procedure based on the stoichiometric decarboxylation of 2-ketoglutarate to succinate and CO_2 during the hydroxylation of peptidyl proline.

1.7 Assay for prolyl hydroxylase activity

Typically 10–50 µl of each fraction from the DEAE column or the gel filtration column are assayed for prolyl hydroxylase activity in a total of 1 ml reaction volume (*Protocol 5*). The reaction mixture contains 0.1 mg of (Pro–Pro–Gly)$_{10}$ as one substrate and alpha-keto [1-^{14}C]glutarate as the second substrate. The enzyme activity is quantified by trapping the [^{14}C]CO_2 formed from the decarboxylation of alpha-keto [1-^{14}C]glutarate and counting the [^{14}C]CO_2 in a liquid scintillation counter. This method involves suspending 0.5 × 1.0 cm pieces of filter paper (Whatman, number 1) from hooks in rubber serum stoppers used to seal the reaction tubes. At the end of the

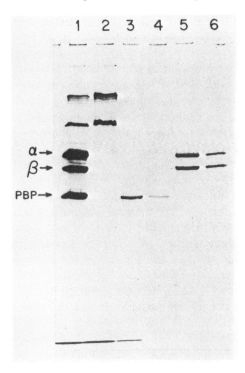

Figure 2. Analysis by SDS-PAGE of aliquots taken from the indicated points along the DEAE-cellulose elution profile. Lane 1 contains pooled affinity eluate material that was applied to the column. Lane 2 breakthrough peak of the DEAE column, fraction 5, showing material which was bound to the affinity column and eluted with poly-L-proline. Lanes 3 and 4 contain PBP, from fractions 33 and 40 of the DEAE column profile, 6 and 3 μg, respectively. The α and β subunits of prolyl hydroxylase are clearly resolved under these conditions. Note that the μg values in Lanes 1 and 2 are approximate due to the difficulty of quantification in the presence of poly-L-proline.

reaction incubation, 0.1 ml of 1.0 M phosphate buffer, pH 5.0, is injected through the stopper into the reaction mixture to liberate the CO_2 which is then trapped on the filter papers saturated with NCS tissue solubilizer (Amersham). The counts are converted to International Units of enzyme activity by determining the μmoles of hydroxyproline per minute represented by a given number of c.p.m. per minute in parallel reaction mixtures of 1 ml with all substrates saturating (1, 5). Hydroxyproline is quantified using an amino acid analyser or a chemical assay for hydroxyproline (10).

An alternative assay involves utilizing alpha-keto [5-^{14}C]glutarate as the substrate and quantifying [1-^{14}C]succinate after precipitating unreacted alpha-keto[5-^{14}C]glutarate with 2,4 dinitrophenyl hydrazine (11). By using 2,4 dinitrophenyl hydrazine, the excess labelled alpha-keto glutarate is precipitated and the labelled succinate remains in the supernatant where it is

removed and counted. Prolyl hydroxylase decarboxylates alpha-keto glutarate in the absence of the prolyl containing substrate, (Pro–Pro–Gly)$_{10}$ in the
reaction mixture, however, a peptidyl prolyl residue is necessary for recycling
of prolyl hydroxylase beyond a single turnover.

The reaction mixture used includes Tris buffer to control the pH, catalase,
DTT, and BSA to scavenge free radicals; and iron and ascorbate to maintain
prolyl hydroxylase in a reduced state. These components are all added in a
master mix for the reaction prior to the addition of labelled alpha-keto
glutarate.

Individual reaction tubes are prepared by adding equal amounts of master
mix and synthetic peptide substrate and varying amounts of prolyl hydroxylase
followed by water to adjust the volume to 1.0 ml (see *Protocol 5*). The
reaction is carried out at 37°C for one hour. After incubation, 25 μl of a
solution of 0.16 M dinitrophenyl hydrazine, 20 mM succinate, 20 mM alpha-
keto glutarate are added in 30% $HClO_4$ to the reaction to stop the reaction to
convert the alpha-keto glutarate to alpha-keto glutarate-2,4-dinitrophenyl
hydrazine with precipitation in acid. After another hour incubation at room
temperature, 50 μl of 1 M alpha-keto glutarate is added to react with any
excess dinitrophenyl hydrazine which reduces the efficiency of scintillation
counting. After an additional 30 minutes incubation, the yellow precipitate is
centrifuged at 3000 *g* for five minutes at room temperature and 50 μl of the
supernatant is removed and added to a scintillation cocktail and counted.

Protocol 5. Assay for 2-oxoglutarate decarboxylating enzymes
using the determination of [^{14}C-1]succinate

Equipment and reagents

- Alpha-keto [1-^{14}C]glutarate (New England Nuclear) Sp. Act. ≃ 50 mCi/mmol
- Alpha-keto [5-^{14}C]glutarate (Amersham) Sp. Act. ≃ mCi/mmol
- Unlabelled α-ketoglutarate (Sigma)
- (Pro–Pro–Gly)$_{10}$ Peninsula Laboratories, Inc.
- Bovine catalase and bovine serum albumin (Sigma)

- 1 M DTT (Sigma)
- NCS tissue solubilizer (Amersham)
- BSA
- 1 mM $FeSO_4$
- 0.5 M Tris–HCl, pH 7.8
- 20 mM Ascorbic acid

Method

1. Prepare enough reaction master mixture for all tubes by adding the
   following in the order listed. The amount indicated is for each reaction
   tube:

   (a) 0.10 ml 0.5 M Tris–HCl buffer, pH 7.8, at 25°C;

   (b) 0.05 ml aqueous solution containing 0.2 mg catalase;

   (c) 0.1 ml of aqueous solution of 1 mM DTT;

   (d) 0.1 ml aqueous solution containing 2 mg BSA;

(e) 0.05 ml aqueous solution of 1 mM $FeSO_4$;

(f) 0.1 ml aqueous solution of 20 mM ascorbic acid.

2. Prepare individual reactions by adding the following, in the order shown, to each reaction tube:

(a) 0.6 ml master mixture;

(b) Water to adjust the final volume to 1.0 ml after enzyme is added;

(c) 0.1 ml (Pro–Pro–Gly)$_{10}$ at a concentration of 1 mg/ml;

(d) Up to 0.3 ml of prolyl-4-hydroxylase to be assayed.

3. Start the reaction by adding 0.1 ml unlabelled alpha-ketoglutarate with 1×10^5 d.p.m. labelled alpha-ketoglutarate, to give a final concentration of 20 mM and add to the reaction mixture. Incubate for one hour at 37 °C.

4. Assay for [^{14}C]CO_2 or [^{14}C]succinate by one of the two methods (see above).

1.8 Purity of prolyl hydroxylase and summary

The purity of prolyl hydroxylase is established by SDS-PAGE (see *Figure 4*, lanes 1 and 2). As shown, purified prolyl hydroxylase is composed of equal amounts of the α and β subunits after being dissociated in SDS (1, 5). Although the enzyme may be contaminated to a small extent by polyproline binding (PBP), these proteins are largely separated from prolyl hydroxylase by the DEAE–cellulose column. Prolyl hydroxylase and profilactin are completely separated from each other by gel filtration chromatography on A-1.5m (*Figure 3*). The recovery of prolyl hydroxylase activity from the agarose column is approximately 90%. Polyacrylamide slab-gel electrophoresis in SDS of the purified enzyme after gel filtration indicates that prolyl hydroxylase prepared in this manner is essentially homogeneous (*Figure 4*). There often appears to be three primary contaminants of prolyl hydroxylase eluted from the affinity column. Two of the contaminants migrate more slowly than prolyl hydroxylase on SDS-PAGE and are separated by DEAE–cellulose chromatography. The third contaminant, profilactin, migrates faster on SDS-PAGE than prolyl hydroxylase and may not be completely separated from prolyl hydroxylase until the gel filtration step.

The overall recovery of prolyl hydroxylase activity starting from the ammonium sulfate precipitate of the chick embryo extract is approximately 60%. The activity of the purified enzyme is greater then 2 U/mg protein where 1 U is expressed as μmoles of hydroxyproline synthesized per minute. Thus the purification from chick embryos is approximately 5000-fold. The enzyme purified using this technique is homogeneous and free of contaminants as tested by SDS-PAGE. This technique allows the preparation of large

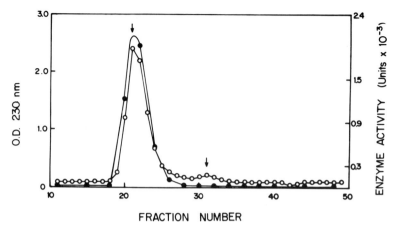

Figure 3. Gel filtration profile of pooled enzyme peak from the DEAE chromatography (peak II, *Figure 1*). Open circles indicate optical density at 230 nm, closed circles indicate enzyme activity. Arrows indicate fractions from which aliquots were taken for SDS-PAGE (see *Figure 4*).

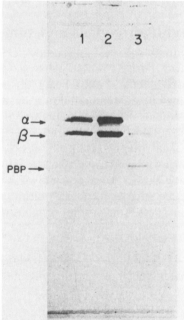

Figure 4. Analysis by SDS-PAGE samples along the gel filtration profile. Lane 1 contains the nearly-pure enzyme applied to the agarose column, 20 μg; Lane 2 contains 24 μg of material from the first peak of the gel filtration column, which appears to be completely pure prolyl hydroxylase. Lane 3 contains 4 μg of the last traces of PBP found in fraction 31 of the gel filtration column as well as some traces of material eluting in the region of the β subunits which may be Cross Reacting Protein (9). These proteins are totally separated from purified prolyl hydroxylase on the gel filtration column.

170

quantities of prolyl hydroxylase using a single chromatography step on DEAE–cellulose, which serves to concentrate the samples from several affinity column runs as well as to remove the polyproline and most of the contaminating proteins. Furthermore, the agarose column used to separate prolyl hydroxylase from traces of PBP can be reused indefinitely unlike the previous method employing a poly-L-proline affinity column (5).

2. Purification of protein disulfide isomerase

2.1 Introduction

Protein disulfide isomerase is a prevalent enzyme found in the endoplasmic reticulum of cells that synthesize secretory proteins. It shares homology with thioredoxin, an enzyme that is more widely distributed in its intracellular location than protein disulfide isomerase.

Protein disulfide isomerase has two functions in collagen biosynthesis. It is non-covalently associated with the α subunits of prolyl hydroxylase to produce an enzyme, containing two subunits of protein disulfide isomerase and two α subunits, that is retained and functions in the endoplasmic reticulum of collagen-producing cells. The second function of protein disulfide isomerase is to catalyse the rearrangement of disulfide bonds in newly synthesized collagen as it is assembled and folded prior to secretion.

The sequence of chicken protein disulfide isomerase (PDI) has been inferred from the studies on the molecular cloning of the β subunit of chicken embryo prolyl 4-hydroxylase, from smooth muscle cDNA libraries (12); from chicken embryo cDNA libraries (3); and from the N-terminal amino acid sequence of the β subunit of chicken embryo prolyl 4-hydroxylase (3, 12, 13). All three of these sequences for the PDI/β subunit from chick embryos match each other and display a 91% level of identity at the amino acid level with rat liver PDI (14).

To purify protein disulfide isomerase, advantage is taken of its solubility in high salt concentration including 60% ammonium sulfate which separates it from most proteins, including native prolyl hydroxylase.

2.2 Preparation of protein disulfide isomerase

Protocol 6. Protein disulfide isomerase preparation and ion-exchange chromatography

Equipment and reagents

- DEAE–52 cellulose (Whatman)
- DEAE–Sephacel (Pharmacia)
- 15-day-old chick embryos
- Agarose (1.5m) (Bio-Rad)

- Polyclonal antibodies against purified protein disulfide isomerase are not commercially available, but can be prepared in rabbits (2, 7) and are used for Western immunoblots to detect PDI in chromatographic column eluates as a faster alternative to testing for enzyme activity of protein disulfide isomerase

---

171

Protocol 6. *Continued*

- Centrificon-30 centrifugal microconcentrators from Amicon or an Amicon ultrafiltrator and PM–30 membranes
- [^{125}I]-Protein A (DuPont/New England Nuclear)
- Buffer 1: 0.01 M Tris–HC1, pH 7.5, 0.2 M NaCl, 0.1 M glycine, 2 mM PMSF
- Buffer 2: 10 mM Tris–HCl, pH 7.8, 100 mM glycine, 50 mM NaCl
- Buffer 3: 25 mM Na citrate, pH 6.2, 100 mM NaCl, 1 mM EDTA
- Buffer 4: 0.05 mM Tris–HCl, pH 7.8, 100 mM glycine, 50 mM NaCl
- Solid $(NH_4)_2SO_4$

Method

1. Homogenize 156 15-day-old chicken embryos in Buffer 1, using an equal volume of buffer to the wet weight (g) of embryos as described in *Protocol 2*.

2. Add ammonium sulfate to 60% saturation (2). Following precipitation of prolyl 4-hydroxylase and other proteins insoluble in 60% ammonium sulfate (354 mg/ml), centrifuge the sample at 4°C for 30 min at 10 000 *g*.

3. Dialyse the supernatant against Buffer 2 at 4°C with four buffer changes.

4. Apply to a DEAE–cellulose anion-exchange chromatography column (2.7 × 30 cm) equilibrated in the same buffer.

5. After washing, elute the column with a gradient from 50 to 350 mM NaCl in Buffer 2. Determine the absorbance at 230 nm for each fraction.

6. Perform Western immunoblot analyses on selected fractions using affinity-purified polyclonal antibodies specific for bovine PDI if available (see below). Pool fractions containing immunoreactive protein, dialyse against Buffer 3 and apply to a DEAE-Sephacel anion-exchange chromatography column (2.7 × 30 cm) equilibrated in the same Buffer 3.

7. Develop the column with a linear gradient from 100 to 600 mM NaCl in the same buffer and analyse as above.

8. Pool peak fractions, rechromatograph on the same column under identical conditions and concentrate in Centricon-30 centrifugal microconcentrators or by using an ultrafiltrator filter with a PM-30 membrane.

9. Fractionate samples on Bio-Gel 1.5m agarose (Bio-Rad) size-exclusion chromatography (1.5 × 75 cm) in Buffer 4. A single peak of protein corresponding to protein disulfide isomerase is obtained.

2.3 Assays of immunoreactive proteins

The α subunit of chicken embryo prolyl 4-hydroxylase was purified using ion-exchange chromatography in 6 M urea (2) and its identity to protein disulfide isomerase demonstrated by sequencing and comparison with sequences of protein disulfide isomerase from humans (13). Affinity-purified polyclonal rabbit immunoglobulins specific for the α subunit of chicken embryo prolyl

4-hydroxylase were prepared using standard procedures (7) and were used to follow the purification of protein disulfide isomerase as an alternative to measuring activity.

Protein samples from various fractions from the DEAE column (*Protocol 6*) were subjected to electrophoresis on polyacrylamide gels containing SDS. The samples were transferred to nitrocellulose membranes, and processed by Western immunoblot analyses as described (15, 16). Bound antigen–immunoglobulin complexes were detected with [^{125}I]Protein A. Autoradiography was performed for 16 h at $-70°C$ using a Kodak intensifying screen.

Protein disulfide isomerase is readily purified from tissues by ammonium sulfate precipitation followed by DEAE chromatography. Protein disulfide isomerase activity is determined by measuring insulin-cleaving activity using assay conditions which ensure that the disulfides of insulin are cleaved exclusively by PDI (17).

Protocol 7. Assay of protein disulfide isomerase activity

Equipment and reagents

- [^{125}I-A Tyr^{14}]insulin (DuPont/New England Nuclear)
- BSA
- Insulin (Boehringer–Mannheim)
- Glutathione (reduced) (Sigma Chemical Company)
- 0.1 M Potassium phosphate, pH 7.5
- 5 mM EDTA
- TCA

Method

1. Perform reactions in 1.0 ml volumes containing 1 μM insulin, 1 μCi of [^{125}I-A Tyr^{14}]insulin, Sp. Act. = 2000 Ci/mmol, 1 mM reduced glutathione, 0.1 M potassium phosphate, pH 7.5, 5 mM EDTA, 3 mg/ml bovine serum albumin.

2. After 15 min at 37 °C, terminate the reaction by adding 1 ml of 10% TCA.

3. Collect the precipitate by centrifuging for 5 min at room temperature at 1000 *g* and wash once with 1 ml of 5% TCA.

4. Determine the radioactivity of the combined supernatants using a liquid scintillation or gamma counter. All values are corrected for non-enzymatic insulin degradation which is determined in parallel assays without enzyme.

References

1. Berg, R. A. and Prockop, D. J. (1973). *J. Biol. Chem.*, **248**, 1175–82.
2. Berg, R. A., Kedersha, N. L., and Guzman, N. A. (1979). *J. Biol. Chem.*, **354**, 3111–17.

3. Parkonnen, T., Kivirikko, K. I., and Pihlajaniemi, T. (1988). *Biochem. J.*, **256**, 1005–11.
4. Cuatrecasas, P. (1970). *J. Biol. Chem.*, **245**, 3059–65.
5. Tuderman, L., Knuuti, E.-R., and Kivirikko, K. I. (1975). *Eur. J. Biochem.*, **52**, 9–16.
6. Tanaka, M. and Shibata, H. (1985). *Eur. J. Biochem.*, **151**, 291–7.
7. Kedersha, N. L., Broek, D., and Berg, R. A. (1986). *Biochem. J.*, **238**, 561–70.
8. Kedersha, N. L. and Berg, R. A. (1981). *Coll. and Rel. Res.*, **1**, 345–53.
9. Berg, R. A., Kao, W. W.-Y., and Kedersha, N. L. (1980). *Biochem. J.*, **189**, 491–9.
10. Berg, R. A. (1982). In *Methods in enzymology*, Vol. 82 (ed. L. W. Cunningham and D. W. Fredricksen), pp. 372–98. Academic Press, New York.
11. Kaule, G. and Günzler, V. (1990). *Anal. Biochem.*, **184**, 291–7.
12. Kao, W. W.-Y., Nakazawa, M., Aida, T., Everson, W. V., Kao, C. W.-C., Seyer, J. M., and Hughes, S. H. (1988). *Connect. Tiss. Res.*, **18**, 157–74.
13. Bassuk, J. S. and Berg, R. A. (1989). *Matrix: Coll. Rel. Res.*, **9**, 244–58.
14. Edman, J. C., Ellis, L., Blacher, R. W., Roth, R. A., and Rutter, W. J. (1985). *Nature*, **317**, 267–70.
15. Bassuk, J. A., Tsichlis, P. N., and Sorof, S. (1987). *Proc. Natl. Acad. Sci. USA*, **84**, 7547–51.
16. Bassuk, J. A., Capodici, C., and Berg, R. A. (1990). *J. Cell Phys.*, **144**, 280–6.
17. Varandani, P. T., and Natz, M. A. (1976). *Biochim. Biophys. Acta*, **438**, 358–69.

Fibronectin and fibronectin fragments

STEVEN K. AKIYAMA and KENNETH M. YAMADA

1. Introduction

The adhesive glycoprotein fibronectin has provided an important model system for investigating mechanisms of cell-adhesive interactions. Fibronectin binds to a number of biological macromolecules including heparin, collagen, fibrin, and cell surface receptors. These binding activities are contained in distinct, protease-resistant domains that can be released from the intact molecule by proteolysis and purified by chromatographic techniques. Although studies are often carried out using intact fibronectin, it is often necessary to use purified fragments of defined binding activities in order to derive molecular mechanisms (1, 2).

For most cells, adhesion to fibronectin is mediated by the central cell-binding domain of fibronectin through a (Gly)–Arg–Gly–Asp–(Ser) (RGD) sequence and a second, distinct region that acts synergistically with the RGD sequence (1–4). The major fibronectin cell–adhesive fragments that contain the RGD sequence in current use in our laboratory are listed in *Table 1*. Two protocols, each yielding a variety of biologically-active fibronectin fragments including the 110 kDa and 75 kDa cell-binding fragments, are given in this chapter. Although the 37 kDa and the 11.5 kDa cell-binding fragments are listed in *Table 1*, their use presents potential difficulties. Whereas the 37 kDa fragment is sufficiently large to retain cell-adhesive activity comparable to that of intact fibronectin, it is too small to adsorb appropriately to plastic or glass and requires covalent coupling with a spacer arm in order to form an adhesive substrate (4). The 11.5 kDa fragment also does not adsorb well to plastic and glass. Furthermore, it is too small to contain the entire cell-adhesive region of fibronectin and retains only relatively low biological activity (5). Although procedures have been developed for the purification of these smaller fragments (4–6) they will not be discussed further here.

The fibronectin fragments that bind to easily purified ligands, such as denatured collagen (gelatin), heparin, and fibrin, can be assayed by using the appropriate ligand immobilized on agarose beads or plastic dishes. The

Table 1. Summary of the cell-adhesive characteristics of fibronectin and major fibronectin fragments[a]

	Protease used	Relative activity[b]	Estimated K_d (μM)	Functions well on plastic?	Contains synergistic region?
Fibronectin	none	++++	0.8	yes	yes
110 kDa	thermolysin	++++	ND	yes	yes
75 kDa	trypsin	++++	0.4	yes	yes
37 kDa	chymotrypsin	+++	ND	no	yes
11.5 kDa	pepsin	+	1–10	no	no

[a] Data taken from references 4–7
[b] The relative activity of fibronectin or fragments in cell spreading assays using baby hamster kidney cells on protein-coated plastic
ND = Not determined.

cell-adhesive fragments present more of a problem because direct binding assays of soluble fragments to cells are neither simple nor rapid (7). It is more practical to use *in vitro* assays for cell–substrate adhesion that serve as simple models for the more complex cell-adhesive processes that occur *in vivo*. Two of the most useful adhesion assays are presented in this chapter—one for simple cell attachment to adhesive substrates and one for the more complex process of cell spreading. Either of these is suitable for rapidly and easily assaying the biological activity of proteins or fragments that have potential cell-adhesive activities. Assays for even more complex cell-adhesive processes such as migration have also been described (8, 9).

Some reagents and methods that are used in common among the various protocols given in this chapter are most conveniently discussed first. All dialysis steps are performed at 4°C, except where noted. All of the chromatography columns are run at room temperature (approximately 22°C) requiring the addition of 0.02% sodium azide to all buffers to inhibit microbial growth. In general, column fractions should not exceed 5–10% of the column bed volume. The protein content of all fractions is determined by the absorbance at 280 nm (A_{280}) and confirmed by sodium dodecylsulfate-polyacrylamide gel electrophoresis (SDS-PAGE) with either Coomassie blue R-250 or silver staining.

In order to ensure maximal recovery of fibronectin and its fragments, only polypropylene or siliconized glass beakers and tubes should be used. In virtually every case, the final products of these protocols are obtained at relatively low concentrations, usually 1 mg/ml or less. The most convenient way to increase the concentrations of fibronectin or its fragments is to precipitate them with ammonium sulfate using the procedure given in the chapter on tenascin (Chapter 8). Fibronectin can be precipitated with 50% saturated ammonium sulfate (29.1 g/100 ml) and the fragments are usually precipitated using 70% saturated ammonium sulfate (43.6 g/100 ml).

The protocols contained in this chapter usually require the addition of the serine protease inhibitor, phenylmethylsulfonyl fluoride (PMSF). Solid PMSF is only slightly soluble when added to aqueous buffers. Therefore, it must be prepared as a 200 mM stock solution in isopropanol, which is then diluted into the aqueous buffer. PMSF dissolved in isopropanol is stable for several weeks at room temperature.

2. Fibronectin

Fibronectin can be purified from plasma, tissue, and cultured cells. Due to its relatively high content of fibronectin in soluble form, plasma is the most convenient source, although protocols for the purification of fibronectin from cultured cells have been developed (10). The protocol given here for the purification of plasma fibronectin was first described by Miekka *et al.* (11). This method is based on the binding of fibronectin out of human plasma on to immobilized gelatin (denatured collagen) first described by Engvall and Ruoslahti (12), but it uses a pH 5.5 buffer to specifically elute the bound fibronectin without causing its denaturation.

Protocol 1. Purification of human plasma fibronectin

Equipment and reagents (see also p. 000)

- EDTA
- PMSF
- 6-aminohexanoic acid
- Sepharose 4B pre-column
- Gelatin–Sepharose (Pharmacia-LKB) column
- TBS: 0.15 M NaCl, 10 mM Tris–HCl, pH 7.0
- 4 M Urea in TBS

- Buffer A: 50 mM Tris–HCl, 50 mM 6-amino-hexanoic acid, 20 mM Na citrate, pH 7.6
- Buffer B: 0.1 M NaCl, 50 mM Na citrate, pH 5.5
- 1 M Na phosphate
- Ammonium sulfate
- 0.02% Sodium azide added to all chromatography buffers to inhibit microbial growth

Method

Note: Dialyse at 4 °C, run chromatography columns at room temperature (approximately 22 °C).

1. Warm plasma to room temperature and bring to 5 mM EDTA, 1 mM PMSF, and 50 mM 6-aminohexanoic acid.

2. Centrifuge at room temperature at 10 000 *g* for 15 min and save supernatant solution.

3. Pack a pre-column of Sepharose 4B and a gelatin–Sepharose column. The bed volume of each should be 1/6 the volume of plasma. Wash the gelatin–Sepharose with 4 bed volumes of urea/TBS. Equilibrate both columns with Buffer A.

Steven K. Akiyama and Kenneth M. Yamada

Protocol 1. *Continued*

4. Apply the plasma to the pre-column as fast as possible (2 bed vols/hr) and apply the flow-through to the gelatin–Sepharose column at a slower flow rate (approx. 20 ml/min). Discard the flow-through fractions of the gelatin–Sepharose column.

5. Wash the gelatin-bound material as rapidly as possible with 4 bed volumes of Buffer A supplemented with 1 M NaCl, then with 2 bed volumes of Buffer A.

6. Elute the fibronectin with Buffer B. Save and pool the fractions containing eluted fibronectin as judged by A_{280} and neutralize immediately by adding 1/20 volume 1 M Na_2PO_4.

7. If necessary, precipitate the fibronectin with 50% saturated ammonium sulfate. Resuspend and dialyse for several days to remove any residual ammonium sulfate.

A wide range of common buffers can be used to solubilize and dialyse the final fibronectin product. Because most of our protocols to generate fibronectin fragments call for the use of Tris as a buffer, it is usually most useful to end up with the fibronectin in TBS. Fibronectin is generally soluble at concentrations up to approximately 5 mg/ml in most physiological buffers. If higher concentrations are needed, fibronectin is soluble at over 10 mg/ml in 0.15 M NaCl, 10 mM 3-[cyclohexylamino]-1-propanesulfonic acid, pH 11. The concentration of fibronectin is most easily determined by measuring the A_{280} and using an extinction coefficient of 1.28 ml mg^{-1} cm^{-1} (13).

3. Preparation of fibronectin fragments

Although a wide variety of proteolytic enzymes have been used to generate biologically active fragments of fibronectin, we have found the most useful to be trypsin and thermolysin. Either of these two enzymes can be used to produce most of the useful fragments in a single digest. To obtain reproducible digestions, proteolytic enzymes should be prepared as stocks of 1–3 mg/ml, frozen as small aliquots on powdered dry ice, and stored at −80°C. Trypsin should be solubilized in 1 mM HCl and thermolysin in TBS supplemented with 1 mM $CaCl_2$. Each aliquot of enzyme is subsequently used for only a single digestion, with any remainder discarded rather than refrozen to ensure that no enzymatic activity is lost as a result of repeated freezing and thawing.

3.1 Preparation of tryptic fragments of human plasma fibronectin

The method described by Hayashi and Yamada (14) can be routinely used for the preparation of the 75 kDa RGD-containing cell-adhesive domain as well

178

Table 2. Biologically active fragments obtainable after typsin digestion of fibronectin[a]

Size (kDa)	Binding activity	Comments
31	heparin/fibrin	NH_2-terminal fragment. Contains the low-affinity heparin-binding site, high-affinity fibrin-binding site
34	fibrin	COOH-terminal fragment
38/24	fibrin heparin	High-affinity heparin binding. The larger fragment is derived from the A chain and the smaller from the B chain of plasma fibronectin
75	cells	Contains both the RGD sequence and synergistic sites
113	cells/heparin	Derived from the A chain of plasma fibronectin. Contains the IIICS and RGD plus synergistic cell-adhesive region and the high-affinity heparin-binding site
146	cells/heparin	Derived from the B chain of plasma fibronectin. Contains the RGD plus synergistic cell-adhesive region and the high-affinity heparin-binding site but not the IIICS site

[a] Data taken from reference 14.

as all the major biologically active domains of human plasma fibronectin, except for the collagen-binding domain (*Table 2*). The most reproducible results are obtained if trypsin previously treated with the chymotrypsin inhibitor L-1-tosylamido-2-phenyl chloromethylketone (TPCK-trypsin, Worthington Biochemicals, approximately 240 units/mg) is used. This protocol relies on the elution of specific fragments from ion-exchange columns using step gradients. Although these are very work-intensive, experience has shown that linear gradients do not resolve the fragments as well. Except where noted, all Tris–HCl buffers described below are adjusted to pH 7.0.

Protocol 2. Preparation of tryptic fragments of human fibronectin

Equipment and reagent (see also text, all Tris–HCl buffers to pH 7.0)

- Buffer A: 30 mM NaCl, 50 mM Tris–HCl, 1 mM $CaCl_2$
- TPCK-trypsin
- PMSF
- Buffer B: 10 mM Tris–HCl, 0.5 M PMSF
- DE-52 cellulose ion-exchange resin (Whatman) column (1.5 × 20 cm)
- Buffer C: 1 M NaCl, 10 mM Tris–HCl
- 10 mM Tris–HCl
- Buffer D: 0.5 M NaCl, 10 mM Tris–HCl
- Buffer E: 10 mM Tris–HCl, 1 mM $MgCl_2$
- Gelatin–Sepharose (Pharmacia-LKB) column (1.5 × 30 cm)
- Buffer F: 4 M urea, 10 mM Tris–HCl
- Heparin–Sepharose (Pharmacia-LKB) column (1.5 × 30 cm)
- Buffer G: 0.13 M NaCl, 10 mM Tris–HCl
- DE-52 column (1.5 × 6 cm)

Method

Note: Dialyse at 4 °C, run chromatography columns at room temperature (~ 22 °C), unless otherwise stated.

179

Protocol 2. *Continued*

1. Dilute 330 mg of human plasma fibronectin to 110 ml with Buffer A, and warm to 30 °C.

2. Add 0.2% TPCK–trypsin (0.66 mg) and incubate for 30 min at 30 °C.

3. Quench the digest by adding PMSF to 1 mM and dialyse against Buffer B for 3 h at room temperature. This particular dialysis step is performed at room temperature rather than 4 °C to prevent the amino-terminal fragment (which is relatively insoluble) from precipitating.

4. Prewash a 1.5 × 20 cm column of DE-52 cellulose ion-exchange resin with 60 ml of Buffer C and equilibrate with 10 mM Tris–HCl.

5. Clarify the dialysed digest by centrifuging at 45 000 *g* for 20 min and apply to the DE-52 column (step 4). The flow-through fractions contain the 31 kDa amino-terminal domain.

6. Wash the DE-52 column with 10 mM Tris–HCl until the A_{280} returns to baseline. Elute the bound material with Buffer D and dialyse overnight against Buffer E.

7. Prewash a 1.5 × 30 cm gelatin-Sepharose column. Wash with 90 ml Buffer F and equilibrate with Buffer E. Apply the dialysed material from step 6 to the gelatin–Sepharose. Collect and pool the flow-through fractions. The gelatin-binding fraction is recovered by eluting with Buffer F.

8. Prewash a 1.5 × 30 cm heparin–Sepharose column with 60 ml Buffer D. Adjust the flow-through fractions from step 7 to 2 mM EDTA and apply to the heparin–Sepharose column. Collect the flow-through and wash the heparin-bound material with 10 mM Tris–HCl until the A_{280} returns to baseline.

9. Elute the 75 kDa cell-binding domain with Buffer G. This material is usually > 80% pure and is often > 95% pure, as judged by densitometer scans of Coomassie blue R-250 stained gels.

10. Elute the heparin-bound material with Buffer D. Dialyse against 10 mM Tris–HCl.

11. Wash and equilibrate a 1.5 × 6 cm DE-52 column as described in step 4. Load the dialysed heparin-bound material and elute with a step gradient of NaCl in 10 mM Tris–HCl starting at 60 mM NaCl, using 80 ml at each step, and increasing the NaCl concentration in 20 mM increments. The 113 kDa fragment elutes at 80 mM NaCl and the 146 kDa fragment elutes at 100–120 mM NaCl (*Table 2*).

12. The 34 kDa fibrin-binding domain is present in the heparin flow-through fractions from step 8. It can be purified on a DE-52 column with a step gradient as described in step 11, eluting at 100–120 mM NaCl.

If only the central cell-adhesive fragment of fibronectin is needed, the yield of the 75 kDa fragment can be increased by digesting fibronectin with 1:50 (w/w) enzyme:substrate ratio for 15 min. Under these conditions, the amount of the 75 kDa fragment is increased while the amounts of the 146 kDa and the 113 kDa fragments described in *Table 2*, and also the gelatin-retarded fragments, are all decreased. In this case, the protocol can be terminated after step 10.

3.2 Preparation of thermolysin fragments of human plasma fibronectin

One of the most versatile and useful methods for the production of specific fibronectin fragments is that developed by Zardi and co-workers (15, 16). This procedure is simple and produces all of the major domains of fibronectin (*Table 3*) in relatively pure form after only a single hydroxyapatite chromatography column (HACC). It is the easiest way to obtain the cell-adhesive domain of fibronectin in the form of a 120 kDa fragment, which has similar cell-adhesive properties as the 75 kDa tryptic fragment. The only way in which we have modified the original procedure has been to purify heparin-, fibrin-, and collagen-binding domains by affinity chromatography using the appropriate immobilized ligand rather than to attempt rechromatography on hydroxyapatite. Except where noted, all phosphate buffers used in this protocol are adjusted to pH 6.8 with NaOH.

Table 3. Biologically active fragments obtainable after thermolysin digestion of fibronectin[a]

Size (kDa)	Binding activity	mM NaPO$_4$[b]	Comments
20	fibrin	18	COOH-terminal fragment. Contains the low-affinity fibrin-binding site.
28	heparin	100	NH$_2$-terminal fragment. Contains the low-affinity heparin-binding site.
29	heparin	130–190	High-affinity heparin-binding fragment derived from the B chain of plasma fibronectin.
38	heparin	120–140	High-affinity heparin-binding fragment derived from the A chain of plasma fibronectin.
40	collagen	7	Binds to denatured collagen (gelatin).
110	cells	55	Contains both the RGD sequence and the synergistic region

[a] Data taken from reference 14.
[b] Concentration of sodium phosphate required to elute each fragment from a HACC at pH 6.8.

Protocol 3. Preparation of thermolysin fragments of fibronectin

Equipment and reagents (see also text, all phosphate buffers to pH 6.8)

- HACC (Bio-Gel HT, Bio-Rad) 5 × 16.5 cm
- Buffer A: 25 mM Tris–HCl, 0.5 mM EDTA, 50 mM NaCl, 2.5 mM CaCl$_2$, pH 7.6
- PMSF
- Thermolysin (Protease type X, Sigma)

- 9 mM Sodium phosphate
- 9 mM Sodium phosphate, 1 mM EDTA
- 1.2 mM Potassium phosphate, 1 M NaCl
- Linear gradient 9–220 mM sodium phosphate

Method

1. Pack a 5 × 16.5 cm HACC. The most convenient form of hydroxyapatite is Bio-Gel HT (Bio-Rad). Wash the column with 2 ml of 1.2 M potassium phosphate, 1 M NaCl and equilibrate with 9 mM sodium phosphate.

2. Dilute fibronectin to 1–1.2 mg/ml with Buffer A. Add PMSF to a final concentration of 0.5 mM and incubate at 25 °C.

3. Add thermolysin to a final concentration of 5 μg/ml and incubate for 4 h at 25 °C.

4. Stop the digestion by adding EDTA to a final concentration of 5 mM and dialysing at 4 °C against 9 mM sodium phosphate, 1 mM EDTA.

5. Warm the dialysed digest to room temperature and clarify by centrifuging at 45 000 g for 20 min.

6. Apply to the HACC and wash with 9 mM sodium phosphate, until the A_{280} returns to baseline. The gelatin-binding fragment is found in the flow-through. For all steps using the HACC, the flow rate should be 80–120 ml/h.

7. Elute the HACC with a linear gradient of 9–220 mM sodium phosphate using a total volume of 2500 ml. The concentration of sodium phosphate that elutes each fragment is given in *Table 3*. To ensure identification of the correct fractions, analyse the appropriate fractions by SDS-PAGE.

8. If necessary, further purify the fibrin-, heparin-, and collagen-binding fragments on their respective affinity columns (see reference 14). The 110 kDa cell-binding domain is usually obtained at > 90% purity. If necessary, further purify on an Ultrogel Ac44 sizing column as described (16).

---

4. Assays for cell-adhesive activity

To test the cell-adhesive properties of fibronectin and its fragments, we use assays which quantitate cell attachment and spreading. The steps involved in the preparation of the adhesive substrates for both assays are essentially the same, so this aspect will be described first. The simple attachment of cells to fibronectin is analysed using the assay described by Nagata *et al.* (18) and Akiyama *et al.* (9) which uses cells that are radiolabelled by incubating with [^{3}H]thymidine. In the cell attachment assay, the wells are usually coated with fibronectin or fragment solution ranging in concentration from 0.5 μg/ml up to 50 μg/ml. Maximal attachment normally occurs above approximately 20–30 μg/ml, at which point 80–90% of the added cells are attached. The maximal level of attachment can, however, vary. For example, 80–90% attachment of human fibroblasts or fibrosarcoma cells can be routinely achieved (9, 19). However, often no more than 50 or 60% of added carcinoma cells will attach to fibronectin substrates in this assay system (S. K. Akiyama, unpublished data). The reasons for such variations are not yet clear.

The quantitative analysis of cell spreading is performed as described by Yamada and Kennedy (17). In general, lower concentrations of fibronectin or fragments are required to promote cell spreading than are needed for simple attachment. The concentration range of fibronectin should be from 0 to 10 μg/ml, with maximal spreading usually occurring at 0.5–5 μg/ml.

Protocol 4. Preparation of adhesive substrates

Equipment and reagents (see also text)

- Dulbecco's phosphate-buffered saline without divalent cations (DPBS⁻) or TBS
- 96-well TC cluster dish
- BSA, crystalline (Calbiochem)
- Serum-free DMEM

Method

1. Dilute fibronectin or fragments in either DPBS⁻ or TBS to a series of concentrations ranging from 0 to 50 μg/ml.

2. Add to the wells of a 96-well tissue culture cluster dish in 100 μl aliquots and incubate at 4 °C for 16–24 h.

3. Remove the coating solution and block the wells by incubating for 30 min at 4 °C with 100 μl of 10 mg/ml heat-denatured BSA. Prepare heat-denatured BSA by dissolving crystalline BSA in DPBS⁻ at a concentration of 10 mg/ml, holding at 80 °C for 3 min, and then transferring to ice until cool as described (17).

4. Wash each well at least 5 times with 100 μl of DPBS⁻. Then fill with 50 μl of serum-free Dulbecco's modified Eagle's medium (DMEM) and place in a tissue culture incubator.

Protocol 5. Cell-attachment assay

Equipment and reagents

- [³H] Thymidine (5 Ci/mmol; Amersham)
- Trypsin
- Serum-free DMEM
- 2% SDS in 0.01 M NaOH

Method

1. Label cells for attachment assays overnight with 5 μCi/ml [³H] thy-midine.

2. Deplete unincorporated label with a 2 h chase in non-radioactive medium.

3. Harvest cells with the briefest possible trypsinization, collect by centrifugation, then resuspend in 5 ml of regular culture medium.

4. Allow the cells to recover from the trypsinization for 20 min at 37 °C.

5. Count the cells, centrifuge, and resuspend in serum-free DMEM at a concentration of 2×10^5 cells/ml.

6. Add 50 μl of cells to the coated wells from *Protocol 4* and incubate for 30 min in a tissue culture incubator. The amount of radioactivity originally added to each well is determined by counting 50 μl of the original cell suspension.

7. Remove unbound cells by gently washing twice with serum-free DMEM and swirling at 150 r.p.m. on an orbital shaker for 30 sec.

8. Solubilize the bound radioactivity in 100 μl 2% SDS in 0.01 m NaOH and quantitate by liquid scintillation counting after neutralization of the solution with 50 μl 0.05 M HCl.

Protocol 6. Cell-spreading assay

Equipment and reagents

- Serum-free DMEM
- Fixative solution: 2.5% glutaraldehyde, 2.5% formaldehyde in PBS, pH 7.3

Method

1. Harvest cells and incubate in normal culture medium for 20 min to allow them to recover from the trypsinization. Suspend in serum-free DMEM at a concentration of 2×10^5/ml as described in steps 3–5 in *Protocol 5.*

2. Add 50 μl cells to the prepared substrates and incubate in a tissue culture incubator. Most cells require 40–60 minutes for maximal spreading. For example, fibroblasts require 60 minutes for optimal spreading, but HT1080 fibrosarcoma cells require only 40 minutes.

3. Terminate the assay by adding 100 μl of Fixative solution to the cells for 1 hour.

4. Count a predetermined number of cells (usually 600 cells/well) from randomly chosen fields and score as either spread or not spread. Alternatively, 10 or more randomly chosen cells from each well can be traced and their surface areas calculated (20).

References

1. Mosher, D. F. (ed.) (1989). *Fibronectin*. Academic Press, New York.
2. Hynes, R. O. (1990). *Fibronectins*. Springer-Verlag, New York.
3. Obara, M., Kang, M. S., and Yamada, K. M. (1968) *Cell*, **53**, 649.
4. Nagai, T., Yamakawa, N., Aota, S.-I., Yamada, S. S., Akiyama, S. K., Olden, K., and Yamada, K. M. (1991). *J. Cell Biol.*, **114**, 1295.
5. Akiyama, S. K., Hasegawa, E., Hasegawa, T., and Yamada, K. M. (1985). *J. Biol. Chem.*, **260**, 13256.
6. Pierschbacher, M. D., Hayman, E. G., and Ruoslahti, E. (1981). *Cell*, **26**, 259.
7. Akiyama, S. K. and Yamada, K. M. (1985). *J. Biol. Chem.*, **260**, 4492.
8. Varani, J., Orr, W., and Ward, P. A. (1978). *Am. J. Path.*, **90**, 159.
9. Akiyama, S. K., Yamada, S. S., Chen, W.-T., and Yamada, K. M. (1989). *J. Cell Biol.*, **109**, 863.
10. Yamada, K. M. and Akiyama, S. K. (1984). In *Methods for preparation of media, supplements and substrata for serum-free animal cell culture* (ed. D. Barnes, D. Sirbasku and G. Sato), pp. 215–30. Alan R. Liss, Inc., New York.
11. Miekka, S. I., Ingham, K. C., and Menache, D. (1982). *Thromb. Res.*, **27**, 1.
12. Engvall, E. and Ruoslahti, E. (1977). *Int. J. Cancer*, **20**, 1.
13. Mosesson, M. W. and Umfleet, R. A. (1970). *J. Biol. Chem.*, **245**, 5728.
14. Hayashi, M. and Yamada, K. M. (1983). *J. Biol. Chem.*, **258**, 3332.
15. Zardi, L., Carnemolla, B., Balza, E., Borsi, L., Castellani, P., Rocco, M., and Siri, A. (1985). *Eur. J. Biochem.*, **146**, 571.
16. Borsi, L., Castellani, P., Balza, E., Siri, A., Pellachia, C., de Scalzi, F., and Zardi, L. (1986) *Anal. Biochem.*, **155**, 335.
17. Yamada, K. M. and Kennedy, D. W. (1984). *J. Cell Biol.*, **99**, 29.
18. Nagata, K., Humphries, M. J., Olden, K., and Yamada, K. M. (1985). *J. Cell Biol.*, **101**, 386.
19. Yamada, K. M., Kennedy, D. W., Yamada, S. S., Gralnick, H., Chen, W.-T., and Akiyama, S. K. (1990). *Cancer Res.*, **50**, 4485.
20. Akiyama, S. K., Yamada, S. S., and Yamada, K. M. (1986). *J. Cell Biol.*, **102**, 442.

8

Tenascin, laminin, and fibronectin produced by cultured cells

HAROLD P. ERICKSON and GINA BRISCOE

1. Introduction

Tissue culture cells synthesize a variety of extracellular matrix (ECM) molecules and incorporate them into a matrix. In most cases, however, the incorporation into the matrix appears to be inefficient, and by far the majority of the ECM proteins are simply secreted into the medium as soluble molecules. One can harvest the 'conditioned medium' from confluent cell cultures every 2–4 days and purify the molecules of interest. If a monoclonal antibody is available in large supply, affinity purification on an antibody column is frequently the preferred method. A disadvantage of immunoaffinity purification is that the protein must be exposed to denaturing or chaotropic agents to elute it from the column. Alternative biochemical purification may be essential for proteins that are easily denatured, and is desirable whenever one is testing for unknown and potentially sensitive functions. Fortunately, for large ECM molecules a high degree of purification can be effected using the simple procedures of:

(a) $(NH_4)_2SO_4$ precipitation;

(b) gel filtration;

(c) ion-exchange chromatography; and

(d) zone sedimentation through sucrose or glycerol gradients.

We will describe in some detail the procedures we developed to purify tenascin (1), and refer the reader also to the very similar purification of Drosophila laminin from cell cultures, which, in addition, describes the use of lectin- and heparin-affinity columns (2). The *Guide to protein purification* (3) provides a comprehensive presentation of modern procedures and general concepts.

For recent reviews of tenascin see references (4, 5).

2. Protein assays

Our preferred assay is SDS-PAGE, and we typically run gels of all relevant column fractions for every step in the purification. The gels show immediately the subunit distribution, indications of proteolysis, and contaminants. We use the Laemmli procedure with a 5% acrylamide running gel and a 3% stacker. 2% β-Mercaptoethanol is added to the sample (but not the gel) for reducing gels. If non-reduced samples are to be run on the same gel, they are separated by two lanes from the reduced ones. Silver staining is usually the most sensitive, but we have noted that some segments of tenascin, in particular ones containing the fibronectin type III domains, stain poorly with silver. For these we use Coomassie blue or (more sensitive) Western blotting. The 320 kDa tenascin band can be easily seen in silver-stained gels of the conditioned medium (where tenascin is 2–15 μg/ml) and followed throughout the purification.

For some studies it is useful to analyse the non-reduced protein. We have previously used agarose gels to resolve the hexabrachion, trimers and nine-mers (6), but have recently developed a 2.7% acrylamide gel system. The 1.9 kDa hexabrachion migrates several millimetres into this gel and can be distinguished from smaller oligomers (*Figure 1D*). A major advantage of the acrylamide gel over agarose is that it can be silver-stained.

Protocol 1. 2.7% Acrylamide gels for resolving proteins up to 2 million Daltons

Equipment and reagents

- Acrylamide stock: 30% acrylamide, 0.85% piperazine diacrylamide (Bio-Rad)
- 3 M Tris–HCl, pH 8.0
- 20% SDS
- 20% Ammonium persulfate
- TEMED
- 0.5 M Tris–HCl, pH 6.5
- 30% Methanol

Method

1. *2.7% Resolving gel*. Prepare four Hoeffer mini-gels by mixing:
   (a) 3.8 ml acrylamide stock;
   (b) 5.2 ml 3 M Tris–HCl pH 8;
   (c) 210 μl 20% SDS;
   (d) 32.9 ml water.

   Degas for 5 min, then initiate polymerization with 175 μl 20% ammonium persulfate and 14 μl TEMED. Pour resolving gel mix to within 2 cm of top of gel plate. Allow resolving gel to polymerize for approximately one hour.

2. *2.4% stacker*. Mix:
   (a) 1.9 ml acrylamide stock;

(b) 6 ml 0.5 M Tris–HCl, pH 6.5;

(c) 120 μl 20% SDS;

(d) 15.9 ml water.

Degas for 5 min, and initiate polymerization with 96 μl 20% ammonium persulfate and 18 μl TEMED.

3. Run the gel fast. We use 150 volts for the 2.7% gels. If run at 100 volts (which we normally use for most gels) the bands are very diffuse.

Note: *Miscellaneous.* A 10-well comb gives a better looking gel than the 15-well comb. Gels can be stored in the refrigerator (wrapped in plastic and in a Ziploc bag) for up to 3 weeks. The gel will have the consistency of mucus, but with careful handling can be silver stained. These gels tend to shrink or swell after staining, and are best preserved in 30% methanol.

In our previous study (6) we determined the extinction coefficient for purified tenascin to be:

$$\varepsilon = 0.97 \; (A_{277} \text{ for 1 mg/ml, 1 cm path}).$$

We have also calculated the extinction coefficient from the amino acid sequence (7). The value of 0.99–1.07 for the major tenascin splice variants is close to the experimental value. The calculated value for fibronectin is 1.41, somewhat higher than the commonly used experimental value of 1.28 (8). The calculated values are probably more accurate than experimentally determined ones (9); for convenience we generally assume $\varepsilon = 1.0$ for tenascin, and 1.4 for fibronectin. Note that these extinction coefficients ignore the weight of carbohydrate, so they are for the peptide mass only.

We generally use the A_{277} value as our measure of protein concentration, but for some purposes the Bradford assay from Bio-Rad is more convenient. Here it is very important to realize that tenascin and fibronectin produce much less colour than most proteins. We found that, per milligram of protein, fibronectin and tenascin produced 40% as much colour as BSA (6). BSA is an especially poor choice of standard for either tenascin or fibronectin unless these differences are factored in. Fibronectin, which is readily available and produces approximately the same colour as tenascin in the Bradford assay, can be calibrated by its A_{277} value and used as a standard in this assay.

For quantitating tenascin in crude extracts we have developed a sensitive ELISA assay (10) that can detect as little as 20 ng/ml tenascin, with an accuracy of about 25%. This sandwich ELISA requires adjustment for the species examined and the available antibodies, and is not trivial to set up. However, once set up it is practical to assay a large number of samples.

3. Dialysing and concentrating proteins

We have frequently encountered catastrophic losses when attempting to dialyse or concentrate dilute tenascin and other ECM proteins. We have

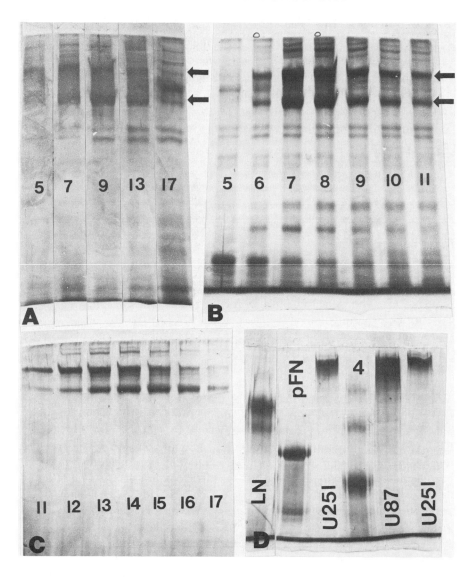

found that soaking the dialysis tubing in 5% Tween-20 for 1 h or more virtually eliminates the loss. We can dialyse samples at 20 µg/ml with complete recovery. When dialysing fibronectin, it is important to eliminate all air bubbles from the dialysis bag. For concentrating samples we sometimes use the Centricon-30 (from Amicon). Treating these with Tween-20 (soaking overnight in 5% Tween, centrifuging 1 ml through the filter, followed by extensive washing with buffer) also dramatically improves the recovery.

Probably the best way to concentrate tenascin (or fibronectin) when it is

Figure 1. A, Fractions from Sephacryl chromatography of the ammonium sulfate precipitate. The 320 and 220 kDa bands of the U-87 glioma tenascin, indicated by arrowheads on the right, are prominent in fractions 7–13. B, Mono-Q chromatography of pooled Sephacryl fractions (7–13). The two tenascin bands (arrowheads on right) are quite concentrated in fractions 7 and 8. C, Glycerol gradient sedimentation of Mono-Q fractions 7 + 8. Note that the 320 and 220 kDa bands are now highly purified (the higher bands are probably cross-linked dimers (6)), and note also that the 320 and 220 kDa splice variants are enriched in the leading and trailing fractions, respectively. D, 2.7% PAGE of non-reduced tenascin and standards. From the left the lanes are: LN = EHS laminin, M_R = 850 kDa; pFN = plasma fibronectin, M_R = 500 kDa; U-251 = tenascin purified from U-251 glioma cultures (the subunit is almost exclusively the 320 kDa splice variant, M_R for the intact hexabrachion is 1900 kDa); lane 4 shows U-251 tenascin that has proteolysed slightly upon storage—arms are cleaved near the central knob, leaving monomers (prominent lower band, $M_R \sim 310$ kDa), dimers and trimers; U-87 is tenascin purified from U-87 glioma cultures. As seen in A–C these hexabrachions consist of equal amounts of the 320 and 220 kDa splice variants, producing a smear of M_R 1900–1300 kDa. U-251 is an independent preparation of the intact, large human hexabrachion (M_R 1900 kDa). Note that this band is several millimetres into the gel.

relatively pure is to use the Mono-Q column. This can be particularly valuable following elution from an antibody-affinity column. The urea-eluted fractions (diluted if necessary to bring the salt concentration below 0.2 M) are pumped on to the Mono-Q column, washed with 0.2 M NaCl in Tris buffer, and eluted in a step to 0.6 M NaCl. This step simultaneously (a) removes the urea, (b) removes proteoglycans that can bind to tenascin, and (c) concentrates the tenascin.

4. Cell cultures

Cells differ enormously in the types and amounts of ECM proteins they secrete. The human glioma line U-251 MG (in particular the clone 3 developed by Dr Darell Bigner, Dept of Pathology, Duke University Medical Center, Durham, NC 27710) is our champion producer of tenascin. Confluent cultures will give 15 μg/ml tenascin in the conditioned medium every three days, about 10 times more than human neonatal skin fibroblasts. The tenascin produced by this cell line is almost exclusively the 320 kDa form containing all the alternatively spliced domains. In the present paper we demonstrate the purification of tenascin from the U-87 human glioma cell line, obtained from the ATCC. Although the level of secretion is much lower, the tenascin produced by this culture is a 50:50 mixture of the large (320 kDa) and small (220/230 kDa) splice variants. Fibroblast cultures from 10-day-old chicken embryo skin give 2–5 μg/ml tenascin. Most other cell lines we have tested produce less than 1 μg/ml.

The best source of mammalian laminin is the EHS mouse tumour, but it can also be obtained from cell culture. Mouse laminin can be obtained in reasonable quantity from F9 teratocarcinoma cells following induction by retinoic acid (11) (note that this laminin is preferentially incorporated in the matrix of cell aggregates, but significant amounts are secreted into the medium). An excellent source of Drosophila laminin is the conditioned medium of the K_c cell line (2).

Fibronectin is secreted in substantial amounts by a number of fibroblast cell lines. Although the classic preparation of cell surface fibronectin was from a urea extract of the cells and matrix (12), in our experience fibronectin is usually much more abundant in the conditioned medium.

For production of secreted proteins, we typically grow cells to confluence in T150 culture flasks or in roller bottles. We generally use 10% fetal bovine serum for the subconfluent cultures, but we then use the minimum amount of bovine serum that the cell line will tolerate (the U-251 MG cells grow well in 1% serum, but the U-87 cells require 10% serum). Conditioned medium is harvested every 3 days when the cultures are fed with fresh medium. Some cell lines start peeling off the substrate after a week or two of confluence, but others (U-251 MG) survive for more than a month, continuously secreting tenascin. Medium is stored at 4°C until we are ready to process it. Tenascin appears to be resistant to most proteases, and we have found that the conditioned medium can be stored at 4°C for weeks with no indication of degradation. (However, once purified the tenascin shows signs of proteolysis within a week or two. We always store purified tenascin at −80°C.) Fibronectin and laminin are much more sensitive to proteases, so media needs to be processed quickly, and treated with protease inhibitors, when purifying these proteins.

All subsequent steps in the purification are done at room temperature.

5. Purification of tenascin

5.1 Initial concentration and purification

Protocol 2. Ammonium sulfate precipitation

Equipment and reagents

- Conditioned medium
- Beckman 45-Ti and type 35 rotors
- Solid ammonium sulfate
- 0–0.2 M Ammonium bicarbonate

Method

1. As a first step this protocol concentrates the protein and achieves a significant purification. Centrifuge conditioned medium at 11 000 g (10 000 r.p.m.) for 30 min in Beckman 45 Ti (higher speeds in this clearing step give somewhat cleaner preparations) to remove cellular and particulate debris.

---

2. Bring the medium to 37% saturation (45% saturation was used for laminin (2), but we find this also precipitates substantial quantities of serum proteins) by adding (slowly and with stirring) 22.1 g solid ammonium sulfate per 100 ml medium. We have not found it necessary to adjust for the small drop in pH.

3. Allow this solution to stir at room temperature for 30–60 min, or overnight at 4°C, and then centrifuge at 120 000 g (32 000 r.p.m. for 30 min in the Beckman type 35 or type 45-Ti rotor (lower speed centrifugation, 10 000–20 000 r.p.m.) in a preparative centrifuge also works, but sometimes results in reduced recovery of tenascin). The same six tubes are used to process up to 1 l or more of medium, accumulating the pellets and discarding the supernatants.

4. Drain the final pellets and resuspend, using a rubber policeman, in 0.2 M ammonium bicarbonate or another buffer. The pellets from 1 l of medium are typically resuspended in 4 ml.

Protocol 3. Hydroxyapatite chromatography (13) as alternative to ammonium sulfate precipitation

Equipment and reagents

- Conditioned medium
- Sodium azide
- Gelatin–agarose matrix column (5 × 20 cm)
- Hydroxyapatite column (2.5 × 6 cm) (Bio-Gel HTP DNA grade, Bio-Rad)
- 4 M urea, 0.15 M NaCl, 0.1 M sodium phosphate, pH 7.0
- 5 mM sodium phosphate, pH 7.0
- Linear gradient 5–300 mM sodium phosphate

Method

1. Add sodium azide (0.1%) to the conditioned medium.

2. Pass medium through two columns connected in series. The first column is a gelatin–agarose matrix which binds and removes fibronectin. The second column is hydroxyapatite. (Note that the binding properties of hydroxyapatite can vary considerably depending on the preparation method. Saginati *et al.* (13) used Bio-Gel HTP DNA grade from Bio-Rad; we have had similar results with this material, but different binding and elution using hydroxyapatite from other sources.) At a flow rate of 100 ml/h, all the tenascin in 2 l of conditioned medium will bind to the hydroxyapatite in 20 h.

3. Separate the two columns and recover the fibronectin by elution of the first column, with 4 M urea (14).

Protocol 3. *Continued*

4. Wash the hydroxyapatite column with 5 mM sodium phosphate, pH 7.0, and then with a linear gradient from 5 to 300 mM sodium phosphate. A peak of partially purified tenascin will elute between 120 and 160 mM sodium phosphate.

5.2 Gel filtration

Because the hexabrachion, laminin, and fibronectin are such large molecules, gel filtration is an ideal purification step to separate them from smaller serum and secreted proteins. As well, tenascin, laminin, and fibronectin are substantially separated from each other. We normally use this as the second step in purification.

Protocol 4. Gel filtration

Equipment and reagents

- Concentrated tenascin from *Protocol 2* or *3*
- 0.22 μm Filter (Millex-GV, Millipore)
- Buffer: 0.15 M NaCl, 0.02 M Tris–HCl, pH 7.9, 0.02% Na azide
- *Either* a Sephacryl S-500 HR column (1.5 × 50 cm) *or* a Sepharose 4B column (3.0 × 40 cm) (both from Pharmacia. Bio-Rad products can also be used and give comparable results)

Method

1. Combine a total of 4–8 ml concentrated tenascin from *Protocol 2* or *3* and filter through a 0.22 μm filter then load on either a Sephacryl S-500HR or a Sepharose 4B column. (Comparable products are available from Bio-Rad, and both companies provide excellent instructions and technical support.)
2. Elute column with Buffer.
3. Monitor the column eluate by UV absorbance and collect 3 ml fractions.
4. Analyse all absorbing fractions by SDS-PAGE (*Figure 1A*).

Tenascin elutes just behind the void volume, followed by and partially overlapping with laminin (not obvious in this preparation) and fibronectin. Proteins of smaller molecular weight in these fractions are either naturally occurring oligomers, partially denatured aggregates, or proteins that bind to tenascin or the other large proteins. Fessler *et al.* (2) included 1 M urea in the column buffer, which should reduce some associations.

5.3 Zone sedimentation through glycerol gradients

Gradient sedimentation can achieve a substantial purification of large molecules from the smaller proteins in conditioned medium, and sometimes is preferred to gel filtration as an early step in purification. For example, certain cell cultures secrete material that clogs the gel filtration columns, and these are much better processed by gradient sedimentation. Gradient sedimentation was used in our earlier work with tenascin (15), and also for Drosophila laminin (2).

We typically sediment our proteins on a 15–40% (v/v) glycerol gradient in 0.2 M ammonium bicarbonate. This is an ideal buffer for preparing rotary shadowed electron microscope samples (16), so gradient fractions can be examined directly. This is also an excellent buffer for storing and freezing proteins. Sucrose gradients give equivalent separation, although centrifugation times will be different. The separation on sedimentation gradients is based on different physical chemistry than that of gel filtration, so this is useful as a final step in purification when gel filtration has been used first (1). The glycerol gradient achieves a potentially important separation of large and small tenascin splice variants. The U-87 tenascin is a 50:50 mix of the 320 and 220 kDa subunits. These are partially separated on the glycerol gradient, providing fractions that are about 75% or more enriched in the small or large splice variants (*Figure 1C*).

Protocol 5. Zone sedimentation

Equipment and reagents

- Ammonium bicarbonate
- Glycerol

- SW 41-Ti rotor C or SW 27 or SW 50-1, depending on sample size

Method

1. Make up standard solution:
   (a) dissolve 1.58 g ammonium bicarbonate in a small volume of water;
   (b) add glycerol by weight:
      i. 18.9 g for 15% (v/v),
      ii. 50.4 g for 40% (v/v);
   (c) add water, with stirring, to 100 ml.

   Note: If the protein sample is a resuspended ammonium sulfate pellet, it should be centrifuged briefly to remove particulate material; in addition, it is frequently necessary to dialyse the sample to remove ammonium sulfate which can make the sample sink through the 15% glycerol.

2. Make a 15–40% glycerol gradient in tubes, load with 500 μl sample per tube for the SW 41-Ti (takes a 12 ml gradient).[a,b]

3. Centrifuge for 18 h at 20°C at 41 000 r.p.m. The 12–13S hexabrachion will migrate about two-thirds of the way to the bottom.

4. Fractionate gradients into 20 fractions, each about the same volume as the sample loaded.

[a] The SW27 can handle larger volumes, add 2–3 ml sample/tube and spin for 48 h at 24 000 r.p.m. at 20°C
[b] The SW 50-1 is good for analytical scale, add 150 μl sample/tube, spin for 15 h at 38 000 r.p.m. at 20°C

5.4 Anion-exchange chromatography

DEAE cellulose is the classic anion-exchange medium and can be used in this purification. However, we have found the Pharmacia FPLC column 'Mono-Q' to be highly reliable and reproducible. For preparations involving up to 2 mg tenascin we use the 1 ml Mono-Q HR 5/5. The Mono-Q HR 10/10 can handle up to 10 mg tenascin. The columns are loaded and run with an LKB HPLC pump and gradient maker. The Sephacryl (or glycerol gradient) fractions rich in tenascin are pooled (typically 10–15 ml), filtered through a 0.22 μm filter, and pumped on to the column.

Protocol 6. Chromatography on Mono-Q

Equipment and reagents

- Mono-Q HR 5/5 column (Pharmacia)
- LKB HPLC pump
- Tenascin-rich (Sephacryl- or glycerol-gradient) fractions
- 0.22 μm Filter (Millipore)
- Buffer A: 0.02 M Tris–HCl, pH 7.9, 0.02% azide
- Buffer B: Buffer A plus 1.0 M NaCl
- Linear gradient: 0.1–0.6 M NaCl in buffer

Method

1. Pool tenascin fractions (~ 10–15 ml) and filter.
2. Load tenascin on the column and wash with 10 ml Buffer A.
3. Apply a 30 ml linear gradient of NaCl (0.1–0.6 M) in the buffer. Tenascin elutes between 0.32 and 0.4 M NaCl, with some material trailing at higher salt concentrations.
4. Wash the column with 10 ml 0.6 M NaCl, and step to Buffer B to elute proteoglycans and some interesting proteins bound to them (1).

Note: The tenascin in the peak Mono-Q fractions is typically 0.2–1.0 mg/ml, and there are only weak contaminating bands on the silver-stained gel (*Figure 1B*).

---

6. Preparation of monomeric fibronectin and hexabrachion arms by reduction of disulfide bonds

The subunits of the fibronectin dimer and the tenascin hexamer are covalently attached by disulfide bonds. In addition, there must be numerous non-covalent, hydrophobic contacts, including the short alpha-helical segment of tenascin and unspecified interface contacts in both molecules. In order to prepare monomeric protein these hydrophobic contacts must be weakened as the disulfide bonds are reduced. Fibronectin appears to require more vigorous denaturation than tenascin for this process.

If fibronectin is treated with dithiothreitol (DTT) at room temperature or in the cold, extensive aggregation occurs before monomers are released. Probably this is due to disulfide exchange catalysed by the DTT. A monomeric fraction can be obtained if the fibronectin is denatured in 4 M urea and the reaction time with DTT is limited to a few minutes.

Protocol 7. Reduced fibronectin monomers by reduction in urea (17)

1. Dialyse fibronectin, 0.2–1 mg/ml, into 0.2 M NaCl, 0.05 M Tris–HCl, pH 8.5. Perform reaction at room temperature. Most sources advise deaerating the solution and maintaining under nitrogen but this is probably not necessary.
2. Add urea to 4 M and DTT to 2 mM. After 10–15 min stop the reaction by adding iodoacetamide to 4 mM. An alternative protocol omitted the urea, increased the DTT to 20 mM, and left the reaction at 37 °C for 4 h (18).
3. Dialyse or pass over a Sephadex G-25 column to remove urea and reagents.
4. To demonstrate the efficiency of the reaction and separate a fraction of pure monomeric fibronectin, sediment the material through a glycerol gradient in a buffer containing 0.2 M NaCl or other salt. The monomeric fibronectin sediments at 7.5S. The yield is rather low (0.1 mg/ml recovered in our experiment), so it may be useful to run several large gradients, combine the peak fractions and concentrate the monomeric fibronectin on a Mono-Q column.

Protocol 8. Reduced tenascin monomers (single hexabrachion arms)

1. Put the tenascin in a buffer at pH 8–8.5, and add DTT to 2 mM for 30 min at room temperature.
2. Stop the reaction by adding iodoacetamide to 4 mM for 30 min. In our earlier work (6) we included 4 M urea during the reduction, but we have found that this is not necessary for tenascin.
3. Layer this reaction mixture directly on to a glycerol gradient for purification by sedimentation. The single hexabrachion arms sediment at 6.5S and recovery is very good. Although the sedimentation co-efficient confirms that the subunits remain as monomers during the centrifugation, we have noted, by electron microscopy, a tendency to reaggregate into small oligomers when removed from the gradient.

These 'monobrachions' should be useful in assays of any activity, to determine whether the hexavalent structure of tenascin is important. There is one report of a naturally occurring monomeric tenascin isolated from adult chicken gizzard (19), but our own similar preparation of gizzard tenascin showed intact hexabrachions (6). The most sensitive site for most proteases is very close to the central nodule (6), so the monomers reported in this gizzard preparation may have been produced by proteolysis or even by accidental exposure to reducing agent.

References

1. Aukhil, I., Slemp, C. A., Lightner, V. A., Nishimura, K., Briscoe, G., and Erickson, H. P. (1990). *Matrix*, **10**, 98–111.
2. Fessler, L. I., Campbell, A. G., Duncan, K. G., and Fessler, J. H. (1987). *J. Cell Biol.*, **105**, 2383–91.
3. Deutscher, M. P. (1990). *Guide to protein purification*. Academic Press, San Diego.
4. Erickson, H. P. and Bourdon, M. A. (1989). *Annu. Rev. Cell Biol.*, **5**, 71–92.
5. Erickson, H. P. (1993). *Curr. Opin. Cell. Biol.*, **5**, 869–76.
6. Taylor, H. C., Lightner, V. A., Beyer, W. F., Jr., McCaslin, D., Briscoe, G., and Erickson, H. P. (1989). *J. Cell. Biochem.*, **41**, 71–90.
7. Aukhil, I., Joshi, P., Yan, Y. Z., and Erickson, H. P. (1992). *J. Biol. Chem.*, **268**, 2542–31.
8. Mosesson, M. W. and Umfleet, R. A. (1970). *J. Biol. Chem.*, **245**, 727–35.
9. Gill, S. C. and Von Hippel, P. H. (1989). *Anal. Biochem.*, **182**, 319–26.
10. Lightner, V. A., Gumkowski, F., Bigner, D. D., and Erickson, H. P. (1989). *J. Cell Biol.*, **108**, 2483–94.
11. Grover, A. and Adamson, E. D. (1985). *J. Biol. Chem.*, **260**, 12252–8.
12. Alexander, S. S., Colonna, G., Yamada, K. M., Pastan, I., and Edelhoch, H. (1978). *J. Biol. Chem.*, **253**, 5820–4.
13. Saginati, M., Siri, A., Balza, E., Ponassi, M., and Zardi, L. (1992). *Eur. J. Biochem.*, **205**, 545–9.
14. Engvall, E. and Ruoslahti, E. (1977). *Int. J. Cancer*, **20**, 1–5.
15. Erickson, H. P. and Taylor, H. C. (1987). *J. Cell Biol.*, **105**, 1387–94.
16. Fowler, W. E. and Erickson, H. P. (1979). *J. Mol. Biol.*, **134**, 241–9.
17. Erickson, H. P. and Carrell, N. A. (1983). *J. Biol. Chem.*, **258**, 14539–44.
18. Odermatt, E., Engel, J., Richter, H., and Hormann, H. (1982). *J. Mol. Biol.*, **159**, 109–23.
19. Chiquet, M., Vrucinic-Filipi, N., Schenk, S., Beck, K., and Chiquet-Ehrismann, R. (1991). *Eur. J. Biochem.*, **199**, 379–88.

Structural analysis of glycosaminoglycans

J. E. TURNBULL, M. LYON, and J. T. GALLAGHER

1. Introduction

1.1 General

Glycosaminoglycans (GAGs) are negatively-charged polysaccharides of differing degrees of complexity which are ubiquitous components of the extracellular matrix and plasma membranes. The GAG chains are largely made up of a repeat disaccharide unit which consists of an amino sugar and a hexuronic acid (Hex A), the exception being keratan sulphate (KS) in which galactose (Gal) replaces Hex A (*Table 1*). The disaccharides may be substituted with ester or amino-linked sulfate groups and the numbers of disaccharides per chain can vary enormously, from about 15 in keratan sulphates, to 50–200 in heparan sulphates/ chondroitin sulphates and over 4000 in hyaluronate (HA). HA has a uniform primary structure and is composed entirely of *N*-acetylated disaccharides (GlcNAc β 1,4 GlcA) which are not sulfated. At the opposite extreme are the heparan sulfates (HS) in which differential sulfation and variations in the Hex A isomers (glucuronate, GlcA or iduronate, IdoA) produce complex domains which represent a considerable problem in qualitative structural analysis.

Sulfated GAGs are covalently-linked to proteins forming proteoglycans (PG). They contain specific short sequences at the reducing ends of the chains that are contiguous with the disaccharide repeat regions and which mediate protein attachment. These protein linkage sequences will not be discussed in detail here, but they are proving to be interesting structures which can be sulfated or phosphorylated in the chondroitin and heparan type chains, or branched in the keratan family (1). KS uniquely contains sialic acid as an end-chain or capping sugar at the reducing terminal, and small amounts of fucose branching from the main polysaccharide backbone (2).

1.2 Disaccharide composition of the sulfated GAGs

1.2.1 Chondroitin sulfate and dermatan sulfate

The chondroitin sulfate (CS) disaccharide is GalNAc (β 1, 4) GlcA which may be sulfated at C-4 or C-6 of GalNAc. In dermatan sulfate (DS) the GlcA is

Table 1. Carbohydrate composition of glycosaminoglycans

Glycosaminoglycan	Disaccharides [a]
Galactosaminoglycans	
Chondroitin sulfate (CS)	A: GalNAc[4S] β 1,4 GlcA
	C: GalNAc[6S] β 1,4 GlcA
	di E: GalNAc[4S, 6S] β 1,4 GlcA
Dermatan sulfate (DS)	B: GalNAc[4S] β 1,4 IdoA
	di B: GalNAc[4S] β 1,4 IdoA[2S]
	di D: GalNAc[6S] β 1,4 IdoA[2S]
Glucosaminoglycans	
Heparan sulfate (HS)	GlcNAc α 1,4 GlcA
	GlcNAc[6S] α 1,4 GlcA
	GlcNSO$_3$ α 1,4 GlcA
	GlcNSO$_3$ α 1,4 IdoA
	GlcNSO$_3$[6S] α 1,4 IdoA
	GlcNSO$_3$[6S] α 1,4 IdoA[2S]
Heparin	GlcNSO$_3$[6S] α 1,4 IdoA[2S]
	GlcNSO$_3$[3S] α 1,4 IdoA[2S]
Keratan sulfate (KS)	GlcNAc[6S] β 1,3 Gal
	GlcNAc[6S] β 1,3 Gal[6S]
Hyaluronate (HA)	GlcNAc β 1,4 GlcA

[a] The disaccharides listed represent the major constituents of each of the GAG types. Abbreviations: GalNAc, *N*-acetylgalatosamine; GlcNAc, *N*-acetyl-glucosamine; GlcNSO$_3$, *N*-sulphoglucosamine; GlcA, glucuronic acid; IdoA, iduronic acid; Gal, galactose; 2S, 3S, 4S, and 6S denote the position of ester-*O*-linked sulfate substituents.

epimerized to IdoA (GalNAc β 1,4 IdoA); C-4 of GalNAc is the most common site of sulfation, but sulfate groups may also be found at C-6 of GalNAc and C-2 of IdoA. DS chains are structures described as co-polymers because they also commonly contain some CS-type disaccharides. In most instances CS and DS chains are quite uniformly sulfated with an average of one sulfate group per disaccharide, but there are notable examples of the occurrence of disulfated disaccharides especially in the CS chains of mucosal mast cells and macrophages. These highly charged units may determine important protein binding properties of the polysaccharides (3).

1.2.2 Heparan sulfate (HS) and heparin

These GAGs are distinguished by the presence of *N*-sulfate groups and α-linked hexosamines. The disaccharide repeat of glucosamine and hexuronate is variable in character because the amino sugar may be *N*-acetylated (GlcNAc) or *N*-sulfated (GlcNSO$_3$), both hexuronate isomers (GlcA and IdoA) are present and *O*-sulfation occurs at various positions in the sugar rings (*Table 1*). HS has about equal proportions of GlcNAc and GlcNSO$_3$, arranged mainly in separate domains within the polymer chains and with the majority of the ester sulfates located in or near the *N*-sulfated regions.

Heparin, the most highly sulfated vertebrate polysaccharide is the unique product of connective tissue mast cells. The major disaccharides are di- and tri-sulfated units of structure GlcNSO$_3$ ($\pm$ 6S) α1,4 IdoA (2S). A small quantity (approximately 5–20% depending on tissue source) of *N*-acetylated disaccharides are also present. Heparin is a potent anticoagulant and the molecular basis of this property is a pentasaccharide sequence:

$$-GlcNAc(6S)-GlcA-GlcNSO_3(3S)-IdoA(2S)-GlcNSO_3(6S)-$$

which interacts specifically with antithrombin III and catalyses the rapid inactivation of thrombin. This important observation suggests that other specific sequences may determine the broad repertoire of protein-binding properties that have been identified in these polysaccharides.

1.2.3 Keratan sulfate (KS)

These short GAG chains consist mainly of mono- and di-sulfated disaccharides of basic structure GlcNAc β 1,3 Gal. The monosulfated units are sulfated at C-6 of GlcNAc and the disulfated structures have an extra sulfate at C-6 of Gal. Some recent evidence suggests that the disulfated disaccharides are present mainly towards the non-reducing end of the chain (4).

2. Glycosaminoglycan extraction and purification

The procedures which can be adopted for the extraction and purification of GAGs will be determined, to a great extent, by two factors: (a) the nature of the source material, i.e. solid tissue, physiological fluid, or cell culture; and (b) whether proteoglycan integrity is to be preserved during the purification process.

Generally applicable procedures for the extraction and purification of proteoglycans have been extensively described in two excellent reviews (5, 6) which deal primarily with connective tissue and cell culture systems, respectively. These procedures may also be successfully adapted for large-scale proteoglycan purification from more highly cellular tissues, for example liver (7).

If intact proteoglycans are not the priority then it can be much simpler to release GAGs in the form of peptidoglycans directly from fragmented tissue or scraped cell layers by digestion with non-specific proteases, such as papain (in 1 M NaCl, 50 mM phosphate, 5 mM cysteine–HCl, 1 mM EDTA, pH 6.8, at 60°C for 24–48 h) or Pronase (0.1 M Tris–acetate, 10 mM calcium acetate, pH 7.8, at 37°C for 24–48 h). Enzyme to substrate ratios of 1:100–200 are recommended. With fatty tissues prior treatment with acetone may enhance the efficiency of proteolytic extraction. Solubilized GAGs may then be fractionated and recovered by ion-exchange chromatography.

In the case of both proteoglycans and protease-released peptidoglycans it may be desirable to generate pure GAG chains before structural analysis can be undertaken. The only available method for this is β-elimination followed by immediate reduction of the reducing xylose residue. The sample, lyophilized or in an unbuffered solution, is treated with 50 mM NaOH, 1 M sodium borohydride at 45 °C for 48 h. This is then neutralized by the careful addition (excessive frothing can occur) of 2 M acetic acid. The GAGs may then be recovered by dialysis and lyophilization. Although the above conditions should not lead to any noticeable depolymerization of the polysaccharide, partial desulfation of HS (probably from ester-sulfate on C-6 of GlcN) has been reported (8) and higher concentrations of alkali and/or longer incubation times should be avoided. In most cases well-purified peptidoglycan samples are adequate for structural analysis and this potential difficulty can thus be avoided.

3. Depolymerization techniques

Structural analysis of GAGs requires reagents for the fragmentation of the chains. The pattern of fragmentation itself provides some information, and also yields products with particular structural characteristics which are amenable to further analysis. A number of methods, both chemical and enzymic, are available for the cleavage of GAG chains at specific linkages. In general, the enzymic methods have the advantage of more restricted linkage specificities, and thus provide more information. They also result in products which retain the sulfation patterns of the original sequences and are, therefore, well suited to further analysis.

3.1 Analytical strategies

A strategy for determining the structure of a GAG can be divided into three separate phases. First, the overall saccharide composition can be established by complete depolymerization and analysis of the disaccharide products (Section 4). Secondly, the organization of the polysaccharide into ordered domains with particular structural characteristics can be determined (Section 5). Finally, a number of approaches can be used to investigate the actual sequence of sugars from defined reference points (Section 6).

3.2 Enzymic methods

3.2.1 Polysaccharide lyases

The most commonly used, commercially available enzymes for depolymerization of GAGs are the polysaccharide lyases. These are a class of bacterial enzymes which cleave specific glycosidic linkages by an eliminative mechanism (9); this results in the formation of oligosaccharides with 4,5-unsaturated uronic acid residues at the non-reducing end. This UV chromophore provides a convenient method for detection of the reaction products. However, formation of the double bond precludes the possibility of distinguishing GlcA from IdoA on the basis of the difference in configuration of the carboxyl group at the C-5 position. This can be overcome to some extent by knowledge of the specificity of the particular enzyme, but emphasizes the value of applying complementary techniques such as chemical cleavage (see below).

Details of the specificities of the commercially available lyases are given in *Table 2*. A number of heparitinases are available for depolymerization of HS and heparin (10). Heparitinase I shows specificity for GlcA-containing disaccharides irrespective of the sulfation of the glucosamine moiety. In contrast, heparinase scissions linkages in highly sulfated disaccharide units containing an IdoA(2S) moiety. Heparitinase II cleaves linkages containing both GlcA and IdoA residues, but is not efficient at cleaving linkages in sequences of unsulfated disaccharides of the type GlcNAc-GlcA.

Table 2. Polysaccharide lyases—specificities and incubation conditions

Lyase	Linkage specificity		Incubation conditions [a]		
			Buffer [a]	Temperature (°C)	Suppliers [b]
Heparitinases [c]					
Heparitinase I	GlcNR	($\pm$6S)α1-4GlcA	1	37	A
Heparitinase II	GlcNR	($\pm$6S)α1-4GlcA/IdoA	1	37	A
Heparitinase III	GlcNSO$_3$($\pm$6S)α1-4IdoA(2S)		1	37	A
Heparinase I	GlcNSO$_3$($\pm$6S)α1-4IdoA(2S)		1	30	B
Heparinase II	GlcNR	($\pm$6S)α1-4GlcA/IdoA	1	25	B
Heparinase III	GlcNR	($\pm$6S)α1-4GlcA	1	43	B
Chondroitinases [d]					
Chondroitinase ABC	GalNAc	β1-4 GlcA/IdoA	2	37	A,B,C
Chondroitinase ACI	GalNAc	β1-4 GlcA	3	37	A,B
Chondroitinase ACII	GalNAc	β1-4 GlcA	3	37	A,B,
Chondroitinase B	GalNAc (4S)	β1-4 IdoA($\pm$2S)	4	40	A

[a] Buffer and temperature conditions recommended by suppliers: (1) 150 mM sodium acetate, 0.1 mM calcium acetate, pH 7.0; (2) 50 mM Tris–HCl, 50 mM sodium acetate, pH 8.0; (3) 50 mM sodium acetate, pH 6.0; (4) 50 mM Tris–HCl, pH 8.0.
[b] Commercial suppliers of lyases: (A) Seikagaku Kogyo Co. Ltd, Tokyo, Japan; (B) Sigma Chemical Co. Ltd; (C) Boehringer–Mannheim Ltd.
[c] These enzymes are called either heparitinases or heparinases, depending on the commercial supplier: The commonly used heparitinase is designated heparitinase I by Seikagaku and heparinase III by Sigma. Similarly, heparitinase II is known as heparinase II, and heparitinase III as heparinase. These three pairs of enzymes appear to exhibit identical substrate specificities as shown in the table (see reference 10 for further details). For details of resistant sequences, see reference 22. R = NAc or NSO$_3$.
[d] The chondroitinases, with the exception of chondroitinase B, appear to cleave independently of sulfation on GalNAc.

With respect to chondroitinases, the ABC lyase is able to cleave *N*-acetyl-galactosaminidic linkages of all types in the galactosaminoglycans, irrespective of the content and position of *O*-sulfate groups, to yield disaccharide products. In contrast, the AC lyases cannot cleave linkages containing IdoA residues and yield larger oligosaccharide products from substrates containing IdoA-repeat sequences (that is, CS/DS co-polymers). The ACI variety is apparently more efficient at cleaving adjacent to GlcA residues in CS/DS co-polymers which contain both IdoA and GlcA residues. Chondroitinases ACI, ACII, and ABC can also cleave *N*-acetylglucosaminidic linkages in HA. Chondroitinase B, in contrast, cleaves *N*-acetyl galactosaminidic linkages to L-IdoA which are predominant in DS polymers. It appears to have a requirement for 4-*O*-sulfation of the hexosamine unit, as well as a preference for IdoA(2S). The activity of these lyases, and the extent of sample digestion, can be conveniently monitored by the increase in UV absorbance at 232 nm. Conditions for a particular sample under which depolymerization of all susceptible linkages is complete can therefore be established. When cleaving biosynthetically radiolabelled samples it is advisable to include carrier GAG of an appropriate type in the digest; this ensures standard digestion conditions for different samples, and allows the reaction to be readily monitored.

3.2.2 Other enzymic methods

The remaining two groups of commercially available enzymes are those for depolymerizing KS and HA. KS can be degraded by the enzymes keratanase and endo β-galactosidase, both of which cleave at non-sulfated galactosidic linkages, though the former appears to require an adjacent GlcNAc(6S) residue (11). Enzymes are also available for the specific depolymerization of HA (for example. Streptomyces and leech hyaluronidases) though the unvaryingly repetitive structure of the polymer precludes any analytical role for them other than as tools for specific identification of the molecule.

3.3 Chemical methods

3.3.1 Deaminative hydrolysis

Deaminative hydrolysis is an invaluable chemical cleavage method, particularly for the structural analysis of the *N*-sulfated polysaccharides HS and heparin. Low pH nitrous acid (pH 1.5) treatment results in specific and near quantitative cleavage at GlcNSO$_3$ residues, the latter being converted into 2,5-anhydromannose, with concomitant chain cleavage and release of SO$_4$ and N$_2$ (12).

With unlabelled samples it is possible to radiolabel the products by reduction with tritiated sodium borohydride (NaB[^{3}H]H$_4$) (13). Care needs to be taken in the quantitative interpretation of nitrous acid scission data, particularly with heparin, due to the occurrence of a minor ring contraction reaction which precludes chain cleavage (12, 14).

Protocol 1. Deaminative hydrolysis of *N*-sulfated
glycosaminoglycans with low pH nitrous acid

1. Cool equal volumes of 0.5 M H_2SO_4 and 0.5 M barium nitrite $(Ba(NO_2)_2)$ on ice.

2. Mix and centrifuge at 12 000 r.p.m. for 2 min at room temperature to pellet the insoluble barium sulphate.

3. Add the supernatant (1 M HNO_2 solution) to the sample (minimum 50% v/v).

4. Incubate for 15 min at room temperature.

5. Terminate by raising the pH with 1 M sodium carbonate or react with excess 1 M ammonium sulphamate.

An alternative procedure using nitrous acid at pH 4 has also been described for the preferential cleavage of hexosamines with free amino groups (12). Depolymerization of GAGs containing *N*-acetylated hexosamine residues is possible following conversion of the HexNAc residues to HexN residues by hydrazinolysis. This method has been applied to the analysis of the disaccharide compositions of CS, DS, and KS (15, 16). A related approach can also be used to assess the disaccharide composition of HS and heparin. Following hydrazinolysis, the *N*-deacetylated polymer can be treated with nitrous acid at pH 3; under these conditions both *N*-sulfated and *N*-unsubstituted glucosamine residues are susceptible to cleavage (14).

3.3.2 Periodate oxidation

An additional chemical method for the depolymerization of HS, heparin, and DS is periodate oxidation. On treatment of HS or heparin with periodate at pH 3 and 4°C, GlcA associated with GlcNAc is selectively oxidized; in contrast, the GlcA, IdoA, and IdoA(2S) residues from *N*-sulfated regions are unaffected (17, 18). By subsequent treatment with alkali, oxyheparan sulfate is fragmented into oligosaccharides of general structure GlcNR-[HexA-GlcNR]$_n$-R′ where R′ is the remnant of an oxidized and degraded GlcA residue (19). With DS, non-sulfated IdoA can be selectively oxidized and cleaved under conditions which preclude oxidation of GlcA and IdoA(2S) (20).

4. Analysis of disaccharide composition

Analysis of saccharide composition requires complete depolymerization of the parent chains to disaccharide products which can be resolved and identified either by comparison with reference standards or by further structural analysis.

4.1 Polysaccharide depolymerization and preparation of disaccharides

It is important to ensure that the yield of disaccharides is as high as possible. This is generally easy to achieve with polymers such as CS and DS, using chondroitinase ABC digestion (21). In contrast, HS and heparin can present some difficulties. With unlabelled samples yields of 77–94% have been reported using a combination of heparitinases (21). A 97% yield of disaccharides from biosynthetically radiolabelled HS has also been reported (22). Using the combined hydrazinolysis and deaminative hydrolysis approach yields of approximately 90% can be obtained (14).

Disaccharides can be readily separated from larger resistant oligosacchar-

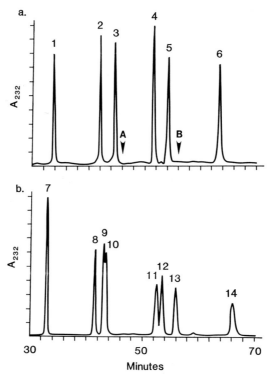

Figure 1. Strong-anion exchange HPLC of GAG disaccharides. Strong-anion exchange (SAX) HPLC of unsaturated GAG disaccharide standards released by polysaccharide lyase scission was carried out as described in *Protocol 2*. The standards separated were: profile (a) HS/heparin disaccharides: 1, UA-GlcNAc; 2, UA-GlcNSO$_3$; 3, UA-GlcNAc(6S); 4, UA-GlcNSO$_3$(6S); 5, UA(2S)-GlcNSO$_3$; 6, UA(2S)-GlcNSO$_3$(6S). The elution positions of the rare disaccharides A [UA(2S)-GlcNAc] and B [UA(2S)-GlcNAc(6S)] are also shown. Profile (b) CS/DS disaccharides: 7, UA-GalNAc; 8, UA-GalNAc(6S); 9, UA-GalNAc(4S); 10, UA(2S)-GalNAc; 11, UA(2S)-GalNAc(6S); 12, UA-GalNAc(4S,6S); 13, UA(2S)-GalNAc(4S); 14, UA(2S)-GalNAc(4S,6S).

ides by gel filtration on Bio-Gel P2 in ammonium bicarbonate eluent (13, 22). This step provides both purification of the disaccharides and an assessment of their yield. Column eluants can be readily monitored for radioactivity or absorbance at 232 nm, and the fractions corresponding to disaccharides can then be pooled and freeze-dried for further analysis.

4.2 Separation of disaccharides

A number of techniques can be used to resolve disaccharides derived from different GAGs. These include paper chromatography (23), high-voltage electrophoresis (24), thin-layer chromatography (25), and HPLC (13, 21). The latter technique is probably the most useful because of its rapidity, reproducibility, and high resolution. With unsaturated disaccharides detection can be achieved using absorbance at 232 nm. Alternatively, in samples which are biosynthetically radiolabelled or have had radiolabel introduced following depolymerization (see Section 3.3.1), radioactivity can be monitored either by in-line monitoring (for example Radiomatic Flo-one/Beta A-200 detector, Canberra Packard) or by scintillation counting of collected fractions.

A weak anion-exchange HPLC method using a Lichrosorb NH_2 column has been described for the separation of all the major GAG disaccharides (21). Reference standards corresponding to these disaccharides can be used to calibrate HPLC columns, and are available commercially from Seikagaku Kogyo Limited (Tokyo, Japan). These disaccharides can also be separated by strong anion-exchange (SAX) HPLC as described in *Protocol 2*. *Figure 1* shows the HPLC profiles obtained for the separation of disaccharide standards derived from HS, CS, and DS.

Protocol 2. SAX HPLC of unsaturated disaccharides from HS, CS, and DS

Equipment and reagents

- Gradient HPLC system
- Analytical SAX Spherisorb columns (5 micron particle size; 4.6 mm × 25 cm resolving column, 4.6 mm × 5 cm guard column).
- Mobile phase (double-distilled water adjusted to pH 3.5 with HCl)

Method

1. Equilibrate column in mobile phase.

2. Inject disaccharide sample (in distilled water) and run mobile phase at a flow rate of 0.5 ml/min (5 minute wash).

3. Elute disaccharides with a linear gradient of sodium chloride (0–0.5 M over 60 minutes) in the same mobile phase.

Protocol 2. *Continued*

4. Detect elution of disaccharides by monitoring UV absorbance or radioactivity.

5. Identify and quantitate the disaccharides by comparison with reference standards.

4.3 Interpretation of data

When calculating the overall disaccharide composition, the level of resistant oligosaccharides (if any), and the presence of unknown components should be taken into account. In addition, the latter may prove to represent important but minor or rare species. With disaccharides derived by lyase scission the original GlcA and IdoA residues cannot be distinguished (see Section 3.2.1). In contrast, disaccharides derived by deaminative hydrolysis have lost their *N*-sulfate groups, but retain asymmetry around C-5 on uronic acid residues, thus allowing separation and identification of GlcA and IdoA-containing disaccharides. These two methods thus provide complementary approaches for compositional analysis.

5. Mapping of oligosaccharide domains

Information on the distribution of constituent disaccharides can be obtained by characterizing their arrangement into domains with particular structural features. Depolymerization of GAG chains with a single specific reagent generates oligosaccharides which contain end-groups derived from the susceptible linkages, and internal sequences defined by their resistance to the reagent. Separation of these products provides a type of analysis called oligosaccharide mapping (22, 26, 27) which gives details of the:

(a) structural complexity of oligosaccharide mixtures;

(b) content of susceptible linkages;

(c) relative arrangement of these linkages in the parent polysaccharide;

(d) relative distribution of resistant linkages.

5.1 Chromatographic resolution of GAG oligosaccharides

A widely used method for assessing the size and distribution of GAG oligosaccharides is gel filtration chromatography on Bio-Gel P6 or P10 (3, 22, 24, 27). Resolution of fragments up to approximately 8–10 disaccharides in length can be obtained, with individual peaks corresponding to oligosaccharides differing in size by one disaccharide unit (*Figure 2*). Anion-exchange HPLC methods can also be used to map oligosaccharide mixtures derived by enzymic scission of KS (28), and heparin (29, 30).

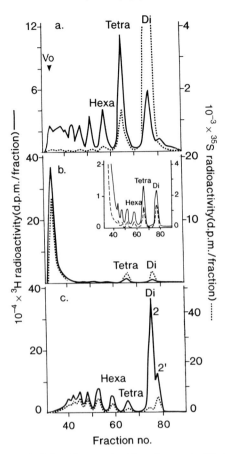

Figure 2. Separation of HS oligosaccharides by gel filtration on Bio-Gel P6. [^{3}H]Glucosamine-labelled HS from human skin fibroblasts in cell culture was treated with (a) low pH HNO$_2$, (b) heparinase, or (c) heparitinase, and the resulting oligosaccharides fractionated on a Bio-Gel P6 column (1 cm × 120 cm) eluted with 0.5 M NH$_4$HCO$_3$ at 4 ml/ h. Fractions (1 ml) were collected for scintillation counting. In each case a series of oligosaccharides ranging from disaccharides (Di) and tetrasaccharides (tetra) to larger oligosaccharides (hexa and above) was observed, the precise pattern depending on the content and distribution of susceptible linkages in the intact molecule. Reproduced from reference 22, with permission.

5.2 Oligosaccharide mapping by gradient PAGE

5.2.1 Gradient PAGE

Gradient polyacrylamide gel electrophoresis (PAGE) is a high-resolution technique for the separation of GAG oligosaccharides of variable sulfate content and disposition. It provides a level of resolution for oligosaccharides larger than tetrasaccharides which is superior to gel filtration or anion-

Table 3. Gradient PAGE—composition of solutions for gel preparation[a]

	Resolving gel concentration[b]		
	Upper limit (T30%/C5%)	Lower limit (T20%/C0.5%)	Stacking gel (T5%/C0.5%)
Acrylamide (Stock A)[c]	9.6 ml	—	—
Acrylamide (Stock B)[c]	—	10.65 ml	3.33 ml
2 M Tris–HCl, pH 8.8	3.0 ml	3.0 ml	—
1 M Tris–HCl, pH 6.8	—	—	2.5 ml
Glycerol	2.5 ml	—	—
Double-distilled water	0.9 ml	2.3 ml	13.95 ml
Ammonium persulfate (10% w/v)	12.5 µl	38 µl	200 µl
TEMED	5 µl	5 µl	20 µl
Total volume	16 ml	16 ml	20 ml

[a] The recipe given is for a single 20–30% resolving gel approximately 27 cm long, 14 cm wide, and 0.75 mm thick, suitable for casting in a 32 cm total length gel system. This allows space for 2–3 cm of stacking gel between the well bottoms and the top of the resolving gel. Recipes for gels of different dimensions can be adapted accordingly.
[b] %T refers to the total concentration (w/v) of acrylamide monomer (i.e., acrylamide plus methylene-*bis*acrylamide); %C refers to the concentration of cross-linker (methylene*bis*acrylamide) relative to total monomer.
[c] Stock A (T50%/C5%): 47.5 g acrylamide and 2.5 g methylene*bis*acrylamide made up to 100 ml in distilled water. Stock B (T30%/C0.5%): 29.85 g acrylamide and 0.15 g methylene*bis*acrylamide made up to 100 ml in distilled water.

exchange HPLC, due to the combination of a discontinuous buffer system with a pore-size gradient polyacrylamide resolving gel (26, 31). Use of the gradient results in far higher resolution than isocratic PAGE which is sufficient for GAG oligosaccharides of relatively uniform charge/mass ratio (for example DS, CS, HA oligosaccharides). Oligosaccharide mapping by gradient PAGE is a rapid and reproducible method for the simultaneous comparison of multiple samples, allowing structural characteristics to be elucidated and compared (22, 26; see *Protocol 3* and *Figure 3*).

Protocol 3. Gradient PAGE oligosaccharide mapping

Equipment and reagents

- Vertical slab gel electrophoresis system and power supply
- Electrophoresis buffer (25 mM Tris–HCl, 192 mM glycine, pH 8.3)
- Acrylamide and buffer stock solutions (see *Table 3*)
- Low volume gradient mixing apparatus and peristaltic pump

A. *Preparing the gradient PAGE gel*

1. Assemble the gel unit (consisting of glass plates and spacers, etc.).

2. Prepare and degas the upper and lower limit resolving gel acrylamide

---

solutions without ammonium persulfate or TEMED. (*Table 3* gives the details for a 20–30% gradient resolving gel.)

3. Add ammonium persulfate and TEMED to the upper and lower limit solutions, mix well, and place separately in the mixing and reservoir chambers, respectively, of the gradient mixing apparatus (or vice versa if the gradient is to be introduced from the bottom of the apparatus).

4. Begin pumping the solution from the mixing chamber into the top of the gel unit (50–100 ml/h).

5. When all the gradient has been transferred, overlay the unpolymerized gel with resolving gel buffer or water-saturated butanol. Polymerization should occur (from the top of the gel downwards) in approximately 1 hour. The gel can then be used immediately or stored at 4 °C for 1–2 weeks.

B. *Electrophoresis*

1. Immediately before electrophoresis, rinse the resolving gel surface with stacking gel buffer (0.125 M Tris–HCl buffer, pH 6.8).

2. Prepare and degas the stacking gel solution, add ammonium persulphate and TEMED. Immediately pour on to the top of the resolving gel and insert a well-forming comb.

3. After polymerization (approximately 15 min) remove the comb and rinse the wells with electrophoresis buffer.

4. Place the gel unit into the electrophoresis tank and fill the buffer chambers with electrophoresis buffer.

5. Load the oligosaccharide samples (5–50 μl containing approximately 10% (v/v) glycerol) carefully into the wells with a microsyringe. Marker samples containing Bromophenol blue and Phenol red in 10% (v/v) glycerol should also be loaded into separate tracks.

6. Run the samples into the stacking gel at 150–200 V for 30 min, followed by electrophoresis at 300–400 V for approximately 16 h. Heat generated during the run should be dissipated using a heat exchanger with circulating tap water.

7. Electrophoresis should be terminated when the Phenol red marker dye is 2–3 cm from the bottom of the gel. (At this point, disaccharides should be 1–2 cm from the bottom of the gel.)

The concentration range of the gradient resolving gel can be altered to optimize the separation of oligosaccharides of different sizes. A 20–30% gradient is good for general purpose mapping, but improved separation of smaller or larger oligosaccharides can be obtained using gel concentrations within the ranges 25–40% and 20–25%, respectively. Different voltage

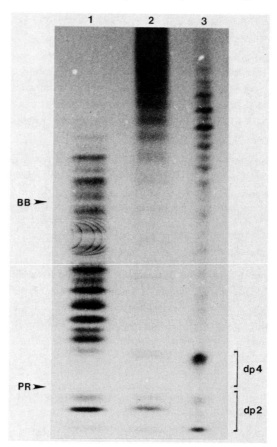

Figure 3. Oligosaccharide mapping of HS oligosaccharides by gradient PAGE. [³H]Glucosamine-labelled HS from human skin fibroblasts in cell culture was treated with heparitinase (track 1), heparinase (track 2), or low pH HNO₂ (track 3), and the resulting oligosaccharides separated by gradient PAGE (20–30% gel; see *Table 3* and *Protocol 3*), transferred to nylon membrane (see *Protocol 4*), and detected by fluorography. The migration positions of Bromophenol blue (BB), Phenol red (PR), disaccharides (dp2), and tetrasaccharides (dp4) were as indicated, and oligosaccharides approximately dp12–14 in size migrated to a similar position to that of BB. Note that only the bottom 19 cm of the gel is shown (23–30% portion of the gradient; see reference 22).

conditions and running times are required for different gradients, and can be established by trial and error.

5.2.2 Electrotransfer and fluorography

GAG oligosaccharides resolved by gradient PAGE can be subsequently transferred on to a positively-charged nylon membrane using electrotransfer methods. They can then be readily detected by fluorography in the case of radiolabelled material or by cationic dye staining for unlabelled samples (see

Section 5.2.3). Electrotransfer can be achieved using standard commercially available 'wet' or 'semi-dry' blotting equipment (22, 26, 32). Wet transfer is carried out as described in *Protocol 4* (see also *Figure 3*).

Protocol 4. Electrotransfer of GAG oligosaccharides from gradient PAGE gels to positively-charged nylon membrane

Equipment and reagents

- Electrotransfer tank and transfer cassette
- Transfer buffer (10 mM Tris–acetate, pH 7.9 containing 0.5 mm EDTA)
- Positively charged nylon-66 membrane (e.g. Zetaprobe, Biotrace RP, etc.)
- Whatman 3MM filter paper

Method

1. After termination of electrophoresis remove the stacking gel and equili-brate the resolving gel in transfer buffer for 10 min.

2. Place the gel on to a sheet of filter paper (Whatman 3MM) soaked in transfer buffer, superimpose a sheet of positively charged nylon-66 membrane, and then overlay with a further sheet of pre-soaked filter paper. Ensure good contact between the gel and the nylon membrane by gently rolling a glass tube over the top of the layered assembly.

3. Sandwich the whole assembly in a transfer cassette and insert into the transfer tank containing transfer buffer precooled to 4°C. The nylon should be on the anodic side of the gel.

4. Transfer at 1.5 V/cm for 3–4 h at 4°C.

5. Remove the nylon membrane and detect oligosaccharides either by fluorography using for example, En³Hance spray (New England Nuc-lear) or by cationic dye staining.

An alternative is the semi-dry electrotransfer technique recently reported for GAG oligosaccharides (for experimental details, see reference 32). Efficient transfer can be achieved in approximately 1–2 h, although small oligosaccharides can transfer very rapidly (for example within 10–15 min). With both transfer methods oligosaccharides with low levels of sulfation bind to the membrane with low efficiency (25), necessitating care in the quantita-tive interpretation of data (for example compare *Figures 2* and *3*). Although additional layers of nylon can be used to capture fragments which transfer through the first layer, the procedure should be assessed and standardized for the particular type of oligosaccharide mixtures under investigation.

Oligosaccharides transferred to a nylon membrane can be recovered prep-aratively by cutting out specific bands (located by staining or fluorography of parallel strips cut from the membrane edge) and eluting with 2 M sodium chloride. In addition, the semi-dry transfer technique can be used to purify a major oligosaccharide from minor contaminants; using multiple layers of nylon, oligosaccharide purity increases proportionally with transfer depth (32).

5.2.3 Detection of non-radiolabelled GAG oligosaccharides

Detection of non-radiolabelled oligosaccharides resolved by gradient PAGE can readily be achieved with cationic dyes if sufficient material is available (generally 50–100 μg of a complex mixture). After removal from the glass plates the gel can be stained for 30 minutes in either 0.5% (w/v) Alcian Blue in 2% (v/v) acetic acid or in 0.08% (w/v) aqueous Azure A, followed by destaining in 2% (v/v) acetic acid or distilled water, respectively. Azure A staining is 2-fold more sensitive than Alcian Blue, but increased sensitivity is obtained by a dual-staining method of first staining with Alcian Blue, followed by Azure A staining and destaining (31). Oligosaccharides with a low charge density or small size may not be visualized using cationic dye staining (see below). Oligosaccharides transferred to a nylon membrane can also be detected with Alcian Blue, by staining for 2 minutes and destaining under the conditions described above (31).

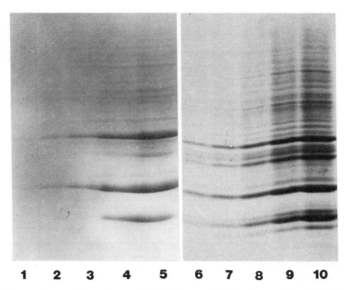

1 2 3 4 5 6 7 8 9 10

Figure 4. Comparison of detection of GAG oligosaccharides in gradient PAGE gels using Azure A alone or in combination with silver staining. Porcine mucosal heparin was digested to completion with heparinase, and various sample loadings were run on a gradient PAGE gel. The gel was cut into two: one half was stained with Azure A alone (tracks 1–5), and the other was stained with Azure A followed by ammoniacal silver (tracks 6–10), as described in *Protocol 5*. The sample loadings were: track 1, 5 μg; 2, 10 μg; 3, 20 μg; 4, 50 μg; 5, 100 μg; 6, 0.5 μg; 7, 1 μg; 8, 2 μg; 9, 5 μg; 10, 10 μg (NB tracks 6–10 contain one-tenth of the loading of tracks 1–5 respectively). Reproduced from reference 33, with permission.

A silver-staining method which is 25- to 50-fold more sensitive than cationic dye staining alone has been developed. The method involves fixation with the dye Azure A followed by staining with ammoniacal silver (reference 33; see *Protocol 5*). This procedure is particularly valuable in situations where the availability of GAG is very limited and/or where radiolabelling is impractical or undesirable (such as clinical tissue/fluid samples). Detection of as little as 1–2 ng of a single oligosaccharide species is possible with this method, which can thus be used for mapping a few micrograms of a complex mixture (*Figure 4*).

Protocol 5. Detection of GAG oligosaccharides on PAGE gels by silver staining

Note: In order to minimize background and artifactual staining all electrophoresis/staining reagents and solutions should be of a high quality. Gel plates and the staining dish (preferably glass) should be very clean, and gels should only be handled using clean gloves.

Equipment and reagents

- Polyacrylamide gel (see *Protocol 3*) 14 cm × 14 cm × 0.75 mm
- Bromophenol blue
- Phenol red
- Azure A

- Ammoniacal silver solution: prepare immediately before use; 10 ml fresh 10% (w/v) silver nitrate solution to mixture comprising 187 ml degassed, distilled water, 1 ml 11.3% (w/v) NaOH solution, 1.73 ml 35% (v/v) aqueous ammonia solution (add silver nitrate gradually with shaking)

A. *Electrophoresis*

1. Electrophorese non-radiolabelled oligosaccharides (preferably derived from 5–10 μg of GAG) on a gradient polyacrylamide gel essentially as described in *Protocol 3*. A gel with dimensions of 14 cm × 14 cm × 0.75 mm (i.e. a standard protein system) is recommended for ease of subsequent gel handling.

2. Load Bromophenol blue and Phenol red marker dyes separately from the oligosaccharide samples as they give rise to a number of stained bands.

B. *Staining*

1. Immediately after electrophoresis fix the gel in 0.08 % (w/v) aqueous Azure A for 15 min with gentle agitation.

2. Destain in frequent changes of distilled water. Very slight acidification of the water with acetic acid (1 μl glacial acetic acid in 200 ml water) will hasten the process.

3. Immerse the destained gel in ammoniacal silver solution, cover and agitate gently for 1 h at room temperature.

Protocol 5. *Continued*

4. Wash the gel three times for 5 min each in degassed distilled water.

5. Develop the silver stain by the addition of 200 ml of freshly prepared 0.01% (w/v) citric acid, 0.038% (v/v) formaldehyde in degassed distilled water. Keep agitated at all times. Development can be more rapid than commonly occurs with proteins.

6. When a suitable level of development has almost been reached decant off the solution and terminate development by the addition of 200 ml of 2.5% (v/v) acetic acid for 30 sec.

7. Wash the gel extensively in water (tap water will suffice) until all cloudiness has disappeared. The stain is stable and will last indefinitely.

The method is applicable to oligosaccharides derived from any GAG species. However, a limitation of this staining technique which must be borne in mind is the oligosaccharide cut-off imposed by the dye fixation step (NB direct staining of the gel does not work). Oligosaccharides below a minimum size and/or charge content are not quantitatively fixed within the gel (33). In general, quantitative analysis of smaller species is best served by gel chromatographic or HPLC techniques.

5.3 Interpretation of data and further analyses

Data from gel filtration profiles can be used quantitatively to calculate the proportion of linkages susceptible to a particular scission technique, and to give information on their distribution within the intact GAG chain (that is contiguous, alternating, or spaced apart by non-susceptible sequences; for details see reference 22). The size distribution of resistant oligosaccharides can also provide useful information on the arrangement of particular types of resistant sequences (for instance heparitinase 1 resistant oligosaccharides in HS represent contiguous sequences of IdoA-containing disaccharides (22, 27)).

Oligosaccharide mapping by both gradient PAGE and HPLC are particularly useful for revealing the structural complexity of oligosaccharide mixtures, and provide a rapid and reproducible means of making detailed comparisons between different species of a particular GAG type. Gradient PAGE mapping provides the most powerful technique available for resolving GAG oligosaccharides, and although it should be used mainly as a comparative method, it can be usefully applied to the detailed resolution of oligosaccharides which have been quantitated and broadly sized by an initial gel filtration step.

Following primary cleavage of the chains by a particular scission method further analyses of resistant sequences (prepared by gel filtration, SAX HPLC, or preparative gradient PAGE methods) can be made by the use of secondary reagents which cleave at different linkages. The distribution of these linkages within the initial resistant oligosaccharides of defined size can be established, thus providing more detailed information on the disaccharide composition and sequence of specific domains of the GAG chain. For example, the distribution of IdoA(2S) residues within IdoA-repeat (heparitinase-resistant) sequences in HS has been investigated using subsequent heparinase treatment (27).

6. Approaches for the sequence analysis of GAGs

The presence of structurally-ordered domains within certain types of GAG chains has prompted the development of approaches for analysing the precise sequence of sugars along the polysaccharide chains. The distribution of sugars relative to a defined reference point, usually the reducing terminal xylose residue, can be assessed by either direct or indirect approaches. In the former, the reducing end of the chains can be radiolabelled specifically at either the xylitol residue (by treatment of PG or peptidoglycan with alkaline NaB^3H_4 (25)) or at the serine residue remaining after exhaustive proteolysis of PG (20).

The indirect approach involves end-referencing of the chains by coupling of the core protein or peptide moieties of PG or peptidoglycan, respectively, to activated Sepharose (34–36). This allows specific degradation of the GAG chain with retention of those oligosaccharides originating at the point of attachment to the protein core (that is the xylose residue). The latter can then be released by treatment with alkaline borohydride. Alternatively, a hydrophobic moiety can be introduced on to the reducing terminal xylose, allowing separation of end-referenced fragments by hydrophobic chromatography (37).

Using the above approaches, the GAG chains can be treated with reagents that cleave at specific linkages. Analysis of the size distribution of the resulting radiolabelled or Sepharose-retained oligosaccharides by gel filtration (20, 25, 34–37) or gradient PAGE (20, 36) gives information on the distance of the susceptible linkage 'downstream' (that is towards the non-reducing end of the chain) from the reducing terminus. Complete digestion defines the position of the first susceptible downstream linkage, whereas partial scission should, in principal, allow the position of linkages further downstream to be obtained. Assessment of the disaccharide composition or biological activity of particular sizes of fragments derived by partial scission can provide further information on the distribution of sugars along the chain (35, 37). Studies using the sequencing strategies described above have shown that the distribution of sugars in HS (25, 34, 36, 38), heparin (35), CS (37), and DS (20) are clearly non-random, and the application and further development of these approaches should prove to be an active area of future research on the structural analysis of GAGs.

Acknowledgements

Thanks are due to Dr Keilichi Yoshida for provision of heparitinases and unsaturated disaccharide standards, and to Drs Robert Linhardt, Hui-Ming Wang, and D. Loganathan at the University of Iowa for a productive collaboration on the lyase specificities. The generous support by the Cancer Research Campaign and the Christie Hospital Endowment Fund are acknowledged. We also thank Mrs P. Jones for preparation of the manuscript.

References

1. Gallagher, J. T. (1989). *Curr. Opin. Cell Biol.*, **1**, 1201–18.
2. Nieduszynski, I. A., Huckerby, T. N., Dickenson, J. M., Brown, G. M., Tai, G., Morris, H. G., and Eady, S. (1990). *Biochem. J.*, **271**, 243–5.
3. Maimone, M. M. and Tollefsen, D. M. (1990). *J. Biol. Chem.*, **265**, 18263–71.
4. Oeben, M., Keller, R., Stuhlsatz, H. W., and Greiling, H. (1987). *Biochem. J.*, **248**, 85–93.
5. Heinegard, D. and Sommarin, Y. (1987). In *Methods in enzymology*, Vol. 144 (ed. L. W. Cunningham), pp. 319–72. Academic Press, New York.
6. Yanagishita, M., Midura, R. J., and Hascall, V. C. (1987). In *Methods in enzymology*, Vol. 138 (ed. V. Ginsberg), pp. 279–89. Academic Press, New York.
7. Lyon, M. and Gallagher, J. T. (1991). *Biochem. J.*, **273**, 415–22.
8. Yanagishita, M. and Hascall, V. C. (1983). *J. Biol. Chem.*, **258**, 12857–64.
9. Linhardt, R. J., Galliher, P. M., and Cooney, C. L. (1986). *Applied Biochem. Biotech.*, **12**, 135–76.
10. Linhardt, R. J., Turnbull, J. E., Wang, H. M., Loganathan, D., and Gallagher, J. T. (1990). *Biochemistry*, **29**, 2611–17.
11. Nakazawa, K., Ito, M., Yamagata, T., and Suzuki, S. (1989). In *Keratan sulphate: chemistry, biology, chemical pathology* (ed. H. Greiling and J. E. Scott), pp. 99–110. The Biochemical Society, London.
12. Shively, J. E. and Conrad, H. E. (1976). *Biochemistry*, **15**, 3932–42.
13. Bienkowski, M. J. and Conrad, H. E. (1985). *J. Biol. Chem.*, **260**, 356–65.
14. Guo, Y. and Conrad, H. E. (1989). *Anal. Biochem.*, **176**, 96–104.
15. Shaklee, P. N. and Conrad, H. E. (1984). *Biochem. J.*, **217**, 187–97.
16. Edge, A. S. B. and Spiro, R. G. (1985). *Arch. Biochem. Biophys.*, **240**, 560–72.
17. Sjoberg, I. and Fransson, L.-A. (1980). *Biochem. J.*, **191**, 103–10.
18. Casu, B. (1989). In *Heparin—chemical and biological properties and clinical applications* (ed. D. A. Lane and U. Lindahl), pp. 25–49. Edward Arnold Press, London.
19. Fransson, L.-A., Sjoberg, I., and Havsmark, B. (1980). *Eur. J. Biochem.*, **106**, 59–69.
20. Fransson, L.-A., Havsmark, B., and Silverberg, I. (1990). *Biochem. J.*, **269**, 381.
21. Yoshida, K., Miyauchi, S., Kikuchi, H., Tawada, A., and Tokuyasu, K. (1989). *Anal. Biochem.*, **177**, 327–32.
22. Turnbull, J. E. and Gallagher, J. T. (1990). *Biochem. J.*, **265**, 715–24.
23. Nader, H. B., Dietrich, C. P., Buonassisi, V., and Colburn, P. (1987). *Proc. Natl Acad. Sci. USA*, **84**, 3565–9.

24. Gallagher, J. T. and Walker, A. (1985). *Biochem. J.*, **230**, 665–74.
25. Edge, A. S. B. and Spiro, R. G. (1990). *J. Biol. Chem.*, **265**, 15874–81.
26. Turnbull, J. E. and Gallagher, J. T. (1988). *Biochem. J.*, **251**, 597–608.
27. Turnbull, J. E. and Gallagher, J. T. (1991). *Biochem J.*, **273**, 553–9.
28. Dickenson, J. M., Morris, H. G., Nieduszynski, I. A., and Huckerby, T. N. (1990). *Anal. Biochem.*, **190**, 271–5.
29. Rice, K. G., Kim, Y. S., Grant, A. C., Merchant, Z. M., and Linhardt, R. J. (1985). *Anal. Biochem.*, **150**, 325–31.
30. Linhardt, R. J., Rice, K. G., Kim, Y. S., Lohse, D. L., Wang, H. M., and Loganathan, D. (1988). *Biochem. J.*, **254**, 781–7.
31. Rice, K. G., Rottink, M. R., and Linhardt, R. J. (1987). *Biochem. J.*, **244**, 515–22.
32. Al-Hakim, A. and Linhardt, R. J. (1990). *Electrophoresis*, **11**, 23–8.
33. Lyon, M. and Gallagher, J. T. (1990). *Anal. Biochem.*, **185**, 63–70.
34. Lyon, M., Steward, W. P., Hampson, I. N., and Gallagher, J. T. (1987). *Biochem. J.*, **242**, 493–8.
35. Rosenfeld, L. and Danishefsky, I. (1988). *J. Biol. Chem.*, **263**, 262–6.
36. Turnbull, J. E. and Gallagher, J. T. (1991). *Biochem. J.*, **277**, 297–303.
37. Uchiyama, H., Kikuchi, K., Ogamo, A., and Nagasawa, K. (1987). *Biochem. Biophys. Acta*, **926**, 239–48.
38. Lyon, M., Deakin, J. A., and Gallagher, J. T. (1994). *J. Biol. Chem.*, **269**, 11208–15.

10

Isolation and characterization of proteoglycan core protein

VINCENT C. HASCALL, MASAKI YANAGISHITA,
ANTHONY CALABRO, RONALD MIDURA, JODY A.
RADA, SHUKTI CHAKRAVARTI, and JOHN R. HASSELL

1. Introduction

Proteoglycans have one or more glycosaminoglycan chains covalently bound to a core protein or glycoprotein. Normally, the glycosaminoglycans, with the exception of hyaluronic acid, contain numerous sulfate ester groups (see Chapter 9). This makes proteoglycans highly polyanionic and confers upon them distinct properties, such as large hydrodynamic volumes and, frequently, high buoyant densities. These properties can be used to advantage in purification and characterization steps. The isolation of proteoglycans first requires effective extraction from cells or tissue and then separation from other extracted macromolecules. Because of the wide diversity of core proteins, there is no universally applicable set of procedures that will be satisfactory for every proteoglycan. Thus the protocols described below, while generally useful, may require modification to adapt them for use in any particular problem. They are chosen to illustrate general principles which can help investigators choose optimal combinations for their particular needs.

2. Metabolic labelling

Presently a large proportion of proteoglycan research utilizes biosynthetic radiolabelling methods. This provides a convenient way for following proteoglycans through purification steps and importantly, for monitoring recoveries at each step along the way. The most convenient precursor for selectively labelling proteoglycans is [^{35}S]sulfate. Typically more than 90% of the incorporated activity with this precursor will be present in proteoglycans. It is frequently used in combination with a carbohydrate precursor such as [^{3}H]glucosamine, which will label the glycosaminoglycans and other oligosaccharides on the core protein as well as the general pool of glycoproteins in the system. Also, [^{3}H]mannose is sometimes used as a fairly specific precursor for asparagine-linked oligosaccharides on the core protein. Once the proteoglycans are purified, such precursors can be used to provide details about the composition, structure, and types of complex carbohydrates on the macromolecules. Radiosulfate is also frequently used in combination with amino acid precursors such as [³H]serine or [³H]leucine, which label the core protein as well as the general pool of proteins and glycoproteins in the system. Such precursors are very useful for monitoring the effectiveness of purification steps, and once the proteoglycans are purified, help define the number and relative sizes of the core proteins in the total proteoglycan population.

Protocol 1. Metabolic Labelling

1. Prepare cell or tissue explant cultures using standard sterile culture techniques and appropriate media for the system.

2. On the day of labelling, replenish the cultures with new medium and preincubate for 1 h to allow the cells to re-equilibrate. At labelling time 0, either replace the medium with fresh medium which contains the radioisotopes or introduce the radioisotopes into the cultures in a small volume, usually less than 2% of the total volume. Take care not to introduce temperature transients in the cultures. Final radioisotope concentrations can be up to 500 μCi/ml for any of the precursors, although, typically, ranges from 50–200 μCi are used. If sulfate-depleted medium is used, the final sulfate concentration should not be less than 100 μM; otherwise the glycosaminoglycans may not be fully sulfated (1). Inorganic sulfate can come from a variety of sources including: the water source, sera (~ 8 μM for each 1% final concentration of serum), antibiotics, and other medium additives.

3. The labelling time can be varied. Typically, sulfate incorporation into proteoglycans will be linear, i.e. in steady state, for 10–24 h. Check this in preliminary experiments by labelling cultures for different time periods. The total labelling time generally should not exceed the time during which proteoglycan synthesis is linear. Note that carbohydrate and amino acid precursors may not show linear incorporation over the same time period because the specific activities of their intracellular precursor pools may change with time from metabolic sources. For example, [³H]glucosamine is usually diluted several hundred fold by glucosamine metabolically synthesized from glucose (2). This is normally not the case with radiosulfate since the metabolic sources of sulfate, namely cysteine and methionine, usually contribute negligibly to the sulfate donor pool, phosphoadenosine-phosphosulfate, and hence the specific activity of the incorporated radiosulfate does not change during the incubation.

4. After the appropriate labelling time, remove the medium, and dilute a small aliquot carefully for counting to determine the precise concentrations of the radioisotopes in the final medium after the labelling period. Briefly wash the culture at 4 °C, typically for 5–10 min, with medium or isotonic saline solution that does not contain radioisotopes. Combine the wash with the labelling medium for further processing as described below. Extract the cell layer with its associated matrix, or the tissue explant as described in *Protocol 2*.

5. Carbohydrate precursors will form covalent bonds with amino groups on proteins by non-enzymatic glycation. Thus, if the labelling medium contained serum or other protein sources, include a control culture without cells or tissues to determine the proportion of total macromolecular radioactivity which is derived from this non-metabolic source.

3. Extraction/solubilization

In most cases, the cell layer or tissue will contain a variety of proteoglycans. Some proteoglycans will be integrated in the extracellular matrix through non-covalent interactions with other matrix molecules such as collagen. Others will be associated with the cell surface either through hydrophobic binding into the plasma membrane via polypeptide intercalation or phosphatidylinositol anchors, or by ionic binding with cell surface molecules. Yet others may be sequestered in intracellular compartments such as in storage or secretory granules. In many experiments, the first objective is to solubilize the maximum amount of proteoglycans in the system independent of the compartment in which they reside. The most effective solvents contain chaotropic reagents, that is those able to denature proteins and dissociate most non-covalent interactions, plus detergents to dissociate hydrophobic interactions. The most widely used chaotropic agent is 4 M guanidine–HCl, and it is frequently used with compatible detergents such as 4% CHAPS or 2% Triton X-100. In general, CHAPS offers the distinct advantage in that it can be removed more easily in subsequent steps, for example by dialysis, whereas Triton X-100, while having somewhat better solubilization capacity, cannot be removed by dialysis due to micelle formation. If sequential extractions are used, solvents with detergents should be used first. Extraction of cell layers with 4 M guanidine–HCl alone should be avoided since such treatment actually facilitates hydrophobic interactions between macromolecules which are difficult to dissociate with detergents in later steps (3).

Because proteoglycans are very susceptible to proteolytic cleavage, a combination of protease inhibitors is usually added to the extraction solvent to inhibit the different broad classes of proteases, and the extractions are usually done at 4°C for the same reason. The protease inhibitors are particularly important if any step involves changing the solvent back to conditions which favour renaturation before the proteoglycans are fully purified. Many proteases can recover activity in such cases. For this reason, proteoglycan purification steps are frequently done in chaotropic solvents to minimize the risk of degradation (see *Protocol 4*).

If tissue explants are to be extracted, extraction efficiency can often be improved if the tissue is frozen and thin slices (less than 1 mm) prepared on a tissue slicer. Otherwise the tissue should be minced finely.

For more extensive discussion of extraction methods see reference 4.

Protocol 2. Extraction/solubilization of proteoglycans

Reagents

- Protease inhibitor cocktail: 10 mM 6-aminohexanoic acid (for cathepsin D-like activity), 5 mM benzamidine–HCl (for trypsin-like activity), 1 mM phenylmethylsulfonyl fluoride (for serine-dependent proteases), 10 mM disodium EDTA (for metalloproteases), 10 mM *N*-ethylmaleimide (for thiol-dependent proteases). Note that NEM will alkylate any exposed sulfhydryl groups on the extracted proteins.

- Solvent: 4 M guanidine–HCl, 4% (w/v) CHAPS, 0.1 M sodium acetate, plus a protease inhibitor cocktail, pH 5.8

Method

1. Prepare the protease inhibitor cocktail just prior to use, from concentrated stock solutions of 6-aminohexanoic acid and disodium EDTA and from dried powders for the other inhibitors. The phenylmethylsulfonyl fluoride and *N*-ethylmaleimide are sparingly soluble in aqueous solutions and can be dissolved at ∼ 100× concentration in methanol first.

2. Extract the cell layer or minced tissue at 4°C overnight with gentle stirring. Use a convenient volume of extractant (typically 1–2 ml for a 35 mm culture dish), but at least 10 volumes per wet weight of minced tissue.

3. Decant the extract and clarify it by low speed centrifugation if necessary. The residue can be re-extracted with another one-half volume of extractant for 3–4 h and the second extract combined with the first after clarification.

4. For most cell culture systems, the residue remaining after the extraction will be negligible. However, solubilize any proteoglycans remaining with the culture dish by papain digestion and then quantitate by molecular-sieve chromatography (*Protocol 3*). If tissue explants were extracted, solubilize the residue by papain digestion and quantitate the proportion of residual proteoglycan in the same way.

This protocol usually solubilizes 80–100% of the proteoglycans associated with the cell layer or the tissue.

---

4. Solvent exchange and quantitation of macromolecular radioactivity

Three compartments have been isolated: the medium fraction, the extract, and the papain-digested residue. Each will contain radiolabel in both macromolecules and unincorporated precursors. *Protocol 3* utilizes a molecular-sieve step to remove the unincorporated radioisotope, to quantitate the incorporated activity, and to exchange the extraction solvent into one which is compatible with subsequent ion-exchange chromatography (*Protocol 4*). Sephadex G-50 (fine) is the molecular-sieve matrix used. The columns are prepared in disposable serological pipettes. The size of the column can be varied conveniently from 1 ml to 24 ml using different size pipettes, and the choice of size depends upon the volume to be applied. Typically the columns give baseline separation between the excluded peak, which contains the incorporated activity, and the totally included volume, which contains the unincorporated precursors, when the volume applied is 25% or less of the total volume. *Protocol 3* is based upon a 2 ml sample volume per 8 ml column. Convenient plastic racks can be made or purchased which will hold 10 columns per rack and which are spaced appropriately for collecting in rows over standard scintillation vial boxes. Thus 10 samples can be processed at a time. The columns are often equilibrated with an 8 M urea solvent or a 10 M formamide solvent. Both are chaotropic, but the stability of the latter is better. Additionally, decomposition products of urea can potentially block N-terminal amino acids, and therefore may be a disadvantage if amino acid sequencing is planned at later steps.

Protocol 3. Sephadex G-50 chromatography

Equipment and reagents

- 10 ml plastic serological pipettes + holder + collecting tray + vials
- 1 ml pipettes
- Glass wool
- Sephadex G-50 (fine)

- Deionized filtered water containing 0.02% sodium azide
- Elution buffer: 10 M formamide, 0.05 M sodium acetate 0.3 M NaCl, 0.5 % CHAPS, containing protease inhibitors (optional)

Method

1. Remove the tops of 10 ml plastic serological pipettes by scoring just below the constriction with a triangular file and snapping them off. Push small glass wool plugs into the point using a 1 ml plastic pipette as a tamp. Place the pipettes in a plastic holder designed to hold 10 pipettes and then place the rack over a collecting tray.

2. Wash and equilibrate Sephadex G-50 (fine) in deionized, filtered water

---

Protocol 3. *Continued*

containing 0.02% sodium azide and suspend this as a 50–70% slurry. Pour the columns and allow the matrix to settle until the resin level is somewhat above the 2 ml mark for a bed volume of 8 ml.

3. Equilibrate each column with elution buffer by allowing about one column volume (8–10 ml) to percolate through the column.

4. Remove sufficient resin from each column to create the top of the bed at the 2 ml mark. These conditions are such that the columns will usually not crack at the top when drained to the surface. Nevertheless, plug the tip of the packed column with a rubber sleeve until the samples are applied.

5. Apply up to 2 ml of sample to each column. If less than 2 ml is applied, after the sample has completely entered into the upper surface of the gel, use an aliquot of buffer to wash the sample into the column such that the sample volume plus the aliquot equals 2 ml.

6. After each column has had a total of 2 ml applied, position the rack over the vials to be used to collect the excluded column volume fractions. The excluded volume is between 2.3–4.5 ml (this can be checked independently by running standards and collecting fractions). Thus, elute the macromolecular fraction from the columns into the collecting vials by applying a total of 3 ml of solvent, using the first 0.1–0.2 ml to wash in the sample. The totally included peak does not begin before 6 ml. Therefore, the 3 ml collected, i.e. between 2–5 ml after beginning to apply the sample, will be free of unincorporated radioactivity.

7. After collecting the excluded volume fractions, discard the columns into the radioactive waste disposal. The discarded columns contain the large majority of radioactivity used in the labelling step and careful disposal minimizes the possibility of contaminating the laboratory.

8. Estimate the volume of each excluded fraction by weighing the collecting vials before and after sample collection and correcting for the density of the solvent, in this case 1.08 g/ml. Count an aliquot to determine total recoveries of incorporated activity in all of the compartments.

The distribution of incorporated radiosulfate provides an estimate of the distribution of proteoglycans in the three compartments because the glycosaminoglycans generated from the residue fraction by papain digestion will still be excluded by the Sephadex G-50 except where GAG chain size is < 15 kDa, in which case a substantial portion will be partially included. The samples are now in a chaotropic solvent which is appropriate for the ion-exchange step used to recover and purify the proteoglycans (*Protocol 4*).

10: Isolation and characterization of proteoglycan core protein

5. Separation of proteoglycans from proteins and glycoproteins

Proteoglycans bind more tightly to anion-exchange matrices than do proteins and glycoproteins. This, plus the development of fast-flow ion-exchange matrices, makes this the method of choice for concentrating proteoglycans and separating them from other macromolecules. There are two potential problems with this technique; non-specific adsorption of contaminating macromolecules and reduced recovery of bound proteoglycans from the resin. The inclusion of detergent and salt in the solvent helps minimize these problems. The use of sufficient, but not excessive amounts of the anion-exchange matrix also helps.

Protocol 4 is particularly useful when the proteoglycans are in relatively large volumes and generally gives better recovery than application of samples to preformed columns.

Protocol 4. Batch ion-exchange method

1. Equilibrate Q-Sepharose with several changes of the 10 M formamide solvent used in *Protocol 3*. The equilibrated slurry can be stored indefinitely. The solvent contains 0.30 M NaCl, adjust this concentration up or down as experience is gained with the proteoglycans in the particular system under study. As a rule, the NaCl concentration should be as high as possible without preventing binding of the proteoglycans, this minimizes non-specific adsorption.

2. Add aliquots of the equilibrated resin to the samples recovered in the excluded volume fractions from the Sephadex G-50 columns (*Protocol 3*). Typically, 1 ml of resin (settled volume) would be sufficient for up to 3–5 mg of proteoglycan. Make the suspension in a conical centrifuge tube and gently mix a few times over about 60 min.

3. Centrifuge the suspension at low speed, approximately 100 *g*. Count an aliquot of the supernatant for ^{35}S-radioactivity to estimate the proportion of the total which has bound to the resin. Typically, 90–95% should bind. If the proportion bound is lower than expected, add another aliquot of slurry and repeat the process.

4. After centrifugation, remove the supernatant. Resuspend the pellet in 10–20 volumes of solvent and wash for 10 min. Repeat the centrifugation and wash step twice.

5. Extract the bound proteoglycans with a solvent that contains sufficient salt to reverse their binding to the resin. Resuspend the pellet in 3–5 volumes of an appropriate solvent, for example a 4 M guanidine–HCl

---

Protocol 4. *Continued*

solvent, or a 10 M formamide solvent which contains 1 M NaCl, and mix intermittently over 30 min. Filter the suspension through a small column prepared by putting a glass wool plug in a plastic serological pipette as described in *Protocol 3*. Wash the retained resin with an additional 2 volumes of the extracting solvent. Collect the filtrate and wash in a preweighed container.

6. Determine the volume of the filtrate and count an aliquot to estimate recovery.

7. Concentrations of NaCl in samples and standards can be measured conveniently by diluting small aliquots with water and using a conductivity meter.

Note: When 0.3 M NaCl is used in the 10 M formamide solvent, hyaluronic acid will not bind to Q-Sepharose. The hyaluronic acid fraction can be recovered from the unbound fraction obtained from step 4 above by diluting it with 10 M formamide solvent which does not contain NaCl to give an initial NaCl concentration of 0.15 M. A slurry of Q-Sepharose equilibrated with the same NaCl concentration can then be added to bind hyaluronic acid, and possibly other polyanionic macromolecules present in the extract. Alternatively, the lower NaCl concentration can be used initially to bind both hyaluronic acid and proteoglycans in the original sample. Recovery of hyaluronic acid from the resin can be low in the absence of pretreatment with proteases (5).

6. Further purification and separation of proteoglycans

Separate classes of proteoglycans can have differences in the average properties of their polydisperse populations; namely in their average charge densities, their average buoyant densities, their average hydrodynamic sizes, and the hydrophobicity of their core proteins. Each of these properties can be exploited to identify and separate different classes of proteoglycans. *Protocols 5–8* illustrate examples for each of these properties.

6.1 Charge

The batch ion-exchange process described above (*Protocol 4*) normally gives good recoveries and achieves a reasonably high level of purification. It can also be used to concentrate the proteoglycans into a much smaller volume. However, it is frequently useful to use a second ion-exchange step to achieve better purity and to separate classes of proteoglycans. In this case, small anion-exchange columns can be prepared, and the samples filtered through them to bind the proteoglycans. Alternatively, the proteoglycans are bound to the matrix by adding an aliquot of ion-exchange slurry as described in *Protocol 4*. In this procedure, the slurry with bound proteoglycans is used to make the small column. A continuous salt gradient is then applied to the column via constant flow pumping and the eluant collected with a fraction collector. If the extraction solvent used in the batch step described in *Protocol 4* was 10 M formamide with 1 M NaCl, the sample can be diluted with a 10 M formamide solvent which contains no salt such that the final solution is 0.30 M NaCl. Under these conditions the proteoglycans will once again bind when filtered through the Q-Sepharose column or when the resin is added as a slurry before packing the column.

Protocol 5. Anion-exchange chromatography with a continuous salt gradient

Equipment and reagents

- 25 ml plastic serological pipette (*Protocol 2*) or commercially produced column (Pharmacia, Bio-Rad, etc.)
- Q-Sepharose resin from stock slurry (*Protocol 4*) in starting buffer
- Peristaltic pump + fraction collector + gradient maker

- Starting buffer: 10 M formamide, 0.3 M NaCl, 0.5% CHAPS
- Final buffer: 10 M formamide, 1.2 M NaCl, 0.5% CHAPS
- Linear NaCl gradient: 0.3–1.2 M NaCl

Method

1. Prepare a small column from the bottom third or so of a plastic serological pipette as described in *Protocol 2*. Typically a 25 ml pipette can be used for a column with a bed volume of 1–2 ml. The larger pipette is used because a larger cross-sectional area on the column gives better flow rates and improves proteoglycan recoveries. Alternatively, columns and flow adaptors to regulate bed height can be purchased from a variety of commercial sources.

2. Add an aliquot of Q-Sepharose resin from the stock slurry in Starting buffer to the sample solution to bind the proteoglycans as described in *Protocol 4*. Add another aliquot to the column to form a small support layer of resin, about 100 μl. After the resin settles in the sample solution, most of the supernate can be passed rapidly through the support layer into a container to collect the unbound fraction. Pack the remaining slurry with bound proteoglycans on to the support layer and wash the column with 3–4 bed volumes to collect the remaining unbound and weakly adherent molecules.

3. Attach one end of the column to a fraction collector via a peristaltic pump and attach the other end of the column to an appropriate gradient maker, set to increase the NaCl concentration to about 1.2 M NaCl. For a 1–1.5 ml bed volume, the gradient can be prepared from 40 ml of the Starting buffer and 40 ml of the Final buffer. The total time for the gradient should be set at 2–3 h at a flow rate of 15–20 ml/h. Collect

Protocol 5. *Continued*

> fractions with an appropriate volume per fraction such that the total analysis during the gradient will have about 80 fractions, typically approximately 1 ml per fraction for a bed volume of 1–2 ml.
>
> 4. Count aliquots of the fractions for radioactivity. Backflush the resin in the column into a test tube with buffer, suspend by vortex mixing, and count an aliquot to estimate how much of the total proteoglycan remains bound to the resin. Typical recoveries will be 85–90%.

In some systems, different classes of proteoglycans have different charge densities and will resolve, at least partially, into separate peaks. This is often the case for systems which make both heparan sulfate and dermatan sulfate containing proteoglycans. In such systems, this anion-exchange method should be used early in the experimental strategy so that the different classes of proteoglycans can subsequently be characterized separately.

Recently, many different types of ion-exchange matrices have become available in HPLC and membrane-cartridge forms. These can offer advantages in analysis time and separation of different proteoglycan species. Each can be tested for capacity, recoveries, and separation properties by some of the procedures described above.

6.2 Density

Glycosaminoglycans have much higher buoyant densities in CsCl equilibrium density gradients than proteins for two reasons: carbohydrates have higher intrinsic buoyant densities than amino acids, and the large number of anionic groups are associated primarily with the dense caesium counter ions in the gradient. For this reason, proteoglycans with higher proportions of glycosaminoglycan to protein have higher buoyant densities. In some cases, the differences between two populations are sufficient that they can be at least partially resolved in the gradients. In *Protocol 6* described below, the gradients are formed in the presence of chaotropic solvents, and are therefore referred to as dissociative gradients. Gradients developed in the absence of chaotropic solvents have been particularly useful for studying cartilage proteoglycan aggregation with hyaluronic acid and are referred to as associative gradients. Dissociative gradients are often established in the presence of 4 M guanidine–HCl as described in *Protocol 6*. Other chaotropic agents such as 8 M urea or 10 M formamide may be used and the same general principles apply. The presence of 0.5% CHAPS improves recoveries, particularly when hydrophobic proteoglycans such as cell-surface heparan sulfate proteoglycans are being analysed. See reference 4 for a more detailed discussion of this methodology.

Protocol 6. Isopycnic CsCl density gradient

1. Equilibrate the radiolabelled sample with a 4 M guanidine–HCl, 0.5% CHAPS, 50 mM sodium acetate, pH 6 solution. Add 0.6 g CsCl/ml of solution to give an initial density of 1.5 g/ml. Experience with the proteoglycans in a particular system may indicate that the initial density should be somewhat lower for better resolution, therefore, adjust the amount of the CsCl accordingly. Addition of 0.15 g CsCl per g of solution will increase the buoyant density by approximately 0.1 g/ml under these solvent conditions. Always check the initial densities by weighing equal volumes of the solution and of water and calculating the ratio.

2. Add solvent at the same density to fill the centrifuge tubes to the levels specified by the manufacturer.

3. Centrifuge the samples for 48–60 h at 10 °C in an appropriate rotor (the use of vertical rotors shortens the centrifugation time). Typical rotors (Beckman) and speeds (r.p.m.) are as follows:
   - 50.2 Ti   33 000
   - 50 Ti   40 000
   - SW 50.1   35 000

   The final gradients should not contain a CsCl pellet in the bottom of the tube. If one is present, it is necessary in subsequent analyses to either use a lower initial buoyant density or to use a lower rotor speed.

4. Collect each final gradient into 6–12 fractions. This can be accomplished most readily by gently inserting a long, blunt-ended needle attached to tubing linked to a peristaltic pump to the bottom of the tube. The gradient is then pumped into a fraction collector, typically at a rate of ~2 min per fraction.

5. Weigh aliquots of the fractions to determine the density gradient, and count aliquots to determine the distribution of the radiolabel. High-density proteoglycans will be concentrated in the bottom fractions while low-density proteoglycans will be concentrated in the upper fractions. Pool fractions as deemed appropriate from the distribution profile.

6. Dialyse, or ultrafilter (Amicon), pooled samples against an appropriate chaotropic solvent with detergent to remove the CsCl and store at −20 °C until further analyses.

6.3 Size

Molecular-sieve chromatography is widely used to separate and characterize intact and degraded proteoglycans. The choice of support matrix, porosity of the matrix, and elution solvents all depend upon the properties of the proteoglycans and the particular application under investigation. Currently matrices designed for high-pressure and flow rates, such as Superose 6, have distinct advantages in terms of speed of analyses and tolerance for chaotropic solvents. For proteoglycans which have very large hydrodynamic volumes, matrices which are more porous, such as Sephacryl S-500, must be used. The procedure in *Protocol 7* illustrates an application for separating and recovering proteoglycan populations which resolve into more than one peak.

Protocol 7. Proteoglycan elution on Superose 6

Equipment and reagents
- Superose 6 column (1 × 30 cm)
- Starting buffer (*Protocol 5*)

Method

1. Equilibrate a Superose 6 column of 25–30 ml, fitted with a prefilter, with the Starting buffer (*Protocol 5*). If the analysis is for analytical purposes only, a solvent with 4 M guanidine–HCl and 0.5% Triton X-100 can be used effectively. Take care to prepare and filter (0.2 μm) solvents as appropriate for medium- to high-pressure liquid chromatography.

2. Centrifuge the sample containing no more than 5 mg of protein (10 000 g for 5 min) or ultrafilter it (0.2 μm). Inject 100–200 μl (up to 500 μl) into the column and collect 0.4 ml fractions at a rate of 0.4 ml/min.

3. Count aliquots of the fractions for radioactivity and pool peaks as appropriate from the distribution of radiolabel.

4. Concentrate the proteoglycans in each of the pooled samples and recover by the batch anion-exchange method (*Protocol 3*).

5. Use columns with column guards, and only wash and care for according to the manufacturers' recommendations. They should be routinely cleaned with 0.1 M NaOH, especially when large amounts of sample are applied.

6.4 Hydrophobicity

In chaotropic solvents, core proteins of some proteoglycans can be separated on the basis of differences in their hydrophobic properties. For example, the dermatan sulfate and keratan sulfate containing proteoglycans from cornea bind to octyl-Sepharose and elute at different detergent concentrations during a detergent gradient (6). Further, heparan sulfate containing proteoglycans which bind to cell surfaces by polypeptide intercalation or phosphatidylinositol anchors bind tightly to such columns and elute only at the critical micelle concentration for the detergent used in the gradient (7).

Protocol 8. Octyl-Sepharose chromatography

Equipment and reagents

- Stock solvent: 10 M formamide, 0.3 M NaCl, no detergent
- Octyl-Sepharose
- Plastic pipette column (as *Protocol 4*)
- Sephadex G-50 (fine)
- Eluting serum albumin in stock solvent (1 mg/ml)
- CHAPS
- BSA

Method

1. Prepare the stock solvent and treat this with activated charcoal to remove impurities that interfere with the chromatographic steps. Solvents with 4 M guanidine–HCl and 8 M urea can also be used and are treated in the same way.

2. Wash octyl-Sepharose several times with the solvent and prepare a small column (bed volume of 5 ml) in a plastic pipette as described in *Protocol 4*. Apply a thin layer of Sephadex G-50 (fine) in the same solvent on top of the column to provide stability.

3. Some batches of octyl-Sepharose have high-affinity binding sites that bind proteoglycans irreversibly. Block these with eluting serum albumin dissolved at 1 mg/ml in the same solvent and apply 50–100 µg/ml bed volume. Add an aliquot of the BSA to an aliquot of the sample. Then dilute the mixture with sufficient stock solution such that the final CHAPS concentration is less than 0.02%. The final albumin and proteoglycan concentrations should be such that no more than 10 µg and 2 µg, respectively, are applied per millilitre of the column bed volume.

4. After sample application, prepare the gradient with 50 ml of stock solvent (initial solution) and 50 ml of stock solvent with 1.5% CHAPS (final solution). Elute the gradient at 25 ml/h with 2 ml fractions collected. Place an aliquot of concentrated detergent in each fraction tube before the gradient is developed to prevent absorption of eluted materials to the tubes, which is a particular problem at CHAPS concentrations below 0.2%.

5. Determine concentrations of CHAPS in the fractions with the carbazole assay for hexuronic acids as described elsewhere (6). Count aliquots for radioactivity to determine elution profiles for the proteoglycans. The use of other detergents such as Triton X-100 or other non-polar reagents should be considered when CHAPS does not effectively elute or separate proteoglycans.

6. Regenerate the columns by washing with 10 column volumes of 95% ethanol and then 1-butanol before equilibration with 20 column volumes of the initial solvent.

7. Quantitation of chemical amounts of proteoglycans

In many cases, such as extraction of tissue explants or of large numbers of cells, there will be enough mass of proteoglycan such that chemical analyses on micro scales can provide quantitation. Further, if sufficient amounts are present, then the properties of the radiolabelled species can be compared with those of the resident proteoglycans in the culture. Several procedures have been developed which involve dye binding to the anionic groups on the glycosaminoglycan chains under conditions where the dye saturates the available ion sites and precipitates the proteoglycans. Under suitable conditions the amount of the dye precipitated can be measured either by reflectance absorption of the precipitate or by resolubilization of the complex and absorbance spectroscopy at an appropriate wavelength. The limit of resolution for the following protocol is approximately 5 ng of glycosaminoglycan and the assay is linear up to about 100 ng. Among several dye-binding procedures described in the literature, the ones using Safranin O have the best sensitivity with lowest backgrounds (8).

Protocol 9. Quantitation of proteoglycans with Safranin O precipitation

Equipment and reagents

- Proteoglycan or glycosaminoglycan standards
- Polyvinylidene difluoride (Immobilon) membrane sheet

- Hybri-Dot manifold (BRL)
- Safranin O reagent: 0.02% Safranin O, 50 mM sodium acetate, pH 4.75
- Shimadzu CS-9000 densitometer

Method

1. Dilute samples appropriately in water to a final maximum volume of 50 μl.

2. Dilute proteoglycan or glycosaminoglycan standards (10–400 ng) to the same final buffer concentration and volume as the samples.

3. Cut one polyvinylidene difluoride membrane sheet to fit a BRL Hybri-Dot manifold.

4. Wet the membrane in 100% methanol and wash it four times in water (5 minutes/wash).

5. Cut one sheet of filter paper to the same size as the membrane and soak in water.

6. Position the filter paper and membrane into the manifold and apply a vacuum until all the wells are dry.

---

7. Add aliquots (50 μl maximum) of the samples and standards to appropriate wells. The samples and standards are suspended on the walls of their respective wells above the well bottom.

8. Apply vacuum to the manifold. The samples and standards remain in the wells, since none of the wells have been sealed to vacuum by their sample or standard volumes.

9. Inject 400 μl of Safranin O reagent into each well with a repeating pipette. The addition of the Safranin O reagent mixes the samples and standards with the reagent and forms an immediate precipitate. The addition of the reagent also drains the contents of the wells and seals them to vacuum. This protocol eliminates loss of precipitate on pipette tips used for mixing sample and reagent. It also provides a uniform start to sample flow from well-to-well after reagent addition. This prevents bleeding of well contents.

10. After all the wells have drained, fill each well with water. This washes any residual precipitate on to the membrane.

11. After all the wells have drained, release the vacuum and remove the membrane from the manifold.

12. Wash the membrane twice in 100% methanol for 5 minutes with gentle shaking to remove any Safranin O not directly involved in the precipitation reaction.

13. Air-dry the membrane and quantitate the precipitated dye complex via reflectance at 490 nm on a Shimadzu CS-9000 densitometer using the Flying-Spot feature.

8. Proteoglycan concentration by ultrafiltration

During the isolation and characterization of proteoglycans it may be necessary to concentrate samples to microlitre volumes and/or to change solvents between steps. The technique of ultrafiltration can concentrate proteoglycan samples of several millilitres to as little as 20 μl with recoveries in excess of 90%. Ultrafiltration uses centrifugal force to drive solvent and low molecular weight solute molecules through a membrane with a defined molecular weight cut-off. Macromolecules such as proteoglycans are retained by the membrane in the sample reservoir. Membranes with molecular weight cut-offs of 3, 10, 30, and 100 kDa are available from Amicon (Centricon microconcentrators) and are selected based on the size of the proteoglycans in a sample. The design of the sample reservoir prevents the retentate from concentrating below a specific deadstop volume. This ensures that even extended centrifugation does not result in concentration to dryness. A cup at the bottom of the device allows for collection of the filtrate, which may also contain molecules of interest. Recoveries are increased by inclusion of carrier BSA in the sample, pretreatment of the membrane with BSA, and the use of chaotropic agents and detergents. It should be noted that detergent micelles, larger than the molecular weight cut-off of the membrane, are retained and concentrated along with the proteoglycan and other macromolecules in the sample. This concentration of detergent can foul the membrane and adversely affect recoveries. Therefore, when selecting a detergent, its critical micelle concentration and micelle size in various buffers should be considered prior to ultrafiltration.

Protocol 10. Ultrafiltration on Centricon microconcentrators

1. Pre-centrifuge the microconcentrator with 2 ml of 0.1 mg/ml BSA in water and rinse the sample reservoir and filtrate cup with water to remove excess BSA.

2. Add the first 2 ml aliquot of sample to the sample reservoir and add 10–20 μg of BSA as an optional carrier.

3. Centrifuge the assembled microconcentrator according to the manufacturer's specifications.

4. Collect the filtrate from the filtrate cup and add a second 2 ml aliquot of sample to the retentate in the sample reservoir.

5. Vortex the retentate gently and centrifuge as before.

6. Repeat steps 4 and 5 until all the sample has been concentrated to the desired volume (20 μl minimum).

7. If a solvent change is desired, make it at this stage by diluting the retentate in the sample reservoir to 2 ml with the new solvent. Vortex the retentate gently and concentrate it as before.

8. Once the sample has been concentrated to the desired volume, invert the sample reservoir and collect the retentate in a cup by centrifugal force.

9. Determine recoveries by quantitating the amount of proteoglycan in both the retentate and filtrate and comparing these values to that of the original sample.

---

9. Characterization of core proteins

The apparent molecular weights of core proteins of proteoglycans can be estimated by SDS-PAGE after enzymatic removal of the glycosaminoglycan side chains. The core protein can be visualized by Coomassie Blue or silver staining, if milligram amounts of proteoglycan are available; or by autoradiography or fluorography if the core proteins are metabolically radiolabelled with an amino acid precursor or are chemically radiolabelled. Chondroitinase ABC is used to degrade chondroitin/dermatan sulfate, heparitinase/heparanase are used to degrade heparan sulfate, and keratanase (an endo-β galactosidase that requires an adjacent sulfated GlcNAc) or endo-β galactosidase (which does not require adjacent sulfated GlcNAc) are used to degrade keratan sulfate. The commercially available preparations of these enzymes are glycosaminoglycan specific but often contain traces of proteases. Therefore, proteolytic inhibitors must be routinely added to the enzyme to block their activities (9). Also, because proteoglycans are present in only small amounts in the initial tissue extract, it is necessary to at least partially purify the proteoglycans and concentrate them sufficiently for detection on SDS-PAGE.

Protocol 11. Detection of proteoglycan core protein by SDS-PAGE

Equipment and reagents

- 20 × Cocktail of proteolytic inhibitors from stock solutions (make fresh): 20 μl 50 mM PMSF in 100% methanol, 200 μl 100 mM NEM in 0.1 M Tris–HCl, pH 7.4, 200 μl 36 mM pepstatin A in 0.1 M Tris–HCl, pH 7.4, 200 μl 100 mM EDTA in 0.1 M Tris–HCl, pH 7.4; to 1 ml with 0.1 M Tris–HCl, pH 7.4
- Chondroitinase ABC (Seikagaku America)
- Keratanase (Seikagaku America)
- Heparitinase (Seikagaku America)

Method

1. Purify the proteoglycan fraction(s) by Q-Sepharose chromatography (*Protocol 4 or 5*) and concentrate them by dialysis and lyophilization, or by ultrafiltration, and exchange the solvent for water. If the proteoglycans are present in milligram amounts, dissolve them at 2–5 mg/ml.

2. Dissolve 5 units chondroitinase ABC in 200 μl distilled water, 10 units of keratanase in 200 μl water and/or 0.1 units of heparitinase in 100 μl water. Store frozen in small aliquots at − 80°C and thaw prior to use.

3. Add proteolytic inhibitors to 2–10 μl of proteoglycan fraction, adjust to make the solution 0.05 M Tris–HCl, pH 7.4, and digest with 0.1 units chondroitinase, 0.2 units of keratanase or 0.005 units of heparitinase for 1–3 h at 37°C in a total volume of 50 μl or less. Also include proteoglycan samples without added enzyme and enzyme samples without added proteoglycan because both proteoglycan and enzymes can have proteins that can be confused as core protein.

4. Run the samples, both reduced and unreduced, on SDS-PAGE (7.5% or 4–20%). The core protein can be visualized by Coomassie Blue staining if present in sufficient amounts. If the core proteins are radiolabelled embed the gel with Fluor, dry it, place the gel against X-ray film and put

Protocol 11. *Continued*

it in a − 70 °C freezer. Develop the film after an overnight or over weekend exposure. In some cases it may be necessary to expose the gel to film for 2–3 weeks.

Note: Enzyme digestions will produce a band (core protein) that is not present in either the undigested proteoglycan fraction or in the enzyme controls. Failure to see a band can be due to not having the proteoglycan 'pure' enough to constitute a sufficiently large proportion of the fraction or not having enough radiolabel in the core protein. Either digest more sample or further purify the proteoglycan. Nucleic acids and some proteins co-isolate with proteoglycan purified by Q-Sepharose chromatography or CsCl density gradient centrifugation even under dissociative conditions, and they contribute to the weight of lyophilized samples. Also, some proteoglycans have two different kinds of glycosaminoglycan side chains and the core protein will not appear as a band unless both types are removed. For this reason, it may be best initially to digest one proteoglycan sample with all three enzyme types combined.

10. Identification of precursor proteins

The core proteins of proteoglycans are initially translation products in the rough endoplasmic reticulum (RER). N-linked oligosaccharides are transferred to the asparagines that serve as acceptors for oligosaccharides as soon as the protein is made in the RER. The protein, or precursor protein in the case of proteoglycans, then moves to the Golgi apparatus, where the bulk of the post-translational modification takes place. This includes the processing of N-linked oligosaccharides to complex type, the addition of O-linked oligosaccharides, and the addition of the glycosaminoglycan side chains.

The precursor protein has a half-life of 20–120 minutes in the RER before transport to the Golgi and addition of the glycosaminoglycan side chain. Antibodies prepared to the core protein of proteoglycans can be used to isolate radiolabelled precursor protein from lysates of cells in culture that have been radiolabelled with a protein precursor such as [35S]methionine (10). A short radiolabelling time of only 30 min with high levels of isotope is usually used so that most of the incorporated radiolabelled protein is still in the RER and has not passed to the Golgi for glycosaminoglycan addition. A longer labelling time (1–2 h) is also acceptable to get maximum incorporation into the precursor pool, since completed proteoglycan does interfere with the detection of the precursor protein. Electrophoresis of this immunoprecipitated material on SDS-PAGE followed by autoradiography/fluorography will establish the size of the precursor protein. By comparing this size to the size of the core protein, as determined by digestion of the proteoglycan with either heparitinase, chondroitinase ABC, or keratanase, it may be possible to establish, if any, proteolytic clipping of the core protein had occurred during proteoglycan maturation.

Protocol 12. Protein-A–Sepharose immunoprecipitation

Equipment and reagents

- Protein-A–Sepharose resin (Pharmacia)
- Washing buffer: PBS, pH 7.2, 0.5% Tween-20, 0.5% SDS, 0.1% bovine albumin, 0.02% Na azide
- Extraction buffer: 0.1 M Tris–HCl, pH 7.2,
- Methionine-free, serum-free medium
- 1.5 ml Conical microcentrifuge tubes

- Rabbit antiserum against proteoglycan 0.015 M NaCl, 1% Triton X-100, 1% de-oxycholate, 0.1 SDS, 0.1 mg/ml aprotinin (Sigma), 0.02% Na azide
- [^{35}S] Methionine, Sp. Act. >100 Ci/mmol (NEN Dupont—NEG 009T)

Method

1. Prepare matrix beads:

    (a) Swell the Protein-A–Sepharose overnight in washing buffer.

    (b) Suspend 1 ml (swollen) resin in 10 ml washing buffer, dispense in 0.5 ml aliquots, and store at − 70 °C.

2. Harvest cells:

    (a) Label a confluent layer of cells in a 35 mm dish with [^{35}S]-methionine (500 μCi/ml) in methionine-free and serum-free medium for 30 min. Remove the medium and extract the cell layer with 1 ml of extraction buffer.

    (b) Allow the extraction buffer to sit on the cells for 2–3 min, then scrape off the cells with a cell scraper.

    (c) Briefly sonicate the cell lysates (3–5 sec) to disrupt the DNA and reduce viscosity. Centrifuge at 15000 *g* at 4 °C for 5 min to remove insoluble material. Save the supernatants for step 4.

    (d) Store the lysates frozen. Centrifuge at 15000 *g* at 4 °C for 5 min before each use.

3. Bind antisera to beads as follows:

    (a) Thaw the swollen resin (Protein-A–Sepharose) and resuspend the beads by vortex mixing.

    (b) Combine 20 μl antiserum raised against the proteoglycan with 100 μl of the suspended Protein-A–Sepharose. Mix for 1–2 h at 4 °C with rotation. At least 0.6 ml, but no more than 1.2 ml, of fluid is needed to ensure proper mixing in a 1.5 ml Eppendorf tube. The volume can be brought up to 0.5 ml with washing buffer.

    (c) Wash the resin (which has the antibody coupled to it) with 1 ml washing buffer and centrifuge at 1000 *g* at 4 °C for 5 min. Wash a total of 3 times.

    (d) Suspend the beads in 100 μl of washing buffer and save for step 5.

4. Pre-absorb lysate (absorbing the lysate will reduce the background in the subsequent immunoprecipitation):

Protocol 12. *Continued*

(a) Add 50 μl of suspended Protein-A–Sepharose and 2.5 μl preimmune-serum to 100 μl of the cell lysate. Mix by rotating for 1 h at 4 °C.

(b) Centrifuge at 1000 g at 4 °C for 5 min. Save the supernatant for step 5 (discard beads). Always save a few microlitres of the lysate to run on the gel.

5. Precipite as follows:

(a) Take the equivalent of 100 μl of the absorbed cell lysate from step 4 (take into account any dilution factor in extraction buffer) and combine it with 100 μl of the suspended Protein-A–Sepharose-antibody (from step 1). Mix for 2 h at 4 °C with rotation.

(b) Wash with 1 ml washing buffer, 3–5 times. Centrifuge at 1000 g at 4 °C for 5 min each to recover the beads.

(c) Store the washed beads at −70 °C.

(d) Extract the beads with 75 μl sample buffer (2% SDS) in a hot-water bath (100 °C) for 3 min.

(e) Centrifuge the Sepharose beads and save the supernatant.

(f) Count 5 μl before applying the sample to an SDS–PAGE gel as in *Protocol 11.*

References

1. Ito, K., Kimata, K., Sobue, M., and Suzuki, S. (1982). *J. Biol. Chem.*, **257**, 917–23.
2. Yanagishita, M., Salustri, A., and Hascall, V. C. (1989). *Methods in enzymology: complex carbohydrates, Part F, Glycosaminoglycan precursors* (ed. V. Ginsburg), Vol. 179, pp. 435–45. Academic Press, New York.
3. Yanagishita, M. and Hascall, V. C. (1984). *J. Biol. Chem.*, **259**, 10260–9.
4. Hascall, V. C. and Kimura, J. H. (1982). In *Methods in enzymology, Part A* (ed. V. Ginsburg) pp. 769–800. Academic Press, New York.
5. Salustri, A., Yanagishita, M., and Hascall, V. C. (1987). *J. Biol. Chem.*, **264**, 13840–7.
6. Yanagishita, M., Midura, R., and Hascall, V. C. (1987). In *Methods in enzymology: complex carbohydrates, Part E* (ed. V. Ginsburg), vol. 138, pp. 279–89. Academic Press, New York.
7. Yanagishita, M. and McQuillan, D. J. (1989). *J. Biol. Chem.*, **264**, 17551–8.
8. Lammi, M. and Tammi, M. (1988). *Anal. Biochem.*, **168**, 352–7.
9. Oike, Y., Kimata, K., Shinomura, T., Nakazawa, K., and Suzuki, S. (1980). *Biochem. J.*, **191**, 193–7.
10. Ledbetter, S. R., Tyree, B., Hassell, J. R., and Horigan, E. A. (1985). *J. Biol. Chem.*, **260**, 8106–13.

11

Elastin

JEFFREY M. DAVIDSON, MARIA GABRIELLA GIRO, and
ROBERT P. MECHAM

1. Introduction

Elastin, when mature, is the most insoluble protein in the vertebrate body. It is synthesized as an $M_r = 70\,000$ precursor from a 3.5 kb mRNA encoded by a single gene into a protein termed *tropoelastin*, which is cross-linked subsequent to secretion by the copper-dependent enzyme, lysyl oxidase (1). Therefore, for most purposes elastin purification simply requires the solubilization and removal of other protein components from connective tissue to leave elastin as the remaining residue. The elastic *fibre*, however, consists of insoluble elastin plus other structural macromolecules, collectively referred to as microfibrillar components (2). Since these molecules are also relatively insoluble, their isolation is also accomplished by selective extraction of the elastic fibre under denaturing and reducing conditions. This chapter will not detail the isolation of these other components.

Elastin is distinguished by its amino acid composition and the presence of unique intermolecular cross-links (3). Elastin is extremely rich in hydrophobic amino acids, and the elastic properties of the polymer are dependent on hydrophobicity. Regularly interspersed among the hydrophobic regions are distinctive cross-link domains, coded by separate exons, that frequently consist of the sequence: Ala–Ala–Lys–Ala–Ala–Ala–Lys–Ala/Phe/Tyr. The pairs of lysines in these polyalanine tracts are oxidatively deaminated by lysyl oxidase to aldehydes which then condense via Schiff base formation into the distinctive and characteristic tetrafunctional residues, desmosine or isodesmosine (*Figure 1*). The lysyl residue adjacent to phenylalanine or tyrosine is thought to be spared from oxidation.

2. Isolation of insoluble elastin

2.1 Introduction

Elastin is generally isolated from tissues by removing all other connective tissue components by denaturation or degradation. The most successful purification procedures rely upon elastin's extreme inertness to protein solvents

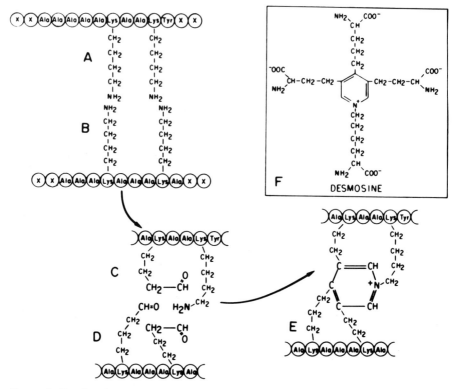

Figure 1. Details of the formation of the cross-linked desmosine in the passage from soluble elastin (tropoelastin) to insoluble elastin. A and B show the apposition of two cross-linking sites on different tropoelastin chains, with the probable position of the lysine side chains before the oxidative removal of the ε-amino groups. Tropoelastin chains spontaneously condense (coacervate) under physiologic conditions due to their hydrophobicity. In A, two alanine residues separate the two lysine residues and a tyrosine follows the second lysine. In B there are three alanine residues between the lysines. The action of lysyl oxidase brings about the oxidative deamination of lysines not adjacent to tyrosine (or phenylalanine). In C and D, aldehydes have replaced three or four ε-amino groups of the lysine side chains. The chains are folded to indicate their probable contribution to the desmosine ring structure. In E, the pyridinium ring of desmosine has been formed with its resonating double bonds. F shows a desmosine molecule free from the peptide linkages to the elastin chains. With isodesmosine, the lysine derived side chain opposite the nitrogen (*para*) is moved to the *ortho* position. From reference 3, with permission.

and its resistance to hydrolysis during treatment with dilute acids or alkali. Lansing *et al.* (4) described a purification of elastin from aortic tissue that included exposure to hot alkali for 45 minutes. Partridge *et al.* (5), utilized an exhaustive autoclave procedure for purifying elastin from ligamentum nuchae that included successive one hour autoclave periods until no further protein appeared in the supernatant. To avoid extensive peptide bond cleavage that

occurs with hot alkali and repeated autoclaving, several alternative purification methods have been proposed, including treatment with a combination of
proteases, chaotropic and reducing agents, and cyanogen bromide. An assessment of the efficacy of the different extraction protocols can be found in
Soskel *et al.* (6). These milder extraction methods are successful to varying
degrees, depending on the starting tissue. Generally speaking, however, hot
alkali and autoclave techniques yield the purest elastin.

The best sources of elastin-rich tissue are the thoracic aorta and, in many
grazing animals, the ligamentum nuchae, a large, yellowish elastic ligament
that runs along the dorsal surface of the spine from the head to the lumbar
region. Tissues may be stored frozen at $-20°C$ after isolation and before
subsequent procedures. For all procedures, it is best to start with material
that is dissected free of adventitious tissue and washed with saline until no
protein is detected in the supernatants. The tissue is then minced with a
Polytron or ground to powder in liquid nitrogen. Fat can be removed by
extraction first with ethanol followed by chloroform:methanol (2:1, v/v).

The purity of isolated elastin can best be assessed by amino acid analysis.
Pure elastin should not contain hydroxylysine (indicating contamination with
collagen), methionine, histidine, tryptophan, or an enrichment in acidic amino
acids (7). The amino acid compositions of human, bovine, porcine, chicken,
and rat elastin are given in *Table 1*.

2.2 Isolation of insoluble elastin

Protocol 1. Isolation of insoluble elastin using hot alkali

1. Suspend minced, defatted tissue in ≥ 9 volumes of 0.1 M NaOH.

2. Heat, with stirring, in a boiling-water bath for 10 min.

3. Cool samples to room temperature.

4. Collect the insoluble residue by centrifugation (20 000 *g*, 20 min, 20 °C),
   or by filtration on a sintered-glass funnel.

5. Repeat the extraction of insoluble material four times.

6. Wash the insoluble residue with 20 volumes of cold water and lyophilize.

Protocol 2. Isolation of insoluble elastin by autoclaving

1. Place minced tissue in a container containing a loose-fitting gauze plug.

2. Autoclave tissue in 20 volumes of distilled water at 25 p.s.i. (121.1 °C)
   for 45 min.

3. Collect the sample by centrifugation (20 000 *g*, 20 min, 20 °C) or on a
   sintered-glass funnel.

Protocol 2. *Continued*

4. Repeat the autoclave procedure using fresh water until no further protein is detected in the supernatant (by frothiness of shaken liquid; usually after the 3rd or 4th extraction).

Protocol 3. Isolation of elastin by the non-degradative procedure of Starcher (8)

Equipment and reagents

- Solution A: 0.05 M Na$_2$HPO$_4$, pH 7.6, 1% (w/v) NaCl, 0.1% (w/v) EDTA
- Solution B: 0.1 M Tris–HCl, pH 8.2, 0.2 M CaCl$_2$
- Solution C: 0.05 M Tris–HCl, pH 8, 6 M urea, 0.05% (v/v) 2-mercaptoethanol
- 0.1 M Tris–HCl, pH 8.2, 0.2 M CaCl$_2$
- Trypsin (Sigma, 2 × crystallized)
- 97% formic acid
- CNBr
- SDS

Method

1. Extract powdered tissue for 72 h at 4 °C with multiple changes of Solution A, twice with water, and then lyophilize. If very small amounts of tissue are available, or if radioactive tracers are being used to monitor synthesis of insoluble elastin, homogenize tissues or cell layer extracts in phosphate-buffered saline containing 1% SDS. Collect the insoluble residue by centrifugation (20 000 *g*, 20 min, 20 °C) and wash once with water. Proceed to step 4.

2. Suspend the lyophilized protein in water (150 ml/g) and autoclave for 45 min at 25 p.s.i. (121.1 °C).

3. Centrifuge at 20 000 *g*, for 20 min, at 20 °C to collect the insoluble residue.

4. Discard the supernatant and wash the insoluble residue twice with water by resuspending and recentrifugation.

5. Suspend the insoluble residue in Solution B (150 ml/g of residue) containing 4 mg/ml of trypsin and incubate at 37 °C for 18 h.

6. Collect the insoluble residue by centrifugation (20 000 *g*, 20 min, 20 °C), wash twice with water, and place in 70% formic acid, containing 20 mg/ml CNBr (50 ml per g of tissue). **Note**: *Perform all handling of CNBr under a fume hood. Do not take the container of CNBr over to the balance to weigh out crystals, as the concentration is not critical and the fumes are extremely noxious.*

7. After shaking at room temperature for 5 h, collect the residue by centrifugation (20 000 *g*, 20 min, 20 °C), and wash twice with water as in step 4.

8. Suspend the residue in Solution C (100 ml/g of residue) and stir overnight at room temperature.

9. Collect the insoluble residue by centrifugation as in step 7, wash successively with distilled water (> 20 volumes), ethanol (10 volumes), and then acetone (5 volumes). Dry *in vacuo* over P_2O_5 or lyophilize after a final wash with water.

Table 1. Amino acid composition of mature (insoluble) elastin [a]

Amino acid	Human aorta [b]	Bovine ligamentum nuchae [b]	Porcine aorta [c]	Chicken aorta [d]	Rat aorta [e]
Hydroxyproline	8	6.48	8	13.9	8.5
Aspartic acid	6.4	4.6	5	3.9	5.2
Threonine	5.6	7.4	11	8.9	8
Serine	7.2	7	11	6.5	7.6
Glutamic acid	13.6	12.3	16	13	12.8
Proline	83.2	92.4	90	110	89.6
Glycine	259.2	262.5	256	258.5	271.5
Alanine	179.2	181.6	181	123	172
Valine	114.4	105.3	92	138.6	102.9
Half-cystine	trace	—	trace	—	—
Methionine	—	—	—	0.4	—
Isoleucine	21.6	19.1	14	17.8	19.3
Leucine	52	47.5	41	47.7	46.5
Tyrosine	7.2	4.7	12	12.4	5.5
Phenylalanine	20.8	23.4	25	18.6	23.2
Hydroxylysine	ND	—	—	—	—
Lysine	4	2.6	5	10.6	3.7
Histidine	0.8	0.4	trace	0.3	0.6
Arginine	4	4.6	5	6.7	5
Lysinonorleucine	ND	1.7	—	1.8	1.6
Isodesmosine	3.2	4.3	—	3.3	4.6
Desmosine	4.8	8.1	—	3.4	7.8
Des + Ides	—	—	3	—	—

[a] Composition is expressed as residues/800 residues.
[b] Data from reference 39.
[c] Data from reference 7.
[d] Data from reference 40.
[e] Data from reference 41.
ND, not determined.

3. Solubilization of insoluble elastin

3.1 Introduction

Because elastin is a covalently cross-linked polymer, solubilization can only be achieved by disruption of peptide bonds by proteolytic enzymes or by acid or base hydrolysis. Three of the most common methods for solubilizing elastin are described below. The reaction products differ in each case, and the original papers should be consulted for a more complete characterization of the solubilized peptides.

3.2 α-Elastin

This procedure utilizes partial hydrolysis of peptide bonds by oxalate, a weak acid, which results in the formation of two soluble components designated α- and β-elastin. α-Elastin was found by Partridge *et al.* (5) to be polydisperse with an average molecular weight between 60 and 85 kDa, while β-elastin was monodisperse and had an average molecular weight of approximately 5 500. Amino terminal analysis suggests that the α-elastin protein contains approximately 17 chains connected by covalent cross-links with an average of 35 amino acid residues per chain. In contrast the β-protein contains two chains with an average of 27 amino acids per chain. The α-elastin fraction undergoes a reversible phase separation (coacervation) and has an amino acid composition similar to that of fibrous elastin. The β-elastin fraction is liberated rapidly at the early phase of hydrolysis and does not coacervate.

Protocol 4. Preparation of α-elastin

Equipment and reagents

- Oxalic acid, 0.25 M
- Steam bath or boiling-water bath
- Autoclave
- 0.01 M NaOAc, pH 5.5, 0.1 M NaCl
- Na phosphate buffer, pH 5.5, containing 0.1 M NaCl
- 0.1 M HOAc

Method

1. Suspend finely milled, purified, insoluble elastin in 0.25 M oxalic acid (1 part elastin to 10 parts acid, w/v).

2. Heat the mixture to ≥ 98 °C on a steam bath or in a boiling-water bath for 1 h and then centrifuge at 16 000 *g*, for 20 min, at 35 °C.

3. Pour off the supernatant, resuspend the insoluble residue in 0.25 M oxalic acid, and repeat the cycle of heating followed by centrifugation until the insoluble residue is completely dissolved, giving a clear yellow solution (at least four to five times).

4. Pool the supernatants from the third and successive extractions.

5. Dialyse the solubilized elastin against distilled water, at 4 °C, until free of oxalate, and then lyophilize (use low molecular weight cut-off dialysis tubing if greater retention of β-elastin is desired).

6. Dissolve the lyophilized protein in 0.01 M NaOAc or Na phosphate buffer (pH 5.5), both containing 0.1 M NaCl and heat at 37 °C for 15–30 min.

7. Centrifuge the solution at the same temperature (20 000 *g*, 20 min,

---

37 °C). The coacervate, representing α-elastin, forms at the bottom of the tube as a viscous syrup. Remove the supernatant by aspiration. Coacervation is reversible so that α-elastin quickly redissolves in buffer upon cooling. Protein in the non-coacervated supernatant is β-elastin. Use several cycles of coacervation to yield a more homogeneous α-elastin fraction.

8. Dialyse vs three changes of 10 volumes of 0.1 M HOAc for ≥ 6 h at 4 °C. Use low molecular weight cut-off dialysis tubing to retain β-elastin.

9. Lyophilize the dialysed α-elastin pellet and β-elastin supernatant.

10. Store the final preparations lyophilized or as a 10 mg/ml solution in 0.1 M HOAc at − 20 °C.

3.3 κ-Elastin (9)

Unlike hydrolysis with oxalic acid, alkaline degradation of fibrous elastin in aqueous organic solvents can be realized at a relatively low temperature. In these mixtures, the organic solvent accelerates the hydrolytic reaction by decreasing hydrophobic interactions within the elastin molecule thereby improving the efficiency of the nucleophilic attack of the hydroxonium ions. The average molecular weight of the peptides released in the first 3 h of hydrolysis is about 12 000 and their desmosine content increases with progressing hydrolysis. Peptides released during the first 3 h are dialysable and, hence, relatively small. A rapid release of peptides occurs after 4 h which, when separated on Sephadex G-100, resolves into an excluded and retarded fraction. The retarded peak does not form a coacervate and contains a major band at 16 000 molecular weight on SDS-PAGE. The excluded peak forms a coacervate and contains two major components of 120 000 and 80 000 molecular weight. The maximal yields of the excluded fractions are obtained after hydrolysis for 6 h when the yield of the retarded peak is at a minimum. On further hydrolysis, the amount of the high molecular weight peak decreases rapidly.

Protocol 5. Preparation of κ-elastin

Equipment and reagents

- 1 M KOH–ethanol (1:4, v/v)
- Perchloric acid, concentrated (Fisher)
- HOAc, glacial
- Chromatographic column, Sephadex G-200, 0.8 × 50 cm, equilibrated with 0.02 M HOAc

Method

1. Suspend elastin powder (1 g per 25 ml), with stirring, in 1 M KOH–ethanol (1:4, v/v) at 37 °C for 6 h.

Protocol 5. *Continued*

2. Centrifuge the mixture (16 000 g, 15 min, 20°C) or filter.

3. Neutralize the soluble fraction with perchloric acid and discard any resulting precipitate.

4. Lyophilize the neutralized supernatant (κ-elastin).

5. Apply 1–5 mg of material to a Sephadex G-200 that has been equilibrated with 0.02 M HOAc and chromatograph in the same solvent. Monitor the elution profile at 230 nm.

6. Pool and lyophilize the initial, high molecular weight peaks (κ_2-elastin).

3.4 Enzymic digestion of elastin

Because elastin contains a paucity of polar amino acids, enzymes with broad-ranging specificity must be used to affect solubilization. The proteases that have been used most frequently include pancreatic and leukocyte elastase (10), although chymotrypsin and trypsin have been noted to solubilize elastin under some conditions (11). Several other enzymes possess elastolytic activity, including subtilisin (12), thermolysin (13), pepsin (14), papain (15), ficin (10), bromelain (10), and nagarase (16). The original papers should be consulted for specific reaction conditions needed when employing these enzymes.

Optimal digestion of elastin with leukocyte elastase occurs at pH 8 in buffers containing 0.1 M NaCl (17). Leukocyte and pancreatic elastases have different peptide bond specificities and produce different peptides from elastin. Pancreatic elastase has a strong specificity for alanyl peptide bonds, whereas the leukocyte enzyme prefers leucyl peptide bonds.

Protocol 6. Digestion of insoluble elastin with pancreatic and leukocyte elastase

Equipment and reagents

- 0.05 M Tris–HCl, pH 8.8
- 0.05 M Tris–HCl, pH 8, 0.1 M NaCl
- Pancreatic elastase (Sigma, type I)
- Leukocyte elastase (Elastin Products Co.)
- PMSF

Method

1. Add pancreatic elastase to insoluble elastin suspended in 0.05 M Tris–HCl, pH 8.8, at an enzyme to substrate ratio of 1:100 (w/w).

2. Incubate the mixture at 37°C with constant stirring until the elastin is completely solubilized (usually between 4 and 12 h, depending upon the specific activity of the elastase).

3. Destroy enzymatic activity by heating to 100°C for 15 min or by adding the protease inhibitor, PMSF, to a final concentration of 2 mM.

4. Identification of lysine-derived cross-links in elastin

4.1 Introduction

Elastin contains several lysine derivatives that serve as covalent cross-linkages between protein monomers. These cross-links, desmosine and isodesmosine (*Figure 1*), as pointed out below, are important in imparting rubber-like properties to elastin. The initial step in their formation is the deamination of the ε-amino group of lysine side chains by the copper-requiring enzyme lysyl oxidase (18). This appears to be the only enzymatic step involved in elastin cross-linking. Under normal conditions, the cross-linking reaction is so efficient that no pool of soluble elastin can be detected in the extracellular matrix. The reactive aldehyde that is formed by the action of lysyl oxidase (α-amino adipic δ-semi-aldehyde, or allysine) is thought to interact with a second aldehyde residue to form allysine aldol or with an ε-amino group on lysine to form dehydrolysinonorleucine. Allysine aldol and dehydrolysinonorleucine can then condense to form the pyridinium cross-links desmosine and iso-desmosine. These structures are unique to elastin and can be used as distinctive markers for the protein. A more thorough discussion of elastin cross-links can be found in Robins (19) and Paz *et al.* (20). Papers by Lent *et al.* (21), Franzblau *et al.* (22), and Mecham and Foster (10, 13) should be consulted as examples of methods and strategies for cross-link analysis.

Obtaining a complete cross-linking profile of elastin is difficult and tedious. A full analysis of the cross-links in elastin requires stabilization of the labile aldimine forms with borohydride since dehydrolysinonorleucine and dehydro-merodesmosine are destroyed by acid hydrolysis. Even then, the reduced forms of allysine and allysine aldol are sensitive to acid hydrolysis and can only be detected following hydrolysis of elastin with NaOH. Since both reduced and non-reduced forms of the cross-links are naturally present in elastin, the ratio of each can only be determined by difference analysis in hydrolysates of reduced and unreduced protein.

The most commonly used methods for quantifying cross-links employ either separation on an amino acid analyser or a dedicated ion-exchange column chromatography system designed to utilize buffer systems that separate cross-linking amino acids along with total amino acids. Such systems, however, require extensive 'fine tuning' to obtain resolution of cross-linking amino acids. For this reason, it is difficult to recommend any one system. Recently, several promising HPLC techniques have been described for quantifying desmosine in tissue hydrolysates (23, 24).

If the target tissue is relatively rich in elastin, gravimetric determination of the dry weight of insoluble elastin is simple and acceptable. Otherwise, desmosine concentration should be used as an index of insoluble elastin content. An RIA and an ELISA for desmosine are available, but the antisera

Jeffrey M. Davidson et al.

may not be totally specific for elastin-derived cross-links (25). We have developed a chromatographic method, modified from that of James *et al.* (26). This is the only approach we have found for adequately resolving the collagen-derived pyridinium cross-links from desmosine and isodesmosine without prior fractionation of collagenous and elastin proteins. The method has the further advantage of being an isocratic procedure, requiring no delay time for re-equilibration of the column and needing only a single HPLC pump.

Protocol 7. Identification of desmosine/isodesmosine by HPLC (Figure 2)

Equipment and reagents

- *n*-Butanol
- HOAc, glacial
- Whatman CF-1 cellulose
- 1-Octanesulfonic acid
- EDTA
- Methanol, HPLC grade
- 6 M HCl
- Isocratic HPLC system with 4.6 × 150 mm C18 column (source not critical), 5 μm particle size, monitored at 271 nm

- Disposable plastic chromatography columns, 10 ml capacity (Bio-Rad)
- Standards of desmosine and isodesmosine (Elastin Products Co.)
- Ultrasonic bath (Branson, Model 1200 or equivalent)
- HPLC buffer: 25 mM Na formate, 5 mM 1-octanesulfonic acid, 1 mM EDTA, adjusted to pH 3.25 with 6 M HCl and made 20% (v/v) with respect to methanol

Method

1. Perform acid hydrolysis of total tissue (dry, fat-free tissue or lyophilized tissue homogenates) or insoluble elastin fractions using 1 ml of 6 M HCl at 110°C for 18–24 h under nitrogen.

2. Isolate the desmosine fraction by the method of Skinner (27). Under the conditions of fractionation, desmosine and isodesmosine as well as the mature collagen cross-links, the pyridinolines, are strongly adsorbed to the support while other amino acid residues are eluted by the mobile phase. As an alternative to commercially-available disposable columns, prepare disposable columns from disposable plastic pipettes which have had the upper part of the bulb cut off with a sharp blade, leaving the remainder of the bulb as a solvent reservoir. Place a glass bead at the bottom outlet to retain the chromatographic medium, and arrange a suitable support for a number of columns equal to the number of hydrolysed samples.

   (a) *Mobile-phase solvent*: *n*-butanol:HOAc:water (4:1:1)

   (b) *Cellulose slurry*: Suspend 10 g of Whatman CF-1 cellulose in 200 ml mobile-phase solvent. Shake to disperse and sonicate briefly in an ultrasonic bath to remove air bubbles.

3. Pour 4–5 ml of cellulose slurry into each column and wash with 5 ml of mobile-phase solvent to eliminate fines.

4. To the hydrolysed sample in 6 M HCl, add 10 ml of glacial HOAc, 1 ml of cellulose slurry, and 4 ml of *n*-butanol.

5. Vortex and apply to the column. Rinse the hydrolysis tube with an additional 1.5 ml of mobile-phase solvent and apply to column.

   (a) Wash the column with 3 × 5 ml aliquots of mobile-phase solvent.

   (b) Elute the cross-link fraction with 5 ml of water into conical centrifuge tubes.

   (c) After low-speed centrifugation (1500 *g*, 5 min, 20°C), carefully aspirate the residual butanol phase overlying the lower, aqueous layer.

6. Lyophilize the aqueous layer and dissolve the residue in 100 μl of 0.05 M HCl, pH 3.

7. Apply the sample (20–50 μl) at a flow rate of 1.5 ml/min to a 4.6 × 150 mm C18 column (5 μm particle size) maintained at room temperature. Perform chromatography by isocratic separation on HPLC (28) for 15 min. Monitor elution at 270–275 nm with a 10 mm flow cell. The elastic cross-links elute after pyridonolines as a well-resolved pair of peaks. This method will easily detect < 10 ng of desmosine.

5. Isolation of tropoelastin

5.1 Introduction

Elastin is secreted as the soluble precursor molecule tropoelastin (29). The precursor is rapidly cross-linked with other tropoelastin molecules to form the insoluble elastic fibre. The cross-linking reaction, catalysed by the enzyme lysyl oxidase, is so efficient that no tropoelastin monomers can be observed in normal tissues. For this reason, tropoelastin can only be isolated from animals maintained on diets that are deficient in copper (a necessary co-factor for lysyl oxidase) or that contain a lysyl oxidase inhibitor such as BAPN. Recently, an *in vitro* technique has been described for isolating tropoelastin from minces of elastin-producing tissues incubated with BAPN (30). This technique avoids the expense and problems of maintaining animals on special diets and yields milligram amounts of purified tropoelastin. The protocol incorporates techniques used previously for isolating tropoelastin from tissues (31), but avoids dialysis which results in extensive losses because of the tendency of tropoelastin to stick to dialysis tubing. Although the procedure outlined below utilizes fetal ligamentum nuchae, other elastin-rich tissues such as neonatal chick, rat, or porcine aorta can be similarly extracted. Because tropoelastin is highly susceptible to proteolysis (32), extreme care should be taken to avoid contamination with proteases. It is advisable to use sterile glassware and solutions during the preparation of the protein.

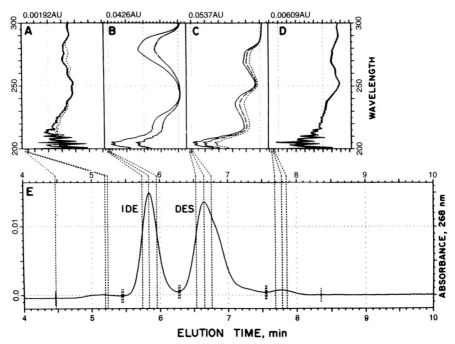

Figure 2. Resolution and identification of elastin cross-links by high-performance liquid chromatography. In this example, purified elastin (50 mg) was hydrolysed and cross-links were fractionated on cellulose as described in the text. The adsorbed isodesmosine (IDE) and desmosine (DES) were resolved by the isocratic method described in the text. The amount represented in this chromatogram is 1/100 of the starting material (500 μg of elastin, or approximately 10 μg of desmosines). Panel E shows the resolution of these compounds, monitored at 268 nm, using a Waters model 960 diode array detector. Spectral identification of the two compounds is confirmed by the absorbance spectra (200–300 nm) shown in panels B and C above.

Protocol 8. Isolation of tropoelastin from elastic tissue

Equipment and reagents

- Dulbecco's modified Eagle's medium (DMEM) containing 10% (v/v) bovine calf serum (HyClone Laboratories Inc.), 100 μg/ml streptomycin, and 100 units/ml penicillin
- β-Aminopropionitrile fumarate (BAPN)
- Penicillamine
- Gentamicin
- Dexamethasone (Sigma)
- N-2-Hydroxy-ethylpiperazine-N-2-ethane-sulfonic acid (Hepes)
- 0.25% HOAc
- 0.5 M HOAc
- Acetone
- n-Butanol
- n-Propanol
- Pepstatin A (Sigma)
- Whatman 50 paper
- Buffer A: 0.5% (v/v) trifluoroacetic acid, aqueous
- Buffer B: 0.5% (v/v) trifluoroacetic acid in n-propanol
- Reverse-phase HPLC column, C3, 4.1 × 150 mm (Hamilton PRP-3)
- Polytron (Brinkman)

252

Method

1. Remove the ligamentum nuchae at a local slaughterhouse from a late-gestation sheep (120–130 day) or bovine (200–270 day) fetus and place immediately on ice in Dulbecco's modified Eagle's medium (DMEM) + 10% bovine calf serum with 2 × antibiotics (see step 3).

2. In the laboratory, remove adherent tissue and mince the ligament finely with a scalpel or, more efficiently, with a McIlwain tissue chopper.

3. Incubate the minced tissue overnight at 37 °C in DMEM supplemented with 100 μg/ml BAPN, 50 μg/ml penicillamine, 80 μg/ml gentamicin, 1 ng/ml dexamethasone, 30 mM Hepes buffer, pH 7.4, and 3% bovine calf serum.

4. After incubation, wash the tissue with ice-cold water to remove serum components.

4a. As an alternative to organ culture, raise young chicks, rats, or swine on a diet containing 0.1% (w/w) α-aminoacetonitrile–HCl and 0.05% ε-aminocaproic acid (33) or with a water source containing the same agents for 7 days. Dissect the thoracic aorta from killed animals and wash as described above.

5. Homogenize in 0.5 M HOAc containing 2.5 μg/ml pepstatin A using a Brinkman Polytron.

6. Stir the tissue mince overnight at 4 °C in 250 ml polyethylene centrifuge bottles and then centrifuge at 16 000 g for 30 min at 4 °C. Discard the insoluble residue.

7. Extract tropoelastin from the supernatant fraction by the sequential, dropwise addition, on ice, of 1.5 volumes of *n*-propanol and 2.5 volumes of *n*-butanol. Remove the precipitate that forms during this step by centrifugation (10 000 g, 20 min, 4 °C) or filtration on Whatman 50 paper at 4 °C.

8. Flash evaporate the supernatant to dryness on a rotary evaporator at ≤ 40 °C.

9. Suspend the residue in a minimal volume of ice-cold, dilute HOAc (0.25% v/v), and precipitate tropoelastin with 3 volumes of ice-cold acetone.

10. Collect the precipitated tropoelastin by centrifugation (50 000 g, 30 min, 4 °C), dissolve in water, and lyophilize.

11. Assess the purity of the isolated tropoelastin by SDS–polyacrylamide gel electrophoresis on an 8% gel (34) and/or by amino acid analysis.

12. If further purification is required, chromatograph the sample on a C3 reverse-phase HPLC column (Hamilton PRP-3, 4.1 × 150 mm) with a linear gradient of A and B buffers to 80% B over 30 min at a flow rate of 0.5 ml/min. Tropoelastin elutes between 50–60% propanol. Alternatively, gel filtration can be used as described below.

Jeffrey M. Davidson et al.

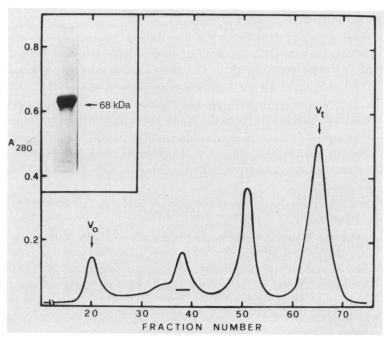

Figure 3. Gel filtration on Bio-Gel A-5m of the propanol/butanol/water (1.5/2.5/1 (v/v)) soluble material from chick aorta. Proteins from the various peaks were analysed by polyacrylamide gel electrophoresis in SDS. Tropoelastin was detected on the underlined peak and its electrophoretic pattern is shown in the inset. A separating gel of 7.5% acrylamide was used and the samples were reduced in 5% 2-mercaptoethanol before application to the gel. The mobility of BSA (68 kDa) is indicated. V_0, void volume; V_t, total volume of the column.

Protocol 9. Isolation of tropoelastin by gel-filtration chromatography (35) (Figure 3)

Equipment and reagents

- Phosphate buffer: 0.01 M phosphate, pH 7.4, 1% SDS, 0.02% NaN_3
- Bio-Gel A-5m (Bio-Rad 200–400 mesh)
- 2-Mercaptoethanol
- 1% SDS
- Glass chromatography column (1.5 × 80 cm)
- Acidified acetone (40 ml of acetone and 1 ml of HCl)

Method

1. Dissolve 10–15 mg of lyophilized material, from *Protocol 8,* step 9, in 2–3 ml of phosphate buffer.

2. Boil for 3 min.

3. Reduce sample with 5% 2-mercaptoethanol.

254

4. Apply solubilized material to the column, previously equilibrated in phosphate buffer at a flow rate of 1–2 ml/h.

5. Combine peak fractions (~ 2 ml) and dialyse against 1% SDS for 24 h at room temperature.

6. Lyophilize and dissolve in a minimal volume of water.

7. Remove SDS by precipitating the protein twice with acidified acetone (1:1 (v/v)), for 24 h, at 20 °C.

6. Direct-binding ELISA for quantitation of tropoelastin

6.1 Introduction

Enzyme-linked immunosorbant assay (ELISA) offers a sensitive assay for quantitation of tropoelastin in cultured cells (smooth muscle cells, fibroblasts, endothelial cells) and biological fluids (lymph, plasma, urine, bronchoalveolar lavage fluid).

Since very little newly synthesized tropoelastin can be detected in the cell layer of cultured fibroblasts and smooth muscle cells (our unpublished results), the protein can be detected in the cell media using specific antisera. Samples of media can be taken after 24–72 h of culture. Two methods are available: direct and indirect ELISA.

Protocol 10. Direct binding ELISA for determining tropoelastin antisera specificity

Note: A direct test is first necessary to determine the titre of antisera.

Equipment and reagents

- Flat-bottom microtitre plates (Immulon-I, Dynatech Lab. Inc.)
- Multi-channel pipettors, 50 and 200 μl capacity
- Voller's buffer (20 mM Na_2CO_3 buffer, pH 9.6): 15.9 g/l of Na_2CO_3 and 29.4 g/l of $NaHCO_3$
- PBS: 8.0 g/l NaCl, 0.2 g/l KCl, 2.14 g/l $Na_2PO_4.7H_2O$, 0.2 g/l KH_2PO_4, pH 7.4
- PBS–TT: 0.05% (v/v) Tween-20, 0.05% (v/v) Triton X-100 in PBS
- α-Elastin, isolated from the species of interest (see *Protocol 4*, this chapter)
- Anti-α-elastin antibodies, raised in rabbit (36)
- Affinity-purified goat anti-rabbit-IgG, peroxidase-labelled (GAR–HRP; Kinkegaard and Perry Labs., Inc.)
- o-Phenylenediamine
- 30% H_2O_2
- Sulfuric acid
- Substrate: dissolve 10 mg of o-phenylenediamine in 1 ml of methanol and add 99 ml of water; add 10 μl of H_2O_2 immediately before use

Method

1. Prepare, on ice in glass tubes, 8 or 12 serial dilutions of α-elastin (or other desired antigen) in Voller's buffer in the concentration range of 1 ng to 10 μg/ml.

Protocol 10. *Continued*

2. Add 200 μl/well of antigen dilutions to flat-bottom microtitre plates to each of a series of wells in rows (12 dilutions) or columns (8 dilutions) to coat the plastic with antigen.

3. Incubate plates overnight at 4 °C in a sealed container lined with wet paper towels to maintain humidity.

4. Serially dilute antisera (from 1:250 to 1:32 000 in PBS–TT) in separate tubes.

5. Wash wells 3 times, soaking with a 3 min soak between washes, with PBS–TT and drain.

6. Add antiserum dilutions to the opposite axis (200 μl/well). Let binding proceed at 4 °C for 30 min.

7. Wash wells 3 times, 3 min each, with PBS–TT.

8. Add GAR–HRP (200 μl/well), diluted 1:2000 in PBS–TT.

9. Let binding reaction go for $\geq$ 90 min at room temperature.

10. Wash wells 3 times, soaking 3 min each, with PBS–TT.

11. Add 200 μl/well of substrate to detect the bound enzyme–antibody conjugate.

12. Stop the reaction after 30–60 min by adding 50 μl of 8 M sulfuric acid to each well with a 50 μl multi-channel pipettor.

13. Determine absorbance at 492 nm using a 96-well microplate reader.

14. Choose an antigen–antibody combination which generates an absorbance in the range of 1.0–1.2 for the indirect test.

6.2 Determination of tropoelastin by direct-binding ELISA

Tropoelastin binds rapidly and irreversibly to the wells of plastic microtitre plates, even in the presence of large amounts of blocking protein. This property is the basis for a direct-binding immunoassay (30) that detects nanogram quantities of tropoelastin directly in cell culture medium. This eliminates sample preparation steps that result in extensive loss of tropoelastin. Another advantage of the direct binding assay is that standard serum-containing culture medium can be assayed directly, since serum proteins block non-specific binding sites in the microtitre plates without affecting tropoelastin binding (30). Additionally the sample requirement for the assay is small (100 μl/well). Thus, serial samples can be taken from the same culture to follow the accumulation of tropoelastin in the medium, allowing temporal analysis of elastin biosynthesis. Measurement of elastin production by a small number of cells is also possible.

Best results are obtained with medium containing 5% fetal bovine serum, although 10% calf serum can be used with only a minimal reduction in assay sensitivity. Pretreatment of microtitre plates by washing in 8 M sulfuric acid enhances tropoelastin binding and reduces plate-to-plate variability.

Protocol 11. Direct-binding ELISA for determining tropoelastin

Equipment and reagents

- DMEM containing 10% bovine serum
- Phosphate-buffered saline, 0.05% (v/v) Tween-20 (PBS–T)
- BAPN
- N-2-hydroxy-ethylpiperazine-N-2-ethane-sulfonic acid (Hepes) 0.25% (v/v) and 0.5 M HOAc.
- 5% calf serum and 30 mM Hepes, pH 7.4
- Flexible polyvinyl chloride 96-well micro-titre plates (Falcon)
- Peroxidase-labelled goat anti-rabbit or goat anti-mouse immunoglobulin (Sigma or Kirkegaard and Perry)
- Antibody to tropoelastin (Elastin Products Co.)
- Tropoelastin, 100 μg/ml in DMEM

Method

1. Acid-wash microtitre plates by soaking for 30 min in 8 M sulfuric acid, rinse well with water, and air dry.

2. Prepare Standard Curve: Dilute the tropoelastin stock solution to a final concentration of 320 ng/ml in tissue culture medium identical to that used for the cell type to be studied. Add serial dilutions of the tropoelastin to one or two columns of the microtitre plate to yield a range of 32–1 ng per well in a volume of 100 μl.

3. Harvest conditioned, serum-containing medium from cells cultured in the presence of 64 μg/ml BAPN, and add 1 volume of 1 M Hepes, pH 7.4 to three volumes of medium.

4. Add three 100 μl aliquots (for triplicate determinations) of the culture medium directly to separate wells of the microtitre plate.

5. Incubate the plate for 3 h at 37 °C.

6. Wash wells three times with PBS–T.

7. Add tropoelastin antibody diluted 1:2000 (can be adjusted to antibody titre) in PBS–T to each well of the plate and incubate for 90 min at 37 °C. Include pre-immune rabbit serum and saline controls on each plate.

8. Wash the wells three times with PBS–T.

9. Incubate for 90 min at 37 °C with peroxidase-labelled GAR–HRP diluted 1:1000 in PBS–T.

10. Wash the wells three times with PBS–T.

11. Develop using the peroxidase substrate (see *Protocol 10*, steps 11–12).

12. Determine absorbance using a microplate reader at 492 nm and construct a standard curve for the determination of unknowns.

6.3 Indirect-binding assay for the determination of tropoelastin

The indirect assay will detect soluble forms of elastin and tropoelastin in the picogram/ml range.

Protocol 12. Determination of tropoelastin by indirect–binding assay (37)

Equipment and reagents

- Same as in *Protocol 10*, plus:
- Round-bottom microtitre plates
- Porcine tropoelastin or tropoelastin from the desired species (see Section 5 this chapter)

- DMEM plus 10% bovine serum or other culture medium used for the experiment

Method

1. On ice, add to round-bottom plates in row A, in duplicate columns:

    (a) 110 μl of a known amount of soluble antigen (for example porcine tropoelastin in culture medium for cell culture evaluation, or α-elastin dissolved in PBS–TT for evaluation of biological fluids, at 20 ng/100 μl).

    (b) 110 μl of culture medium as a possible background contribution from cross-reactive material in serum or 110 μl of PBS–TT as a blank for biological fluids.

    (c) 110 μl of unknown samples.

2. Add to rows A–H, 110 μl PBS–TT and serially dilute (two-fold) from row A–H.

3. Add 110 μl of antiserum, diluted in PBS–TT to 2 × the final concentration as defined by the direct test (1:2000, for example), to each well.

4. Allow the reaction go to equilibrium overnight at 4 °C.

5. Prepare flat-bottomed plates coated with a fixed amount of α-elastin (20 ng/well or as defined by the direct test) and washed with PBS–TT as previously described (3 times, with a 3 min soak between washes).

6. Transfer 200 μl of the contents of each round-bottom well to the corresponding position of the washed, coated plate.

7. Let competitive binding proceed for 30 min at 4 °C.

8. Follow steps 7–13 of the direct test (*Protocol 10*).

One can now construct a standard curve with which to compare unknowns for the quantitation of elastin/tropoelastin. It is important to run standard curves and unknowns in duplicate because of well-to-well variation in binding efficiency. We also subtract the background that can be present in the media due to the presence of elastin peptides in the serum we use for growing cells. For this purpose we run the medium used in the experiment as an unknown sample, and dissolve the antigen (pig tropoelastin) for the standard curve in the same medium. Culture medium is often changed to newborn bovine serum at the start of an accumulation experiment because this serum produces lower background values.

Because of the non-linear nature of antibody–antigen reactivity, a competitive inhibition curve is constructed from the standards and analysed by logit–log regression using a computer program based on the algorithm developed by Rodbard (38) (Microplate Manager, Bio-Rad). Estimates of unknown concentration are based on extrapolation from the fitted data. In general, estimates will only be accurate if values for 2–3 dilutions of the unknown fall on the linear region of the sigmoid inhibition curve. These values are used to determine a mean value for each sample. For critical comparisons, it is best to run assays on the same plate.

In standard practice, antigen concentrations are then normalized to cell number and incubation time (in the case of cell culture), creatinine (in the case of urine), or volume (in the case of other biological fluids).

Acknowledgements

Supported in part by the Department of Veterans Affairs, and NIH grants AG06528, GM37387, and HL26499.

References

1. Reiser, K., McCormick, R. J., and Rucker, R. B. (1992). *FASEB J.*, **6**, 2439.
2. Cleary, E. G., Gibson, M. A., and Fanning J. C. (1989). In *Elastin and elastases* (ed. L. Robert and W. Hornebeck), Vol. 1, p. 31. CRC Press, Boca Raton.
3. Sandberg, L. B., Soskel, N. T., and Leslie, J. G. (1981). *New Engl. J. Med.*, **304**, 566.
4. Lansing, A. I., Rosenthal, T. B., Alex, M., and Dempsey, T. (1952). *Anat. Rec.*, **114**, 555.
5. Partridge, S. M., Davis, H. F., and Adair, G. S. (1955). *Biochem. J.*, **61**, 11.
6. Soskel, N. T., Wolt, T. B., and Sandberg, L. B. (1987). In *Methods in enzymology*, (ed.) Vol. 144 (ed. L. W. Cunningham), p. 196. Academic Press, Orlando.
7. Sandberg, L. B. and Davidson, J. M. (1984). *Peptide and Protein Reviews*, **3**, 169.
8. Starcher, B. C. and Galione, M. J. (1976). *Anal. Biochem.*, **74**, 441.
9. Moczar, M., Moczar, E., and Robert, L. (1979). *Connect. Tiss. Res.*, **6**, 207.
10. Mecham, R. P. and Lange, G. (1982). In *Methods in enzymology*, Vol. 82, (ed. L. W. Cunningham and D. W. Frederiksen), p. 744. Academic Press, New York.
11. Mecham, R. P. and Foster, J. A. (1979). *Biochim. Biophys. Acta*, **577**, 147.
12. Foster, J. A., Rubin, L., Kagan, H. M., Franzblau, C., Bruenger, E., and Sandberg, L. B. (1974). *J. Biol. Chem.*, **249**, 6191.
13. Mecham, R. P. and Foster, J. A. (1978). *Biochem. J.*, **173**, 617.
14. Houle, D. and LaBella, F. (1977). *Connect. Tissue Res.*, **5**, 83.
15. Thomas, J. and Partridge, S. M. (1960). *Biochem. J.*, **74**, 600.

16. Crombie, G., Foster, J. A., and Franzblau, C. (1973). *Biochem. Biophys. Res. Commun.*, **52**, 1228.
17. Senior, R. M., Bielefeld, D. R., and Starcher, B. C. (1976). *Biochem. Biophys. Res. Commun.*, **72**, 1327.
18. Kagan, H. M. (1986). In *Regulation of matrix accumulation* (ed. R. P. Mecham), p. 322. Academic Press, New York.
19. Robins, S. P. (1982). *Methods Biochem. Anal.*, **28**, 329.
20. Paz, M. A. Keith, D. A., and Gallop, P. M. (1982). In *Methods in enzymology*, Vol. 82 (ed. L. W. Cunningham and D. W. Frederiksen), p. 571. Academic Press, New York.
21. Lent, R. W., Smith, B., Salcedo, L. L. Faris, B., and Franzblau, C. (1969). *Biochemistry*, **8**, 2837.
22. Franzblau, C., Faris, B., and Papaioannou, R. (1969). *Biochemistry*, **8**, 2833.
23. Soskel, N. T. (1987). *Anal. Biochem.*, **160**, 98.
24. Yamaguchi, Y., Haginaka, J., Kunitomo, M., Yasuda, H., and Bando, Y. (1987). *J. Chromatogr.*, **422**, 53.
25. Gunja-Smith, Z. L. (1985). *Anal. Biochem.*, **147**, 258.
26. James, I. T., Perrett, D., and Thompson, P. W. (1990). *J. Chromatogr.*, **525**, 43.
27. Skinner, S. J. M. (1982). *J. Chromatogr.*, **229**, 200.
28. Oliver, R. W. A. (ed.) (1989). *HPLC of macromolecules: a practical approach*. Oxford, New York.
29. Davidson, J. M. and Giro, M. G. (1986). In *Biology of the extracellular matrix*, (ed. R. P. Mecham), Vol. 1, p. 177. Academic Press, New York.
30. Prosser, I. W., Whitehouse, L. A., Parks, W. C., Stahle-Bäckdahl, M., Hinek, A., Park, P. W., and Mecham, R. P. (1991). *Connect. Tiss. Res.*, **25**, 265.
31. Sandberg, L. B. and Wolt, T. B. (1982). In *Methods in enzymology*, Vol. 82, (ed. L. W. Cunningham and D. W. Frederiksen), p. 657. Academic Press, New York.
32. Romero, N., Tinker, D., Hyde, D., and Rucker, R. B. (1986). *Arch. Biochem. Biophys.*, **244**, 161.
33. Rich, B. C. and Foster, J. A. (1982). In *Methods in enzymology*, Vol. 82, (ed. L. W. Cunningham and D. W. Frederiksen), p. 665. Academic Press, New York.
34. Hames, B. D. and Rickwood, D. (ed.) (1990). *Gel electrophoresis of proteins: a practical approach*. Oxford, New York.
35. Bressan, G. M., Castellani, I., Giro, M. G., Volpin, D., Fornieri, C., and Pasquali Ronchetti, I. (1983). *J. Ultrastr. Res.*, **82**, 335.
36. Davidson, J. M. and Sephel, G. C. (1987). In *Methods in enzymology*, Vol. 144 (ed. L. W. Cunningham), p. 214. Academic Press, New York.
37. Giro, M. G. and Davidson, J. M. (1988). In *Methods in enzymology*, Vol. 163 (ed. G. diSabato), p. 656. Academic Press, New York.
38. DeLean, A., Munson, P. J. and Rodbard, D. (1978). *Am. J. Physiol.*, **235**, E97.
39. Jacob, M. P. and Robert, L. (1989). In *Elastin and elastase* (ed. L. Robert and W. Hornebeck), Vol. 1, p. 49. CRC Press, Boca Raton.
40. Whe-Yang Kao, W., Bressan, G. M., and Prockop, D. J. (1982). *Conn. Tiss. Res.*, **10**, 263.
41. Deyl, Z., Hora'kova', M., Adam, M., and Macek, K. (1982). *Biochem. Biophys. Res. Commun.*, **106**, 94.

The catabolism of extracellular matrix components

SUSAN J. FISHER and ZENA WERB

1. Introduction

Breaching of the extracellular matrix (ECM) barrier by cells involves proteo-lytic enzymes capable of degrading the ECM components. The most import-ant of these enzymes is the collagenase family of metalloproteinases which have been recently reviewed (1, 2). The members of this multigene family include Pump-1, three types of stromelysin, interstitial collagenase, and the 72 and 92 kDa gelatinases/type IV collagenases. These enzymes possess sequences necessary for maintenance of the zymogen forms, a zinc-binding catalytic domain, and various numbers of structural regions resembling those found in ECM components. The wide range of specificity that the metallo-proteinases exhibit is important in growth, debridement after injury, tissue morphogenesis and remodelling, and invasion. For example, stromelysin degrades fibronectin, elastin, proteoglycan core proteins, and laminin as well as the non-helical regions of collagen types IV, V, VII, and IX and the amino terminus of collagen type I. Interstitial collagenases can degrade collagens of types I, II, III, VIII, and X, but the degradation of basement membrane collagens requires specialized enzymes that degrade type IV collagen. The latter proteinases can also degrade collagens of types V, VII, and X as well as fibronectin, denatured collagens, and casein. Tissue-type (tPA) and urokinase-type (uPA) plasminogen activators, members of the serine pro-teinase family, also play an important role in ECM degradation and these have been recently reviewed (3). They may act directly, cleaving fibronectin and other components. Alternatively, PAs may act as part of a proteinase cascade in which plasmin activates procollagenase and prostromelysin.

Devising an experimental strategy to study ECM-degrading proteinases requires a basic understanding of these specialized enzymes. One important consideration is that proteinase substrate specificities are usually very difficult to define. As a result these enzymes are classified as metallo-, serine, aspartic and cysteine proteinases on the basis of their catalytic mechanisms (reviewed in reference 4). The most commonly used method for determining the class(es) to which a proteolytic activity belongs is to use a battery of class-specific inhibitors (see *Table 1*). Because many of the matrix-degrading enzymes are either metallo- or serine proteinases, determining the class to which an ECM-degrading activity belongs is important for narrowing the scope of the investigation and for targeting certain enzyme systems for future study. Since it is likely that more than one enzyme or enzyme system is involved in a degradative process, inhibitors are also used to determine the relative contribution of various proteinases to the degradative process.

Before more specific assays are attempted, several problems that are consistently encountered when studying proteinases should be considered. Some are inherent to these specialized enzymes. Others are introduced by the complex nature of the tissue fluids and extracts that are the source of proteinases of biological interest. One important consideration is that proteinases are often present as proenzymes; these must be converted to the active form before activity can be detected. Conversely, endogenous activators present in a sample can lead to inconsistent estimates of proteinase activity or degradation of the enzyme. Another important consideration is that proteinases are often complexed with inhibitors that control their activity *in vivo*. Separating these complexes is necessary to obtain an accurate estimate of total proteinase activity.

This chapter emphasizes methods for studying ECM-degrading metalloproteinases and the plasminogen activators. In most cases there are several methods that could be used to determine whether a given proteinase activity is involved in matrix catabolism. We have chosen the most sensitive ones because the source of proteinase is usually limiting. Thus, the following protocols can be used to analyse cultured cells, cell extracts, and/or conditioned medium.

2. Zymography

Zymographic detection on substrate gels is often a good choice for the initial characterization of a proteinase activity. Although not all proteinases renature, thus allowing their detection by this method, many of the ECM-degrading enzymes do (for example collagenases, stromelysins, and plasminogen activators). This technique permits important information, such as estimated molecular weight and proteinase class, to be obtained. In addition, it circumvents two common problems encountered in the assay of proteinases: (a) many proenzymes show activity when assayed in this manner, thus eliminating the requirement for activation; and (b) electrophoresis can result in the separation of non-covalent proteinase–inhibitor complexes. This method also offers excellent sensitivity. Assay of 1–10 μl of unconcentrated conditioned medium or lysates from a wide variety of cultured cells, including macrophages, fibroblasts, endothelial cells, and human tumour cells, is routine.

The substrate gels differ from standard SDS–polyacrylamide gels (5) in two respects. First, the gels are made by incorporating the protein substrate of interest into the polymerized acrylamide matrix. Second, the sample is mixed with a higher concentration of SDS (without reducing agents) and is not boiled (except when proteinase inhibitors are to be detected, as described in Section 2.3.3). Substrate gels using gelatin or casein (*Protocol 1*) are most often used in initial studies (6). More specialized substrates facilitate the detection of specific proteinases. For example, gels containing plasminogen and casein (Section 2.3.1) are used to detect plasminogen activators (7). Gels containing elastin can be used for the specific detection of elastinolytic enzymes (Section 2.3.2). Globin, fibronectin, laminin, α-elastin, carboxymethyl transferrin, and cartilage proteoglycans have also been used successfully. Enzymes that degrade the carbohydrate portions of matrix molecules can also be detected by this method. For example, hyaluronidase-degrading enzymes can be visualized by using substrate gels that contain this glycosaminoglycan (8). Finally, the conditions can be modified so that inhibitors such as the tissue inhibitors of metalloproteinases (TIMPs) can also be detected (Section 2.3.3) (6). In all cases the protein should be reasonably pure ($> 90\%$) and either soluble or capable of forming a non-sedimenting suspension in aqueous solutions.

Protocol 1. Substrate gel zymography

Equipment and reagents

- 4 × Lower gel buffer: 18.15 g Tris base (Sigma), 0.4 g SDS (Bio-Rad), 50 ml double-distilled water, pH adjusted to 8.8 with HCl, final volume to 100 ml with double-distilled water
- 4 × Upper gel buffer: 3 g Tris base, 0.2 g SDS in 20 ml water, pH adjusted to 6.8 with HCl, final volume to 50 ml with double-distilled water
- 4 × Laemmli sample buffer (without Bromophenol blue): 1 g SDS, 2.5 ml glycerol (Bio-Rad), 5 ml 4 × upper gel buffer, double-distilled water to 10 ml
- Stock solutions: 10 × Tris: 30.275 g in 1 litre; 10 × glycine–SDS: 142.25 g glycine (Bio-Rad) + 10 g SDS in 1 litre
- Running buffer: 0.025 M Tris, 0–19 M glycine, 0.1% SDS, pH 8.3 (all prepared from 10 × stock solutions, see above)
- Hoefer electrophoresis apparatus (Hoefer Scientific Instruments)

- Laemmli stacking gel (4% acrylamide): 1.5 ml 30% acrylamide:bisacrylamide (Bio-Rad), 6.1 ml H_2O, 2.5 ml 4 × upper gel buffer, 100 μl fresh 10% ammonium persulfate (Bio-Rad), 5 μl TEMED (Bio-Rad)
- Phosphate-buffered saline (PBS), pH 7.2
- Coomassie Blue stain: 5 g Coomassie Blue in 600 ml H_2O, filter, add 300 ml isopropyl alcohol, 100 ml glacial acetic acid
- Gelatin (Sigma, Type A from porcine skin)
- Casein (Sigma, α-casein)
- Sodium azide (NaN₃)
- Triton X-100
- 50 mM Tris–HCl, pH 7.6
- Pre-stained molecular weight markers (BRL)
- Rubber policeman
- 1 cc syringe with 26-gauge needle

A. *Preparation of samples from cultured cells*

1. Assay the conditioned medium without prior concentration. (**Note:** Medium can be mixed 3:1 (v/v) with Laemmli sample buffer and stored at $-70\,°C$.)

Protocol 1. *Continued*

2. Rinse cultured cells three times with ice-cold PBS, then lyse by adding as small a volume of 4 × Laemmli sample buffer as possible.

3. Scrape the cells from the dish using a rubber policeman.

4. Produce a cell suspension by using a 1 cc syringe fitted with a 26-gauge needle to aspirate the cells in sample buffer, approximately 20 times.

5. Centrifuge to remove any particulate material.

6. Remove a sample and determine the protein concentration by the method of Lowry *et al.* (9).

7. Freeze the sample at $-70\,°C$.

(**Note**: Since proteinases can autoactivate or be cleaved by other proteinases in the sample, initial analyses should be performed with fresh samples. Frozen samples can be used once the electrophoretic pattern has been established.)

B. *Preparation of gels*

1. Dissolve gelatin or casein in H_2O at 3 mg/ml (3 × stock) with 0.02% NaN_3. Proteins may be dissolved in lower gel buffer if necessary. (**Note**: Gelatin is the best substrate for detecting the type IV collagenases and interstitial collagenase, and casein is best for detecting stromelysin.)

2. Prepare 30% acrylamide:bisacrylamide by dissolving 29.2 g of acrylamide and 0.8 g of bisacrylamide in double-distilled H_2O and adjusting to a final volume of 100 ml. Filter and store at 4°C in a dark bottle.

3. Prepare a substrate gel incubation buffer that will provide the conditions necessary for the enzyme. For example, matrix-degrading metalloproteinases, uPA, and tPA are assayed in PBS containing calcium, pH 7.2. For acid proteinases, a lower pH and the addition of EDTA and reducing agents may be better.

4. Prepare 14 ml of 10% acrylamide containing 1 mg/ml gelatin (or casein) by mixing 4.66 ml of 30% acrylamide:bisacrylamide, 3.50 ml of 4 × lower gel buffer, 4.66 ml of 3 mg/ml gelatin (in double-distilled H_2O) containing 0.02% NaN_3, and 1.04 ml H_2O. Degas for 15 min at ambient temperature. Add 50 µl of 10% ammonium persulfate (freshly prepared) and 7 µl of TEMED. (**Note**: 10% acrylamide gels give the best resolution of the enzymes in the range of mol. wt 30 000–100 000.)

5. Pour the acrylamide solution carefully into the mould, avoiding bubbles; overlay with H_2O or butanol-saturated H_2O and allow to polymerize for 120 min. After polymerization the gels can be stored in self-

sealing plastic bags with extra 1 × lower gel buffer at 4 °C for up to 2 weeks.

6. Prepare 10 ml Laemmli stacking gel. This is enough for five mini-gels or two to three regular-gel stackers.

7. Pour Laemmli stacking gel around a 10-, 15-, or 20-well comb, using a Pasteur pipette to fill to the top of the mould. Allow to polymerize for 40–90 min at room temperature.

8. Mix samples 3:1 (v/v) with 4 × sample buffer (do not boil!) and load by underlaying with a Hamilton syringe. When feasible, run the gels at 4 °C to reduce interaction of the enzyme with the substrate and to achieve more accurate estimates of molecular weight. Run at 15 mA/gel while stacking and at 20 mA/gel during the resolving phase. To monitor the progress of the electrophoresis, use prestained molecular weight markers (BRL) as standards. For low molecular weight enzymes, terminate electrophoresis when the dye front reaches the bottom of the gel. For enzymes of mol. wt. 40 000–100 000, terminate electrophoresis when the mol. wt. 25 000 marker reaches the edge of the gel.

9. Remove the SDS and allow the proteins to renature by soaking in 2.5% Triton X-100 (in 50 mM Tris, pH 7.6) with gentle shaking for 60 min at ambient temperature with one change. Rinse the gel three times in the incubation buffer of choice (i.e. one that will permit proteinase activity) and incubate in this buffer at 37 °C for 2–24 h with gentle shaking. For detection of plasminogen activators and metalloproteinases, the routine buffer is 50 mM Tris–HCl, pH 7.8, containing 150 mM NaCl and 5 mM $CaCl_2$.

10. Stain the gel with Coomassie Blue (minimum 30 min) with shaking and destain in either 45% methanol (or isopropyl alcohol) and 10% acetic acid. Optimize the contrast between cleared regions and background by varying the staining and destaining conditions. The areas containing enzyme are pale or clear against a blue background (see *Figure 1*).

11. Transilluminate the wet gel against a light box and photograph it.

12. Air-dry the gel on a Lucite slab overlaid with a single wet-stretched sheet of cellulose nitrate held in place with a Lucite frame and clamps.

13. Scan the gel using a densitometer and/or store in a notebook.

2.1 Using substrate gels to determine proteinase class

By adding a variety of inhibitors to the substrate gel incubation buffer, it is possible to determine the proteinase class of the activity or activities of interest (10). When inhibitors of the appropriate class are used, no enzymatic activity is observed in the molecular weight region that was formerly cleared of substrate. In most cases duplicate samples are run in different lanes of the

Susan J. Fisher and Zena Werb

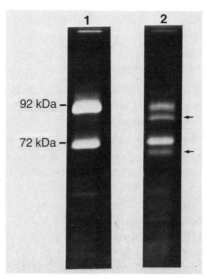

Figure 1. Substrate gel zymography for the detection of proteinases. Samples of conditioned medium from cultures of human placental villi were analysed on 10% polyacrylamide gels containing 1 mg/ml gelatin. (1) Without activation the proenzyme forms of the 72 kDa and 92 kDa type IV collagenases were visible. (2) After activation with APMA the lower molecular weight active forms were also detected (arrowheads).

same gel. One lane is processed as described in *Protocol 1* for detection of proteinase activity (*Figure 1*) and the others are incubated in substrate buffer containing the various inhibitors. *Table 1* lists potentially useful inhibitors and recommended concentrations for their use in this assay. In all cases, the inhibitors are added to the buffer in which the gels are incubated. In some cases either (NPGB) and PMSF are included in the sample buffer used for solubilizing the cells and medium or the samples are incubated with the inhibitor for 30 min (4°C) before electrophoresis. If it is necessary to dissolve the inhibitor in either dimethyl sulfoxide (for instance NPGB) or ethanol (for instance PMSF), control samples and gels are incubated with the solvent alone in the same final concentration as that introduced into either the sample or the incubation buffer. Once the class or classes have been established, the effect of more specific inhibitors can be tested. For example, we have used the TIMPs and transition-state analogues of collagenase to block specific metalloproteinase activity.

2.2 Methods to increase sensitivity

The sensitivity of this assay can be increased by decreasing the substrate concentration. However, under these conditions cellular proteins in the sample may also stain. One variation which may be useful in this circumstance

Table 1. Proteinase inhibitors

Class		Compound	Concentration
Serine		PMSF	2–10 mM
		ε-amino caproic acid	10–100 mM
		p-nitrophenyl-p'-guanidobenzoate	0.1–0.2 mM
	trypsin-like	leupeptin[a]	0.1–10 mM
		benzamidine	0.1–10 mM
	chymotrypsin-like	chymostatin[a]	0.2–2 mM
	elastase-like	elastinal	0.2–2 mM
	uPA and tPA	plasminogen activator inhibitor-1	0.1–1 μM
	uPA and tPA	plasminogen activator inhibitor-2	0.1–1 μM
	uPA	amilioride	1 mM
Metallo-		EDTA	1 mM
		1,10-phenanthroline	0.3 mM
		phosphoramidon	100 μg/ml
	ECM-degrading	TIMPs	10 μg/ml
Aspartic		pepstatin A	0.1–1 mM
Cysteine		trans-epoxy-L-leucyl-amido-(4-guanidino) butane	10–50 μM
		N-ethylmaleimide	2 mM
		organomercurials	1 mM

[a] Also inhibits cysteine proteinases.

is to use radiolabelled protein substrate and autoradiography to detect areas of proteolysis. In addition, overlay assays may be used (11). For example, a substrate for detecting plasminogen activators is mixed at 45°C from 0.8% agar, 1.3% non-fat milk, 40 μg/ml plasminogen in PBS, then cast on a flat glass or plastic surface and allowed to solidify at 20°C. An SDS–polyacrylamide gel electrophoresis (PAGE) gel, from which the SDS has been removed by soaking for 1–2 h in 0.1 M Tris–HCl (pH 8.1) containing 2.5% Triton X-100 then rinsing in double-distilled H_2O, is placed on the substrate. Both are covered with plastic film and incubated at 37°C for 3–30 h. Enzymatic activity is visible as a clearing of the substrate and is visualized by dark-field illumination. Alternatively, SDS-PAGE gels are overlaid with membranes impregnated with MeO–Suc–Ile–Gly–Arg–AFC (7-amino-4-trifluoromethyl coumarin), Val–Leu–Lys–AFC, or Glutaryl–Ala–Ala–Ala–AFC (Enzyme System Products) for the fluorescent detection of plasminogen activators, plasmin, or elastase, respectively.

2.3 Variants for detecting particular proteinases or inhibitors

2.3.1 Substrate gel for detecting plasminogen activators

A substrate gel is prepared, run, incubated, and stained as described in *Protocol 1* except that the running gel contains both casein (1 mg/ml) and plasminogen (13 µg/ml). Under these conditions uPA and/or tPA in the sample cleaves plasminogen, yielding plasmin which then degrades the casein. The areas containing enzyme are pale or clear against a blue background. Purified uPA and tPA, for use as standards, can be purchased from American Diagnostica.

2.3.2 Substrate gel for detecting elastinolytic activity

To demonstrate elastases specifically, substrate gels can be co-polymerized with soluble α-elastin (Elastin Products) at 1 mg/ml and the elastinolytic enzymes can be visualized by using *Protocol 1*. The only difference is that these gels usually need longer incubation periods (24–48 h).

2.3.3 Substrate gel for detecting proteinase inhibitors

An important aspect of the substrate gel technique is that it often results in the separation of inhibitor–proteinase complexes. By performing a variant of the substrate gel technique it is possible to visualize the inhibitors in electrophoretically separated samples. Because the net proteinase activity in any sample depends on the ratio of proteinase to inhibitor, knowing the relative amounts and types of inhibitors present in a sample is also important.

The gel is prepared and run and the SDS removed as described in *Protocol 1*, and the gel is then incubated in medium containing proteinase. To demonstrate the presence of TIMP-1 and/or TIMP-2, gels are incubated in 4-aminophenylmercuric acetate (APMA)-activated fibroblast-conditioned medium for 1–8 h. The gel is then stained as described in *Protocol 1*. Areas of the gel containing no inhibitor and from which the substrate has been removed by proteolytic action are pale blue. In contrast, protein bands that have inhibitory activity are stained and appear a darker blue (*Figure 2*). Similarly,

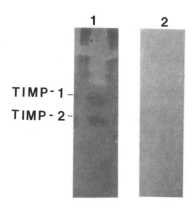

Figure 2. Substrate gel zymography for the detection of proteinase inhibitors. (1) Samples of chondrocyte-conditioned medium contained both TIMP-1 and TIMP-2 at 25 kDa and 21 kDa, respectively. Note that under the conditions used proteinases were also detected above the inhibitor. (2) In contrast, Coomassie Blue staining of an identical gel showed no protein bands.

incorporation of plasminogen activators in the gel can be used to detect plasminogen activator inhibitor PAI-1 and PAI-2.

3. Radiolabelled substrates

Based on preliminary information obtained by using the substrate gel (*Protocol 1*) or other techniques, more specific assays for particular proteinases can be attempted. Because of their relative sensitivity, many employ radiolabelled substrates. In general, these assays are more quantitative than the substrate gel technique. However, interassay comparisons are limited by differences in the efficiency of substrate labelling. Despite the limitations of these assays, an assessment of total relative proteinase activity is important, and the data can be compared with those obtained from other assays that were performed with the same batch of labelled substrate or with those obtained by using proteinase standards.

It is important to note that radiolabelled proteins need not be used if sufficient enzyme and substrate are available to monitor the reaction electrophoretically. When unlabelled collagen, fibronectin, and laminin substrates are used, enzyme–substrate ratios should be between 1:2 and 1:50 (w/w). The reactions are terminated by the addition of SDS-PAGE sample buffer containing reducing agents, and the products are analysed by SDS-PAGE. Proteins are visualized by silver staining.

3.1 Assay of plasminogen activators

Plasminogen activators convert plasminogen, a serum zymogen, to plasmin, a potent fibrinolytic enzyme. The method developed by Unkeless *et al.* (12) offers the sensitivity necessary to measure plasminogen activator secretion by intact cells and also to estimate plasminogen activator concentrations in cell-free conditioned medium and cell lysates (*Protocol 2*). Briefly, tissue culture dishes are coated with a thin film of ^{125}I-labelled fibrinogen, which is dried and converted to fibrin by the addition of thrombin. The generation of plasmin by plasminogen activators is then measured from solubilization of the ^{125}I-labelled fibrin.

Protocol 2A. ^{125}I-labelled fibrin assay of plasminogen activators— purification of fibrinogen

Equipment and reagents

- Bovine fibrinogen (Calbiochem)
- 0.1 M phosphate buffer, pH 6.4
- Ammonium sulfate
- 0.5 M NaCl, pH 7.0
- Dialysis tubing
- 5 mM Na$_2$PO$_4$, 0.12 M lysine, pH 7.0
- 0.6 M NaCl, pH 7.0
- Ethanol
- 10–20 ml of 0.6 M NaCl, pH 7.4
- 0.22 μm Millipore filters

Protocol 2A. *Continued*

(**Note**: Use plastic containers to minimize clotting)

Method

1. Dissolve 4 g of fibrinogen in 400 ml of 0.1 M phosphate buffer, pH 6.4.
2. Add 200 ml of distilled water and leave on ice for 6–18 h.
3. Filter and discard precipitate.
4. Make filtrate one-third of the final concentration with ammonium sulfate, pH 7.0, and stir for 2 h at ambient temperature.
5. Centrifuge the precipitate at 10 000 g for 10 min at 4 °C and dissolve in 50 ml of 0.5 M NaCl, pH 7.0.
6. Dialyse three times for a total of 12 h in the same buffer at ambient temperature.
7. Measure the fibrinogen concentration by determining the protein concentration by the method of Lowry *et al.* (9).
8. Add 5 vol. 5 mM Na_2PO_4, 0.12 M lysine, pH 7.0, to 10 mg/ml fibrinogen in 0.6 M NaCl, pH 7.0. Cool to 0 °C.
9. Add ethanol to 7% final concentration and stir for 30 min at 4 °C. (Repeat ethanol precipitation if significant plasmin concentration remains.)
10. Centrifuge at 10 000 g for 10 min at 4 °C and dissolve the precipitate in 10–20 ml of 0.6 M NaCl, pH 7.4, then dialyse in the same buffer.
11. Pass through filter paper and sterilize by Millipore filtration.
12. Measure the protein concentration and store frozen at −20 °C.

Protocol 2B. ^{125}I-labelled fibrin assay of plasminogen activators—iodination of fibrinogen

Equipment and reagents

- 0.5 mCi ^{125}I-labelled NaI carrier free
- 50 ml of chloramine T (2 mg/ml in 0.3 M PBS, pH 7.4)
- 100–500 μg purified fibrinogen
- 150 ml of a saturated tyrosine solution in water
- 0.22 μm filter
- 200 ml of 5% bovine serum albumin, pH 7.4
- Sephadex G-25
- Bovine serum albumin

Method

1. Add 0.5 mCi ^{125}I-labelled NaI to 50 ml chloramine T and purified fibrinogen and mix at ambient temperature for 2 min.
2. Combine 150 ml of saturated tyrosine and 200 ml of 5% bovine serum

albumin, pH 7.4, then pass the mixture over a small column (2 ml) of Sephadex G-25 to separate the ^{125}I-labelled fibrinogen from the bulk of unincorporated iodine-125.

3. Pool the labelled fibrinogen and filter through a 0.22 μm filter, adding bovine serum albumin (5%) to minimize losses.

4. Determine acid-precipitable radioactivity.

5. The labelled fibrinogen can be stored frozen at − 20°C, although the proportion of acid-precipitable protein may decrease rapidly after several weeks' storage.

Protocol 2C. ^{125}I-labelled fibrin assay of plasminogen activators— preparation and activation of ^{125}I-labelled fibrin plates

Equipment and reagents

- 24-well, 16 mm diameter tissue culture plates or 35 mm tissue culture dishes
- Fibrinogen
- Dulbecco's medium
- 10% fetal bovine serum (FBS)
- Hanks' balanced salt solution (HBSS)
- ^{125}I-labelled fibrinogen from *Protocol 2B*

Method

1. Dilute the ^{125}I-labelled fibrinogen with sterile distilled water so that 0.3 ml of the final solution contains approximately 10^5 acid-precipitable c.p.m. and 20 mg of unlabelled fibrinogen.

2. Keep the solution warm to prevent precipitation of fibrinogen and dispense 0.3 ml samples to each well of the tissue culture dish or plate. (**Note**: Use a glass rod to distribute fibrinogen over the surface of 35 mm dishes.)

3. Dry at 45°C in an incubator under sterile conditions for a minimum of 3 days.

4. The radioactive plates can then be stored for several weeks at ambient temperature or in the cold.

5. Just before use, incubate each well with 1 ml of Dulbecco's medium plus 10% FBS for 2 h or overnight at 37°C to convert fibrinogen to fibrin. (**Note**: Thrombin solutions can also be used.)

6. Wash twice with HBSS.

7. Plates may be kept for 21 days after activation. Wash again before use.

Protocol 2D. ^{125}I-labelled fibrin assay of plasminogen activators—procedure for assay of fibrinolysis: intact cells

Equipment and reagents

- ^{125}I-labelled fibrin-coated wells (*Protocol 2C*)
- HBSS
- Intact cells
- Acid-treated serum

- Acid-treated, plasminogen-free serum or serum-free medium (e.g. Nutridoma)
- Purified plasminogen (Sigma or American Diagnostica)
- Soybean trypsin inhibitor or aprotinin (Sigma)

Method

1. Add various cell concentrations, including the number required to make a nearly confluent monolayer, to ^{125}I-labelled fibrin-coated wells.

2. Initiate the fibrinolytic assay by washing the monolayer three times with HBSS to remove inhibitors and incubate in Dulbecco's medium plus acid-treated FBS. Acid-treated dog serum or serum-free medium supplemented with plasminogen can be used for cells with low levels of activity.

3. Plate appropriate controls in parallel, including:

   (a) wells containing medium alone (no cells);

   (b) cells cultured in acid-treated serum from which plasminogen has been removed by lysine-affinity chromatography or serum-free medium;

   (c) cells cultured as in (b), but in the presence of plasminogen to show plasminogen dependence by reconstitution of the degradative activity;

   (d) cells cultured in the presence of 10% FBS and 60 μg/ml soybean trypsin inhibitor or in serum-free medium plus 10 μg/ml aprotinin, to suppress fibrinolysis.

   (**Note**: (a) and (d) are required; (b) and (c) are optional.)

4. Remove samples at desired intervals and determine the radioactivity in a gamma spectrometer to measure solubilization of fibrin.

5. Monitor cells by phase-contrast microscopy during the course of the assay, which can be extended for up to 24 h to detect low levels of activity.

6. Express results as the percentage of total radioactivity released by trypsin.

Note: One drawback of the ^{125}I-labelled fibrin plate assay is the lack of standardization, which makes it difficult to compare results from different batches of reagents. Arbitrary units can be used to compare results; for instance 1 unit represents 5% of the trypsin-releasable radioactivity solubilized in 1 h at 37°C. Urokinase reference standards can also be useful. The activity of different cell lines can be normalized to the number of cells required to cause release of 50% of available radioactivity under standard conditions.

Protocol 2E. ^{125}I-labelled fibrin assay of plasminogen activators—procedure for assay of fibrinolysis: conditioned medium

Reagents

- Compatible serum-free medium with plasminogen
- Cell monolayers
- HBSS
- 10 mM NH$_4$HCO$_3$

Method

1. Wash cell monolayers carefully at least three times with HBSS to remove inhibitors.
2. Grow cells in compatible serum-free medium with plasminogen for 48 h.
3. Collect the conditioned medium, spin in a microcentrifuge (500 g for 5 min at 4°C) to remove cell debris, and store the supernatant at -20°C until assay. (**Note**: The secreted proteinases are stable for several cycles of freezing and thawing, but the enzymes are sticky and easily lost on precipitates and surfaces when present in dilute solutions.)
4. Measure activity in conditioned medium either immediately or after concentration.
5. To amplify dilute proteinase activities, dialyse conditioned medium for 12 h at 4°C, against three changes of 20 vol. of 10 mM NH$_4$HCO$_3$, a volatile salt, and lyophilize. The lyophilized material is then reconstituted in the appropriate assay buffer.
6. Perform the assay as described in *Protocol 2D* for intact cells.

Protocol 2F. ^{125}I-labelled fibrin assay of plasminogen activators—procedure for assay of fibrinolysis: cell lysates

Reagents

- HBSS
- Cell monolayers
- 0.1–0.2% Triton X-100
- Isotonic saline

Method

1. Wash cell monolayers twice with HBSS, then add 0.5 ml of 0.1–0.2% Triton X-100.
2. Scrape cells from the dish with a rubber policeman, and store lysate at -20°C.
3. Alternatively, harvest cells by scraping in isotonic saline and collecting by centrifugation at 500 g for 5 min at 4°C.

Protocol 2F. *Continued*

4. Remove nuclei by low-speed centrifugation at 5000 g for 10 min at 4°C, then store lysate at -20°C.

5. Perform the assay as described in *Protocol 2D* for intact cells.

3.2 Assay of interstitial collagenase

Interstitial collagenase can degrade a variety of collagens. However, ^{3}H-labelled collagen type I is most often used to assay the activity of this enzyme (13). Radiolabelling is accomplished by using [^{3}H]acetic anhydride to *N*-acetylate the free amino groups of the molecule.

Protocol 3A. ^{3}H-labelled collagen (type I) assay—type I collagen labelling

Equipment and reagents

- Bovine type I collagen (Vitrogen, Collagen Corporation)
- 13 mM HCl
- 5 mCi [^{3}H]acetic anhydride

- *p*-Dioxane (dried by adding Na_2SO_4 crystals)
- Dialysis tubing
- 5 × Phosphate buffer concentrate, pH 7.5: 1 M NaCl, 90 mM Na_2HPO_4/NaH_2PO_4

Method

1. Dissolve bovine type I collagen (3 mg/ml) in 20 ml of 13 mM HCl

2. Centrifuge at 40 000 g for 1 h at 4°C.

3. (**Caution**: [^{3}H]acetic anhydride is volatile and should be handled with extreme care.) Concentrate the [^{3}H]acetic anhydride in the end of the ampoule by submersion in a dry ice/ethanol bath. Open the ampoule under a fume hood and add 0.5 ml dried *p*-dioxane.

4. Rapidly mix the collagen solution with 5 ml of the 5 × phosphate buffer concentrate to yield a pH of 7.5 and a final concentration of 0.2 M NaCl, 18 mM Na_2HPO_4/NaH_2PO_4. Check the pH by using a micropipette to remove a small sample to pH paper and adjust the pH to 7.5 if necessary.

5. Add 0.5 ml, or a total of 5 mCi, of [^{3}H]acetic anhydride over a period of 5–10 min and stir the reaction mixture vigorously during this period. Allow the reaction to proceed under a fume hood for 30 min at 4°C.

6. Dialyse the solution against three changes of 13 mM HCl (total volume of 5000 ml) with changes every day until the dialysable radioactivity falls below 1000 c.p.m./ml.

7. Further purify the labelled collagen by performing at least two cycles of acid solubilization and precipitation.

8. Separate into aliquots of 5000 c.p.m. and store ^{3}H-labelled collagen type I at -20°C.

Protocol 3B. ³H-labelled collagen (type I) assay—the assay

Reagents

- Conditioned medium or cell extract
- ³H-labelled collagen type I
- 250 mM Tris–HCl, pH 7.4, containing 1 M NaCl and 25 mM CaCl₂ (5 × concentrated buffer)
- 10 mM APMA dissolved in 50 mM NaOH (made fresh daily)

Method

1. Bring ³H-labelled collagen type I (5000 c.p.m./tube) to neutral pH by mixing 4:1 (v/v) with 5 × concentrated buffer to yield a final incubation buffer concentration of 50 mM Tris–HCl, pH 7.4, 0.2 M NaCl, 5 mM CaCl₂.

2. Check the pH by using a micropipette to remove a small sample to pH paper and adjust to 7.4 if necessary.

3. Add a sample of the conditioned medium or cell extract to tubes containing ³H-labelled collagen and incubate the sample for 1–16 h at ambient temperature. Assay the sample in the presence of 1 mM APMA to determine the effect of proenzyme activation on collagenase activity and in the presence of 0.3–1 mM 1,10-phenanthroline and 1–10 mM PMSF to determine the contribution of the degradative action of metallo-and serine proteinases, respectively (see *Table 1*). Digest the ³H-labelled collagen with 1 µg/ml trypsin to estimate the per cent of denatured collagen.

4. Stop the reaction by adding 20 mM EDTA or 5 mM 1,10-phenanthroline.

5. Mix the samples with one volume of *p*-dioxane and chill on ice for 10 min.

6. Centrifuge the solutions (10 000 *g*, 10 min, 4 °C).

7. Determine radioactivity in samples of the supernatants in a beta spectrometer.

3.3 Assay of type IV collagenases

The degradation of type IV collagen is carried out by specialized enzymes, of which the most widely studied are the 72 and 92 kDa collagenases. These proteinases have a broader specificity than implied by their names; they can degrade collagens of types V, VII, and X as well as fibronectin, denatured collagens, and casein. However, they are usually assayed by using ³H-labelled collagen type IV (14).

Protocol 4. ³H-labelled collagen (type IV) assay

Reagents

- [³H]propionyl-labelled human type IV collagen (DuPont–New England Nuclear)
- 50 mM Tris–HCl, pH 7.5, containing 0.2 M NaCl and 10 mM CaCl₂ (assay buffer)
- 10% Trichloroacetic acid (TCA) containing 0.5% tannic acid

Method

1. Dissolve the substrate (3000 c.p.m.) in assay buffer and add the sample to be analysed (conditioned medium, cell extracts) to a total volume of 650 μl. Assay the sample in the presence of 1 mM APMA to determine the effect of proenzyme activation on collagenase activity and in the presence of 0.3–1 mM 1,10-phenanthroline and 1–10 mM PMSF to determine the contribution of the degradative action of metallo- and serine proteinases, respectively (*Table 1*). As controls for specificity incubate one sample of the labelled substrate with trypsin and another with interstitial collagenase.

2. Incubate at 30 °C for 18 h or at 37 °C for 4 h.

3. Terminate the assay by adding, in sequence, 20 μl bovine serum albumin (1 mg/ml) and 100 μl 10% trichloroacetic acid containing 0.5% tannic acid.

4. Incubate on ice for 30 min and centrifuge at 5000 *g* for 15 min at 4 °C.

5. Determine the radioactivity in samples of the supernatant in a beta spectrometer.

(**Note**: The products may also be visualized after SDS-PAGE by fluorography.)

---

4. Chromophores

The use of chromophores to assay proteinase activity permits better standardization of the assays. There is less variation in data acquired from different experiments than with radiolabelled substrates, allowing for meaningful comparison of these results. As with radiolabelled substrates, it is possible to assess the ratio of proteinase activity to inhibitor activity, and thereby to determine the net activity present in the sample. The detection limits range from relatively insensitive (azo-conjugated substrates) to extremely sensitive (thiopeptolide substrates).

4.1 Azo-conjugated substrate

Azocoll is a dyed hide powder consisting largely of insoluble gelatin. Azocasein is a dyed milk protein. The former is a better substrate for metalloproteinases, such as the gelatinases and collagenases, and the latter is better for stromelysin and serine proteinases, such as plasmin. The solubilization of peptides is followed by the release of red dye.

Protocol 5. Assays with azocoll or azocasein

Equipment and reagents

- Azocoll (Calbiochem)
- Azocasein (Calbiochem)
- 50 mM Tris–HCl buffer, pH 7.6, containing 200 mM NaCl, 10 mM CaCl$_2$
- APMA
- 1,10-Phenanthroline
- PMSF
- Assay tubes

Method

1. Suspend azocoll or azocasein at 1 mg/ml in 50 mM Tris–HCl buffer, pH 7.6, containing 200 mM NaCl, 10 mM CaCl$_2$.

2. Add 2 ml of the suspension to each assay tube.

3. Add 0.4 ml of enzyme sample (conditioned medium, cell extract).

4. Also assay the sample in the presence of 1 mM APMA to determine the effect of proenzyme activation on collagenase activity and in the presence of 0.3–1 mM 1,10-phenanthroline and 1–10 mM PMSF to determine the contribution of the degradative action of metallo- and serine proteinases, respectively.

5. Measure total available substrate by using trypsin.

6. Incubate for 14–20 h at 37°C with gentle agitation. At the end of the incubation period, add 100 µl of 1 M sodium acetate buffer, pH 5.0, to stop the reaction and to reduce interference due to Phenol red in samples of culture medium.

7. Centrifuge the tubes at 2000 g for 10 min at 4°C and measure the A_{520} of the supernatants.

8. Express activity as units: 1 U of neutral proteinase hydrolyses 1 mg of azocoll or azocasein under the given experimental conditions.

4.2 Thiopeptolide substrate assay of matrix-degrading metalloproteinases

The assay described in *Protocol 6* was adapted from that of Weingarten *et al.* (15). The short, modified peptide thiopeptolide, only recently commercially available, offers excellent sensitivity for the assay of ECM-degrading metalloproteinases. For example, 5–25 µl of conditioned medium from cultured human fibroblasts gives an excellent response. However, there is no way to determine which of the metalloproteinases is responsible for the observed activity since all give a nearly identical response. In addition, serine proteinases, such as the plasminogen activators, do not degrade this substrate.

Protocol 6. Thiopeptolide assay

Reagents

- Thiopeptolide: Ac–Pro–Leu–Gly–
  SCH[CH$_2$CH(CH$_3$)$_2$]CO–Leu–Gly–OC$_2$H$_5$
  (Bachem, Philadelphia, PA)
- Absolute ethanol
- 10 mg/ml trypsin (type I from bovine pan-
  creas; Sigma) dissolved in 1 mM HCl

- Soybean trypsin inhibitor
- 4,4'-Dithiodipyridine
- 50 mM Hepes buffer, pH 7.0, containing
  10 mM CaCl$_2$

Method

1. Assay conditioned medium without concentration. If medium contains serum, include a control of the medium alone. Prepare cell extracts as described in *Protocol 1*. Remove SDS from the sample by dialysing against three changes, 20 vol. each, of PBS at 4°C over 24 h.

2. Activate some samples by incubating 100 μl of conditioned medium or cell extract with 1 μl of trypsin solution (10 mg/ml in 1 mM HCl) for 30 min at 37°C; stop the reaction by adding 20 μl of soybean trypsin inhibitor (5 mg/ml in 0.05 M Tris–HCl with 0.01 M CaCl$_2$, pH 7.5). (**Note**: APMA interferes with the assay but can be used if it is dialysed out before the assay is begun. In this case the sample is activated by incubating with 1 mM APMA for 30 min, then dialysed at 4°C over 24 h against three changes, 20 vol. each, of PBS.)

3. Verify that the activity reflects that of metalloproteinases by pre-incubating (10 min) some samples with 0.3–1 mM 1,10-phenanthroline.

4. Prepare a 1.6 mM stock solution of thiopeptolide by dissolving in absolute ethanol. (**Note:** This stock solution can be stored at − 20°C.)

5. Dilute stock solution 1/10 (v/v) using 50 mM Hepes buffer, pH 7.0.

6. Prepare the reaction mixture consisting of 40 μM thiopeptolide, 800 μM 4, 4'-dithiodipyridine in 50 mM Hepes buffer, pH 7.0, containing 10 mM CaCl$_2$.

7. Start the assay by adding 5–25 μl of conditioned medium or cell extract to 95–75 μl of reaction mixture in a microcuvette at ambient temperature, and start timing the reaction. (**Note:** Volumes may be multiplied proportionately to perform the assay in larger-volume cuvettes if sensitivity is not a consideration.)

8. Read the A_{324} at 5 min intervals for 30 min.

5. Assays using extracellular matrices (ECMs)

Assays that employ actual ECMs create *in vitro* environments that mimic those encountered *in vivo*. For example, cells can be plated on radiolabelled

ECMs and degradative activity can be monitored as a function of the release of labelled components into the medium (see *Protocol 7*). Alternatively, the basement membrane-like matrix secreted by the EHS tumour can also be used to assay invasion (see *Protocol 8*). This matrix offers the added advantage of promoting the differentiation of certain cells, including mammary epithelial cells, keratinocytes, endothelial cells, and placental trophoblasts (16).

5.1 Radiolabelled ECMs

Matrices produced by several cell types have been isolated *in vitro*. The character of the matrix or matrices chosen should resemble that encountered by cells in the biological process under study. The teratocarcinoma (PF-HR9, PYS-2) and bovine corneal endothelial cell matrices resemble basement membranes and contain relatively large amounts of laminin and type IV collagen. In contrast, the R22 vascular smooth muscle cell matrix contains collagens I and III and resembles a blood vessel matrix. Because the composition of these matrices and the degree of cross-linking varies for each cell type, no one isolation method can be uniformly applied.

Protocol 7A. Monitoring extracellular matrix degradation by using ^{3}H-labelled PF-HR9 matrix (17)—Preparation of fibronectin-coated coverslips

Reagents

- 20% H_2SO_4
- 0.1 M NaOH
- Silane (γ-aminopropyltriethoxysilane, Suppelco)
- 0.25% glutaraldehyde in PBS
- Fibronectin (Collaborative Research)
- Ca^{2+}- and Mg^{2+}-free PBS (CMF-PBS)
- 70% ethanol

Method

1. Place coverslips in 20% H_2SO_4 overnight (use forceps to transfer).

2. Wash twice in double-distilled H_2O (1 min per wash).

3. Wash once in 0.1 M NaOH (1 min).

4. Wash twice in distilled H_2O (1 min per wash) and blot dry. (**Note:** Coverslips and holder must be completely dry before going on to the next step.)

5. Place rack in silane (undiluted γ-aminopropyltriethoxysilane). Cover completely and incubate for 4 min at ambient temperature. (**Note:** The silane solution is reusable only if water is excluded. This is easily monitored since the solution turns cloudy if any water is present.)

6. Wash with double-distilled H_2O (1 min).

7. Wash with CMF-PBS (1 min).

Protocol 7A. *Continued*

8. Cover with 0.25% glutaraldehyde in CMF-PBS for 30 min. (**Caution!** Glutaraldehyde should be used only in a fume hood).

9. Wash three times with CMF-PBS (1 min per wash) and blot dry.

10. Incubate in a solution of 15 µg/ml fibronectin in CMF-PBS, pH 7.2, for 1 h.

11. Transfer coverslips to racks.

12. Place racks in a solution of 70% ethanol for 2–18 h to sterilize coverslips.

13. Under sterile conditions, place coverslips, fibronectin side facing up, in 24-well plates, wash with five changes of CMF-PBS and leave 0.5 ml in each well.

14. Wrap the sides of the plates with Parafilm and store at 4°C in CMF-PBS.

(**Note:** Alternatively, tissue culture wells may be coated for 1 h with fibronectin, washed in CMF-PBS and stored as described for coverslips.)

Protocol 7B. Monitoring extracellular matrix degradation by using ^{3}H-labelled PF-HR9 matrix (17)—Plating and radiolabelling PF-HR9 cells

Reagents

- Dulbecco's modified Eagle's medium (DMEM) H-16 containing 5% FBS
- DMEM H-16 containing 2.5% FBS
- 1 mg ascorbic acid in 1 ml double-distilled H_2O
- [^{3}H]glucosamine, [^{3}H]proline, or [^{3}H]leucine

Method

1. Plate PF-HR9 cells in DMEM H-16 containing 5% FBS on fibronectin-coated coverslips at a density of 5×10^5 cells/22 mm^2 well. (**Note:** Cells can also be plated directly on to matrices produced in tissue culture wells.)

2. Grow the cells until they are confluent.

3. After the cells reach confluence, change the medium to DMEM H-16 containing 2.5% FBS.

4. Allow cells to produce matrix for 10 days. During this time, change the medium every other day. Add fresh ascorbic acid (final concentration of 50 µg/ml of medium) every day.

5. To radiolabel the cells, add 10 μCi/ml [^{3}H]glucosamine, [^{3}H]proline, or [^{3}H]leucine to medium on day 6.

6. Change to unlabelled medium on day 8.

7. Harvest the matrix on day 10.

Protocol 7C. Monitoring extracellular matrix degradation by using ^{3}H-labelled PF-HR9 matrix (17)—Harvesting the matrix

Reagents

- Hypo buffer: 10 mM Tris–HCl, pH 7.5, containing 0.1% bovine serum albumin and 0.1 mM CaCl$_2$
- NP-40 hypo buffer: same as hypo buffer except add 0.25% Nonidet P-40 (NP-40)
- Deoxycholate buffer: 50 mM Tris–HCl, pH 7.5, containing 0.1% deoxycholate (1 g/l)
- PBS containing gentamicin (50 μg/ml)

Method

1. Aspirate the culture medium.

2. Wash cells three times with hypo buffer. Let the third wash stand for 2–3 min.

3. Wash twice (2 min each) with NP-40 hypo buffer.

4. Wash twice with deoxycholate buffer.

5. Wash twice with PBS.

6. Cover each well with PBS containing gentamicin. Wrap plates with Parafilm and store at 4 °C for up to 6 weeks.

Protocol 7D. Monitoring extracellular matrix degradation by using ^{3}H-labelled PF-HR9 matrix (17)—Assessment of matrix degradation

Reagents

- PBS
- Tissue culture medium

Method

1. Wash the ^{3}H-labelled matrices with PBS.

2. Plate cells on the matrices.

3. Examine cell cultures daily by using an inverted phase-contrast microscope and keep a record by taking representative photographs.

Protocol 7D. *Continued*

4. Monitor the release into the medium of radiolabelled components by removing a sample (100–200 μl) of medium and determining radioactivity in a beta spectrometer. Replace with an identical volume of fresh medium.

5. To estimate the rate of spontaneous release of radiolabelled matrix components, some wells should contain radiolabelled matrix and tissue culture medium but no cells.

Note: The size of the ^{3}H-labelled components released into the medium can also be estimated by gel filtration chromatography. Acidify the sample and chromatograph on a 10 × 1 cm column of Sephadex G-50 (fine) eluted with 0.1 M pyridine acetate buffer, pH 5.3. Collect 0.5 ml fractions and determine radioactivity in a beta spectrometer.

5.2 Invasion assays using Matrigel

The use of three-dimensional substrates formed from Matrigel to study invasion has several important advantages over the matrices described in the preceding section. First, the EHS matrix (18) is commercially available (Collaborative Research). However, it is important to note that individual lots of this matrix vary widely in protein composition. In general, we use only batches containing at least 10 mg/ml protein. Another problem has been that the growth factors present in the matrix could have important effects on the cultured cells. If this is a significant issue, then these factors (for example epidermal growth factor, transforming growth factor α) can be removed by extraction with ammonium sulfate (19). Second, the EHS matrix does not gel until it is warmed to 37°C. Thus, substances (for instances inhibitors) can be dissolved while the Matrigel is in a liquid state. This solves the problem of delivering reagents to the underside of an invasive cell whose contacts with the matrix may not otherwise be accessible. Third, plugs of Matrigel, like tissues, can be fixed and sectioned to assess the degree of invasion morphologically (see *Protocol 8*). Fourth, by coating nucleopore filters with Matrigel it is possible to quantify the invasion process (see *Protocol 9*). Finally, these assays are easily adapted to assess invasion of other types of three-dimensional matrices, such as those constructed from type I collagen (for example Vitrogen, Collagen Corporation).

Protocol 8. Assessing Matrigel invasion by light microscopy (20)

Reagents

- 00 BEEM capsules (Polysciences Inc.)
- Matrigel
- 3% paraformaldehyde in 0.1 M sodium cacodylate buffer, pH 7.4
- 50%, 95%, and 100% acetone
- JB-4 solutions A and B (Polysciences Inc.)
- 50% acetone/50% JB-4 Solution A
- Benzoyl peroxide

Method

1. Cut off the top portion of the BEEM capsule and sterilize the remainder by ultraviolet irradiation.

2. Pour 325 µl of Matrigel into the capsule and gel for 30 min at 37 °C.

3. Plate cells on the plugs.

4. At various intervals (for example 24, 48, and 72 h), fix the Matrigel plugs with 3% paraformaldehyde in 0.1 M sodium cacodylate buffer, pH 7.4, for 30 min.

5. Gently cut the cap off the bottom using a razor blade, push the plugs out from the capsules with a glass rod and fix for an additional 30 min.

6. Wash the plugs in several changes of 0.1 M sodium cacodylate buffer, pH 7.4.

7. Dehydrate in acetone and infiltrate with JB-4 monomer:

    (a) 50% acetone, 2 × 15 min

    (b) 65% acetone, 2 × 15 min

    (c) 100% acetone, 2 × 30 min

    (d) 50% acetone/50% JB-4 monomer (Solution A), 1 × 30 min

    (e) JB-4 monomer, 1 × 10 min

    (f) JB-4 monomer, overnight

8. Place the plug in moulds containing an embedding medium of JB-4 solution A (0.5 ml), benzoyl peroxide (2.25 mg) and JB-4 Solution B (12.5 µl). Polymerize under vacuum (15–20 mm Hg) at 4 °C for 12 h.

9. Cut 2-µm sections on a JB-4 microtome, float on a water bath, transfer to coverslips, and air dry at ambient temperature.

10. Stain with haematoxylin and eosin to demonstrate morphology or with more specialized reagents if the protocol is compatible with JB-4.

To assess more precisely the ability of various agents (for instance antibodies, protein or chemical inhibitors) to perturb invasion *in vitro*, a system for quantifying this process is necessary (diagrammed in *Figure 3*). A key element of using the method described in *Protocol 9* is in determining when invasive cells are first visible on the underside of the filter. Confining the assay to this initial time point affords the greatest discrimination, since cells that migrate passively through channels formed by the most highly invasive cells are not scored. By choosing times that are less than one cell cycle the problem of scoring dividing cells on the bottom of the filters is also eliminated. Acquiring the data by scanning electron microscopy (SEM) is labour intensive but can be applied in all situations, including the culture of primary cells where non-uniform radiolabelling of cells is a problem. In addition, when

Susan J. Fisher and Zena Werb

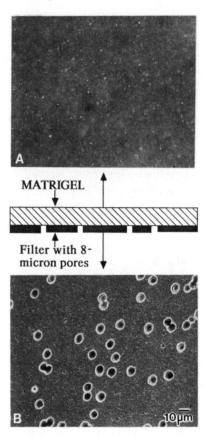

Figure 3. A system for quantifying basement membrane invasion. A polycarbonate filter containing 8 µm pores was coated with Matrigel. SEM of the top of the filter showed a smooth amorphous coating through which no pores were visible. Thus, the Matrigel coating formed a continuous barrier through which the invading cells must pass to reach the underside of the filter. In contrast, SEM of the lower surface of the filter showed the pores, but no Matrigel coating. (Reproduced with permission from the Rockefeller University Press.)

mixed populations of morphologically distinguishable cell types are present, their invasive potential can be distinguished by SEM. If cell lines are used, then radiolabelling is the method of choice.

Protocol 9. Quantification of Matrigel invasion by using scanning electron microscopy (20)

Equipment and reagents

- Matrigel
- 2.5% glutaraldehyde in 0.1 M sodium cacodylate buffer, pH 7.4
- 6.5 mm Transwell filter inserts
- 50%, 70%, 85%, 90%, 95%, and 100% ethanol solutions

284

Method

1. Add 10 μl of Matrigel to 6.5 mm Transwell inserts with polycarbonate filters containing 8 μm pores.

2. Incubate at 37 °C for 30 min to allow the matrix to gel.

3. Plate approximately 2×10^5 cells in 200 μl of medium on to the coated filters, adding medium to cover the bottom of the filter as well.

4. Incubate the cultures for approximately 24 h or until invasion is first evident.

5. Wash twice with medium and fix with 2.5% glutaraldehyde in 0.1 M sodium cacodylate buffer, pH 7.4, for 30 min.

6. Rinse the samples in cacodylate buffer.

7. Pass the filters (5 min each time) through increasing ethanol con-centrations (50%, 70%, 85%, 90%, 95%) up to 100% (two changes, 5 min each).

8. While still attached to the inserts, the samples are critical point dried. At this point the top portion of the insert can be cut off (do not disturb the attached filter!). This allows more samples to be dried at one time.

9. Cut the filters off the inserts with a scalpel blade and mount with either the upper or lower surface exposed for coating.

10. Sputter-coat the filters with a 20–25 nm layer of gold palladium.

11. Examine in a scanning electron microscope (*Figure 4*).

12. Take 10 representative photographs (150 ×) of each filter underside in a systematic fashion from all sectors, avoid the edges of the filters where there may be a meniscus effect on the matrix covering. For each condition tested, examine at least two filters.

13. Overlay the photographs with a grid consisting of points forming equilateral triangles, and count the number of cells that overlap with grid points. (**Note**: A sample grid can be photocopied from Weibel and Bolender (21).)

14. Express the data as the percentage of grid points that overlap with invading cells.

Note: Alternatively, the samples can be fixed as described above, washed in cacodylate buffer, and invasion quantified by wet SEM.

Acknowledgements

This work was supported by grants from the National Institutes of Health (HD 26732, HD 23539, HD 24180, and P50 DE 10306) and by a contract from the Office of Health and Environmental Research, US Department of Energy (DE-AC03-76-SF01012).

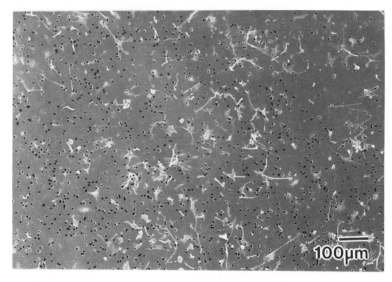

Figure 4. Scanning electron microscopy of invading cells. Human trophoblasts were plated on the top of matrix-coated filters. Invasion was assessed by morphometric analysis of wet scanning electron micrographs of the undersides of the filters (see also *Figure 3*).

References

1. Alexander, C. M. and Werb, Z. (1989). *Curr. Opin. Cell Biol.*, **1**, 974.
2. Matrisian, L. (1990). *Trends Genet.*, **6**, 121.
3. Saksela, O. and Rifkin, D. B. (1988). *Annu. Rev. Cell Biol.*, **4**, 93.
4. Barrett, A. J. (1986). In *Proteinase inhibitors* (ed. A. J. Barrett and G. O. Salvesen), p. 3. Elsevier, Amsterdam.
5. Laemmli, U. K. (1970). *Nature*, **227**, 680.
6. Herron, G. S., Banda, M. J., Clark, E. J., Gavrilovic, J., and Werb, Z. (1986). *J. Biol. Chem.*, **261**, 2814.
7. Heussen, C. and Dowdle, E. B. (1980). *Anal. Biochem.*, **102**, 196.
8. Guntenhoener, M., Pogrell, A., and Stern, B. (1982). *Matrix*, **12**, 388.
9. Lowry, O. H., Rosebrough, N. J., Farr, A., and Randall, R. J. (1951). *J. Biol. Chem.*, **193**, 265.
10. Fisher, S. J., Cui, T.-y., Zhang, L., Hartman, L., Grahl, K. T., Zhang, G.-Y., Tarpey, J. F., and Damsky, C. H. (1989). *J. Cell Biol.*, **109**, 891.
11. Vassalli, J.-D., Dayer, J. M., Wohlwend, A., and Belin, D. (1984). *J. Exp. Med.*, **159**, 1653.
12. Unkeless, J. C., Tobia, A., Ossowski, L., Quigley, J. P., Rifkin, D. B., and Reich, E. (1973). *J. Exp. Med.*, **137**, 85.
13. Birkedal-Hansen, H. (1987). In *Methods in enzymology* (ed. L. W. Cunningham), Vol. 144, p. 140. Academic Press, New York.
14. Stetler-Stevenson, W. G., Krutzsch, H. C., Wacher, M. P., Marguliies, I. M. K., and Liotta, L. A. (1989). *J. Biol. Chem.*, **264**, 1353.

12: The catabolism of extracellular matrix components

12: The catabolism of extracellular matrix components

15. Weingarten, H., Martin, R., and Feder, J. (1985). *Biochemistry*, **24**, 6730.
16. Stoker, A. W., Streuli, C. H., Matrins-Green, M., and Bissell, M. J. (1990). *Curr. Opin. Cell Biol.*, **2**, 864–74.
17. Fisher, S. J., Leitch, M. S., Cantor, M. S., Basbaum, C. B., and Kramer, R. H. (1985). *J. Cell. Biochem.*, **27**, 31.
18. Kleinman, H. K., McGarvey, M. L., Hassell, J. R., Star, V. L., Cannon, F. B., Laurie, G. W., and Martin, G. R. (1986). *Biochemistry*, **25**, 312.
19. Taub, M., Wang, Y., Szczesny, T. M., and Kleinman, H. K. (1990). *Proc. Natl. Acad. Sci. USA*, **87**, 4002.
20. Librach, C. L., Werb, Z., Fitzgerald, M. L., Chiu, K., Corwin, N. M., Esteves, R. A., Grobelny, D., Galardy, R., Damsky, C. H., and Fisher, S. J. (1991). *J. Cell. Biol.*, **113**, 437.
21. Weibel, E. and Bolender, R. (1973). In *Principles and techniques for electron microscopy* (ed. M. A. Hayat), p. 237. Van Nostrand Co., New York.

13

Use of extracellular matrix and its components in culture

HYNDA K. KLEINMAN, MAURA C. KIBBEY,
FRANCES B. CANNON, BENJAMIN S. WEEKS, and
DERRICK S. GRANT

1. Introduction

The interaction of cells with the extracellular matrix is an important determinant of cell behaviour (1). Extracellular matrices not only provide structural support but also elicit biological responses which are mediated by specific components of the matrix through interactions with cell surface receptors. Matrices contain collagens, proteoglycans, and glycoproteins. The type and amount of each component depend on the cell type with which it is associated. Our understanding of the biological activity of the extracellular matrix and its components has been enhanced by the ability to purify the components and to examine their effects on cells in culture. From these studies, it is clear that components such as collagen types I and IV, laminin, and fibronectin influence cell behaviour in a cell specific manner. Cells adhere, grow, migrate, and differentiate on matrix components, and, in many cases, do so better than on plastic or glass surfaces. Specific sites on matrix molecules are being defined for these processes with distinct domains, and synthetic peptides having been shown to possess biological activity (2). Specific cellular receptors for each matrix macromolecule have also been defined. This chapter focuses on the use of basement membrane preparations and the components laminin and collagen in cell culture (3, 4). Many of the techniques described herein can be directly adapted to the other matrix glycoproteins including fibronectin, entactin, vitronectin, thrombospondin, tenascin, osteopontin, and osteonectin.

2. Basement membranes

Basement membranes are specialized extracellular matrices which underlie epithelial and endothelial cells and surround muscle, fat, and the entire nervous system (3). Some of their diverse functions include:

(a) separating the epithelium from the underlying stromal matrix;

(b) supporting epithelium, regulating the passage of macromolecules in the kidney and blood vessels;

(c) promoting the growth and differentiated phenotype of the contacting cells.

Basement membranes are composed of type IV collagen, perlecan (a heparan sulfate proteoglycan), and the glycoproteins laminin, entactin, and sometimes fibronectin (see Chapters 1, 4, 7, 10, and 14). With the exception of fibronectin, these components are usually isolated from the EHS (Engelbreth–Holm–Swarm) tumour grown in C57BL/6 mice and are chemically and antigenically identical to authentic basement membrane components. Although these components can be isolated from tissues such as placenta, heart, or from cell conditioned medium, the yields are generally poor, the purity is questionable, and proteases which modify the molecules are in some cases present. In general, all the commercially available basement membrane components and preparations (Matrigel, laminin, and type IV collagen) are isolated from the EHS tumour.

A reconstituted basement membrane, isolated from the EHS tumour termed Matrigel, has been found to promote the differentiation of many cell types *in vitro* (4, 5). This matrix is composed of laminin, entactin, collagen IV, and perlecan, which are isolated as a crude extract from the EHS tumour. Matrigel contains all these basement membrane components and is a solution at 4°C, but when the temperature is raised to 24–37°C, the components interact with each other and within 1 hour form a polymerized gel. The gel can be used as a substratum for cells or, alternatively, the cells can be mixed at 4°C with the Matrigel and then either plated on the dish or injected into animals. When tumour cells are mixed with Matrigel and injected into mice, tumours rapidly grow. It increases both the incidence and the rate of tumour growth from primary as well as established cell lines.

Various studies have shown that cells are more differentiated when in contact with a basement membrane (*Table 1*). Endothelial cells form capillary-like structures with a lumen within 12 hours when plated on a surface of Matrigel. Mammary epithelial cells not only form ducts but also show an 80-fold increase in casein production, whereas on plastic, both endothelial cells and mammary cells form monolayers. Various other glandular cells such as thyroid, salivary, bile, Sertoli, and oviduct cells, show increased differentiation on Matrigel. Neural cells are very responsive to Matrigel and it has been used in the brain to promote graft survival. *In vitro*, cells such as dorsal root ganglia not only show extensive process outgrowth but also production of myelin within 4 days. Other neuronal cells such as neuroblasts from the olfactory epithelium require Matrigel for survival and differentiation and show significant responses *in vitro*. Thus, epithelial cells which normally contact a basement membrane are more differentiated *in vitro* when in con-

Table 1. Effect of reconstituted basement membrane on cells and organ differentiation

Cell line	Biological effect
Sertoli cells	Become 15- to 20-fold more columnar; germ cell survival and differentiation
Dorsal root ganglion	Outgrowth and myelination
Melanoma cells	Rapid pigmentation
Neuroblast from olfactory epithelium	Odorant response to chemicals
Pancreatic acinar cells	Formation of aggregates with secretory vesicles near lumen
Salivary gland cells	Alignment into networks with cell clusters
Submandibular gland tumour (A253)	Formation of glands and ducts
Oviduct cells	Formation of tubes with a lumen with polarized secretory cells
Mammary epithelial cells	Increased casein production; formation of ducts, ductules, and lumina
Kidney cells	Tubular cells form tubes; glomerular cells grow in clumps
Hair follicle	Shaft formation
Thyroid cells	Thyroglobulin production
Endothelial cells	Formation of capillary networks of tube-like structures
Type II pneumocytes	Cuboidal cells and maintenance of phosphatidylcholine synthesis
Trophoblasts	Invasion with matrix degradation
Rat bile duct cells	Formation of acinar-like structures
Hepatocytes	Maintenance of albumen synthesis and cytochrome P450
Bone cells	Canaliculi-like structures with increased mineral matrix production

tact with Matrigel. In addition, many of the cells have a reduced growth rate. A typical response of kidney epithelial cells to culture on Matrigel is shown in *Figure 1*.

2.1 Growth of EHS tumour

This tumour is maintained by passage in mice. The tumour is routinely grown in various laboratories and can be obtained from ATCC, Rockville, MD, and shipped by overnight delivery in medium. Upon receipt of the tissue, it must be processed immediately and injected into mice. All manipulations must be performed at 4°C.

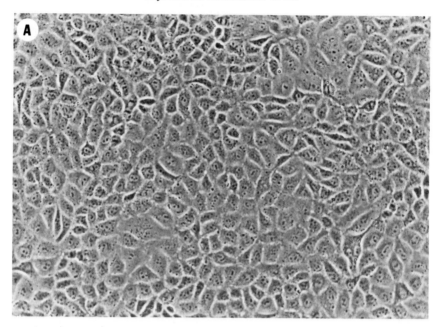

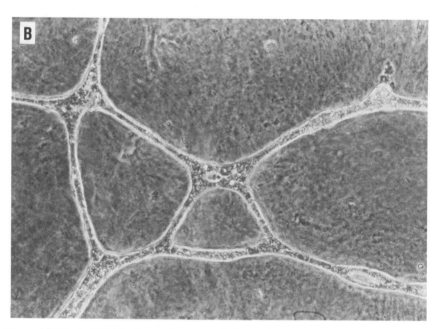

Figure 1. Effect of Matrigel on culture morphology of normal rat kidney epithelial cells. Normal rat kidney epithelial cells, clone NRK52E (11) were plated and grown on plastic (Panel A) or Matrigel (Panel B) using the procedure detailed in *Protocol 4.* Magnifications: Panel A, 125 ×; Panel B, 250 ×.

Protocol 1. Transplantation of EHS sarcoma in mice

1. Rinse tumour tissue with phosphate-buffered saline (PBS) containing penicillin (50 μg/ml) and streptomycin (50 units/ml) (Gibco).
2. Aspirate and flush tumour initially through a syringe and subsequently through a syringe with a 16-gauge needle to disperse the cells.
3. Allow the tissue to settle or pellet at low speed (100 g for 1 min at room temperature) and decant supernatant fraction.
4. Inject using a 16-gauge needle 0.5 ml of settled volume in to each site subcutaneously into the backs or intramuscularly into the hind limbs of 6-8 week-old C57BL/6 mice.
5. After 2–3 weeks, the mice develop large tumours and begin to die. Tumours should then be harvested and either used immediately, frozen in liquid nitrogen and stored at − 70 °C, or passaged into mice for continuous growth.

2.1.1 Additional considerations in tumour growth

It may be necessary to prepare lathyritic tissue for the isolation of type IV collagen or collagen-enriched Matrigel. One week after tumour injection, mice are fed ground chow diet containing 20 g/kg BAPN and the monoamine oxidase inhibitors (which prevent the breakdown of the BAPN) iproniazid phosphate (0.4 g/kg) and pargyline hydrochloride (0.04 g/kg) (from Sigma). All weighed materials are dissolved in 300 ml H_2O and mixed in a diet mixer for 15–30 min. Other facts concerning tumour growth are:

(a) One tumour generally is passaged into 10 mice.
(b) Unused cells can be frozen in medium plus 5% DMSO.
(c) Generally, the frozen material does not grow well.
(d) If tumours get too large (that is greater than 4 g), necrotic foci and infections develop.

2.2 Matrigel
2.2.1 Preparation of Matrigel

Protocol 2. Preparation of Matrigel

Equipment and reagents

- Polytron homogenizer
- Chloroform
- High-salt buffer with protease inhibitors: 0.05 M Tris–HCl, pH 7.4, containing 3.4 M NaCl, 4 μm EDTA, 2 μm NEM
- Urea buffer: 0.15 M NaCl, 0.05 M Tris–HCl, pH 7.4
- Tris-buffered saline (TBS): 0.05 M Tris–HCl, pH 7.4, 0.15 M NaCl

---

Protocol 2. *Continued*

Method

1. Perform all manipulations at 4 °C.

2. Homogenize or Polytron the tumour (1 g/3 ml high salt buffer with protease inhibitors) until dispersed in high-salt buffer containing protease inhibitors.

3. Centrifuge the mixture at 7000 *g* for 15 min. Discard the supernate.

4. Suspend Polytron pellets in the high-salt buffer with protease inhibitors (approx. the same total volume of buffer as step 2).

5. Centrifuge the mixture at 7000 *g* for 15 min. Discard the supernate.

6. Repeat steps 4 and 5.

7. Suspend Polytron pellets in 1 g/ml of urea buffer and stir overnight at 4 °C.

8. Centrifuge mixture at 14 000 *g* for 20 min. Save supernate.

9. Homogenize pellets in 1 g/0.5 ml urea buffer.

10. Centrifuge mixture at 14 000 *g* for 20 min. Save supernate.

11. Combine supernate from steps 8 and 10 and dialyse against at least 10 volumes of Tris-buffered saline (TBS) containing 5 ml chloroform per litre. A glass container or cylinder must be used. (This is a sterilization step.)

12. After 2–10 h, remove dialysis bags, rotate to disperse contents, and continue dialysis for at least 2 h against at least 10 volumes of TBS lacking chloroform for at least 3 changes.

13. Perform the final dialysis against at least 10 volumes of media salts such as MEM, DMEM, or medium 199, this is usually done overnight.

14. The contents inside of the dialysis bag are sterile, therefore empty in a sterile hood using sterile instruments and a sterile receptacle. Rinse the outside of the bag with alcohol, cut open at one end (held with forceps or a haemostat), and pour the contents carefully into a large vessel on ice.

15. Perform a protein determination and dilute Matrigel with additional medium to a final protein concentration 10–15 mg/ml.

16. Aliquot and store the suspension at − 20 °C. The preparation is stable for at least 1 year.

2.2.2 Additional considerations

(a) The procedure takes 3 days.

(b) Approximately 5 g of tumour (equal to one mouse) yields 7 ml of 10 mg/ml Matrigel.

(c) Matrigel is commercially available from Collaborative Biomedical Products.

(d) It can be refrozen if all the aliquot is not needed.

(e) Matrigel should not be used dried. Too much salt accumulates on the dish.

2.2.3 Preparation of growth factor-reduced Matrigel

Protocol 3. Preparation of growth factor-reduced Matrigel

1. Perform the steps for the preparation of regular Matrigel as described in *Protocol 2* to step 11. Dialyse for at least 2 h against at least 10 volumes of TBS for 3 changes. No chloroform is needed.

2. Empty the dialysis bags and add 20% (w/v) $(NH_4)_2SO_4$ (16.4 g/100 ml) slowly with stirring until all the $(NH_4)_2SO_4$ is dissolved. Stir for 1 h.

3. Collect the precipitate by centrifugation at 7000 *g* for 15 min. Discard supernate.

4. Homogenize pellets with a Polytron in the original volume with TBS and stir until in solution.

5. Dialyse the mixture for at least 2 h against 2 changes of 10 volumes of TBS.

6. Repeat steps 2 and 3.

7. Polytron pellets in half of the original volume with TBS and stir until in solution.

8. Dialyse the mixture for at least 2 h against at least 10 volumes of TBS containing chloroform (5.0 ml/l).

9. Treat dialysed material exactly as described in steps 11–15 in *Protocol 2*.

2.3 Use of Matrigel in cell culture

2.3.1 General

Because Matrigel polymerizes at 24–37°C, it must be thawed slowly at 4°C. This can be done by constantly rotating it in your hands or by placing it in a refrigerator overnight. Matrigel must be kept on ice at all times before use. The following two protocols describe the major ways Matrigel can be used as a culture substrate.

Protocol 4. Standard method for using Matrigel as a culture substrate

1. If using a 35 mm culture dish, place it in a sterile hood.

2. Swirl 500–700 µl of Matrigel immediately to cover the bottom of the dish.

Protocol 4. *Continued*

3. Cover and place in a 37 °C incubator and allow to polymerize for 30–60 min. Transport gently at all times.
4. Add medium and cells as usual for regular cell culture. For endothelial cells, usually 0.25 ml Matrigel/16 mm dish and 4×10^4 cells in 0.5 ml are added. Cultures can be observed within 6–12 h.

Alternative methods:

(a) *Thin coat*—pour Matrigel on to a precooled dish, swirl, and immediately pour off excess. Excess can be used on another dish. Generally, 3 ml of a 10 mg/ml solution can be used to coat 5–10 150 mm dishes.
(b) *Superthin*—add dilute Matrigel (2–3 mg/ml) and pour on to dish. Immediately pour excess from dish. Excess can be used on several other dishes.
(c) *Matribeach*—add Matrigel (1 ml/35 mm dish), swirl, and let it polymerize with dish raised at one end by placing an object under the dish. This will give Matrigel of different thicknesses.

Protocol 5. Cell culture within Matrigel

1. Mix cells and Matrigel together on ice. Cells should be in as small a volume of cold medium salts (no serum) as possible to prevent dilution of the Matrigel.
2. Add 1 ml of mixture/35 mm culture dish.
3. Cover and place in 37 °C incubator for 30–60 min. Transport gently at all times.
4. Add medium and incubate as usual. Structures will be inside the Matrigel and may have to be viewed by histological section.

Protocol 6. Matrigel *in vivo*

1. For each recipient mouse, mix freshly isolated cells (5×10^5 or greater) and Matrigel together in a final volume of 0.5 ml on ice. Cells should be in as small a volume as possible to prevent dilution of the Matrigel. Typically 250 μl of cold medium containing 2×10^6 cells/ml is mixed with 250 μl of 10 mg/ml Matrigel.
2. Inject mixture subcutaneously, using a 19-gauge needle for tissue samples and a 23-gauge for cultured cells, in the dorsal flank of nude (for human cells) or syngeneic mice. The injections should be done quickly to prevent the Matrigel from solidifying.
3. Rotate syringe when withdrawing to prevent leakage. The needles may need to be changed frequently because of blockages developing.

Protocol 7. Cell recovery from Matrigel

1. Remove medium and rinse gently with PBS at least twice to remove serum components.

2. Add 2.4% dispase (Boehringer–Mannheim) in medium salts and incubate at 37 °C until cells release. Typically 5–20 min is required.

3. Gently collect cells, add medium plus serum to inhibit the enzyme, and centrifuge for 5 min at 100 *g* at room temperature.

---

3. Use of laminin in cell cultures

Although all the components of basement membranes—laminin, entactin, type IV collagen, and heparan sulfate proteoglycan—have biological activity, the most intensely studied has been laminin (6, 7; see Chapters 1, 4, and 8). Initially described in 1979 as a basement membrane component, laminin has been subsequently shown to have numerous biological activities (*Table 2*),

---

Table 2. Biological activities of laminin

Cell adhesion
    Cells attachment to collagen IV
    Cell–cell aggregation
    Reduces fibroblast attachment

Cell morphology
    Induces Sertoli cell polarity
    Suppression of blebs
    Induces Schwann cell elongation
    Stimulates endothelial cell flattening and spreading

Cell growth
    Increased proliferation of epithelial cells, Schwann
      cells, and hepatocytes

Cell migration
    Chemotactic and haptotactic

Differentiation
    Induces embryoid body formation by F9 cells
    Teratocarcinoma cell differentiation
    Melanoma cell pigmentation
    Increases tyrosine hydroxylase
    Increases neural process formation

Metastases
    Promotes metastasis *in vivo*
    Promotes metastatic phenotype *in vitro*
    Stimulates collagenase IV activity

Phagocytosis
    Promotes phagocytosis

Table 3. Active peptides from laminin

Chain	Peptide	Residues	Activity
A	SIKVAV	2099–2105	Adhesion, migration, spreading, collagenase IV, metastases, neurite outgrowth
A	ALRGDNP	1118–1128	Adhesion, spreading, migration, metastases
B1	RYVVLPRPVCFEKGMNYTVR(F9)	641–660	Adhesion, heparin binding
B1	PDSGR	902–906	Adhesion, migration, ↓ metastases
B1	YIGSR	929–933	Adhesion, migration, ↓ differentiation, ↓ metastases
B2	RNIAEIIKDA	1542–1551	Neurite outgrowth

including promotion of cell adhesion, migration, growth, neurite outgrowth, differentiation, and metastases. Laminin is composed of three chains designated A (M_r = 400 000), B1 (M_r = 210 000), and B2 (M_r = 200 000) which are held together by disulfide bonds (see Chapters 1, 4). At least six active sites on laminin have been identified by synthetic peptides based on their biological activity in culture (*Table 3*). The activity of these peptides has been documented when the peptides are tested either as a solid substratum or in solution. Depending on the cell type, the peptides show different activities.

3.1 Use of laminin

Laminin is commercially available from several companies, including Collaborative Biomedical Products and Sigma. It is generally sold at 1 mg/ml in a sterile solution and can be used directly in culture (7). When first thawed it should be aliquotted and then stored at −70°C. It should not be lyophilized.

Protocol 8. Use of laminin as a culture substrate

1. Add laminin (1–10 μl/16 mm dish) to a dish containing 200 μl of serum-free medium. Incubate for 30–60 min to allow the laminin to bind.
2. Remove unbound material from the dish.
3. Add cells in medium and culture as usual.

3.1.1 Additional considerations

(a) Laminin can be used dried on the dish. It is usually diluted in water and dried to avoid salt build-up or rinsed before use with serum-free media to remove excess salts.

(b) For solution studies, laminin is added directly to the culture medium along with the cells and/or after the cells have been plated.

(c) This method can be used for coating dishes with other adhesion glyco-proteins including fibronectin, vitronectin, thrombospondin (see Chapters 1, 5, 7, and 8).

3.2 Use of laminin peptides in culture

3.2.1 General

Laminin peptides are commercially available from several companies, includ-ing Peninsular Labs and Bachem, Inc. They should appear as a white powder. All peptides should be in the amide form. The non-amide peptides are less active. If peptides are not readily soluble in water, a drop of acetic acid can be added. Peptides can be stored frozen at $-20°C$ or as a dry powder at room temperature.

Protocol 9. Use of laminin peptides as culture substrates

1. Add peptide (10–50 μg/16 mm dish in 100 μl serum-free medium) to culture dishes. Shake for several minutes and air dry with or without gentle shaking overnight. Falcon plastic works well.

2. Sterilize the plate by either exposure to UV for 2–12 h or rinsing in 70% alcohol.

3. Incubate dish for 15–30 min with 3% BSA in medium salts to block unoccupied sites.

4. Remove medium plus BSA and replace with regular medium contain-ing cells. Subsequent culture is performed using standard procedures.

3.2.2 Use of soluble laminin peptides in culture

Peptides are generally dissolved directly into medium at a concentration of 1–5 mg/ml. The pH of the medium should be checked and adjusted if necessary. Sonication can be used to solubilize peptide. Peptides can be filter-sterilized using a low protein-binding filter (for instance Millex GV; Milli-pore). Peptide solution is then added to the cells at the time of plating or at any subsequent time.

3.2.3 Additional considerations

(a) Peptides are very stable in the powdered form and relatively stable in solution. Precaution dictates that peptides in solution be stored frozen.

(b) Peptide-coated plates can be prepared in advance and stored until use at 4°C.

4. Collagen-coated dishes

4.1 Introduction

Collagen substrates have been used for many cells, and, in general, promote adhesion, growth, and/or differentiation (8–10). Type I collagen has most often been used because it is the most abundant (see Chapters 1 and 2). It can be used as a dried substrate, as a three-dimensional gel, or as a floating gel. Fibroblastic cells actually contract the gel to a very small area. Type I and V collagens are commercially available.

4.2 General considerations in using type I collagen

Type I collagen is sold as a 3–4 mg/ml solution (in dilute acetic acid) by Gibco, Collaborative Biomedical Products and Vitrogen. Sigma sells type I collagen as a lyophilized product. The lyophilized material can be solubilized at 5 mg/ml in 0.5 M HOAc at 4°C by stirring overnight. It can be sterilized by chloroform dialysis, as described for Matrigel, or by UV irradiation after drying on the dish. An alcohol rinse of the dried material on the culture dish is also effective. Collagen is generally not used as a wet substratum because of its acidic pH. Type V collagen can be obtained from Collaborative Biomedical Products or the Heyltex Corp.

Protocol 10. Use of collagen as a culture substrate

A. *Dried substrate method*

1. Add collagen (any type) to the dish (10–50 µg/35 mm dish) in the presence of water and air dry.
2. Sterilize as described above, and rinse with medium to neutralize pH.
3. Add cells as usual.

B. *Collagen gel method*

1. Neutralize collagen (type I only) at 3–5 mg/ml with 10 × PBS and NaOH while keeping on ice and place in a culture dish at 500–1000 µl/ 35 mm dish.
2. Warm to 37°C. Gelation occurs in 30–60 min.
3. Add cells and culture as usual.

C. *Floating gel method*

1. Allow type I collagen to gel.
2. Either mix the cells with the collagen after neutralization or add after the gel forms.
3. Release the gel from the dish by mechanical agitation.

4.3 Additional considerations

Sigma designates its collagen preparations by roman numerals. This is confusing because Roman numerals are used to distinguish the collagen types. Most of the collagens sold are actually type I. *One must read the descriptions carefully.*

5. Conclusions

In vivo cells are not normally in contact with plastic or glass. Extracellular matrix components have been found to promote better cell growth and to maintain the differentiated phenotype. While this chapter has suggested some substrata that work well, the conditions may vary for different cell types. The researcher is encouraged to vary the amount and type (mixtures work well) of matrix used to obtain the best desired effect.

References

1. Hay, E. D. (1991). In *Cell biology of the extracellular matrix* (2nd ed.), p. 419. Plenum Press, New York.
2. Yamada, K. M. (1991). *J. Biol. Chem.*, **266**, 12809.
3. Martin, G. R. and Timpl, R. (1987). *Annu. Rev. Cell. Biol.*, **3**, 57.
4. Kleinman, H. K., Graf, J., Iwamoto, Y., Kitten, G. T., Ogle, R. C., Sasaki, M., Yamada, Y., Martin, G. R., and Luckenbill-Edds, L. (1987). *NY Acad. Sci.*, **513**, 134.
5. Kleinman, H. K., McGarvey, M. L., Hassell, J. R., Star, V. L., Cannon, F. B., Laurie, G. W., and Martin, G. R. (1986). *Biochemistry*, **25**, 312.
6. Kleinman, H. K., Cannon, F. B., Laurie, G. W., Hassell, J. R., Aumailley, M., Terranova, V. P., Martin, G.R., and Dubois-Dalcq, M. (1985). *J. Cell. Biochem.*, **27**, 317.
7. Beck, K., Hunter, I., and Engel, J. (1990). *FASEB J.*, **4**, 148.
8. Kleinman, H. K., Luckenbill-Edds, L. L., Cannon, F. B., and Sephel, G. (1987). *Anal. Biochem.*, **166**, 1.
9. Kleinman, H. K., Klebe, R. J., and Martin, G. R. (1981). *J. Cell Biol.*, **88**, 473.
10. Vuorio, E. and de Crombrugghe, B. (1990). *Annu. Rev. Biochem.*, **59**, 837.
11. DeLarco, J. E. and Todao, G. J. (1978). *J. Cell Physiol.*, **94**, 335.

14

Immunological identification of extracellular matrix components in tissues and cultured cells

PETER S. AMENTA and ANTONIO MARTINEZ-HERNANDEZ

1. Introduction

Within the last two decades considerable advances have been made in the knowledge of extracellular matrix (ECM) composition and function. The initial concept of the ECM was that of an inert, support structure morphologically characterized by the presence of banded and beaded fibres, fibrils, filaments, granular material, and specialized matrices—the basement membranes. The isolation of the individual ECM components has led to the development of monospecific antibodies. One application of these antibodies has been for immunohistochemical studies. This approach provides information, unattainable by biochemical means, concerning the distribution of the individual ECM components. Immunohistochemistry can be performed at both the light and electron microscopic level. Light microscopy immunohistochemistry provides a preliminary *road map* of matrix distribution in tissues and organs. However, only electron microscopy with its superior resolution, can define the exact matrix–matrix and cell–matrix relationships. Cell culture studies have provided data on the synthetic capacity of individual cell types in culture; however, *in vitro* cell behaviour may not reflect the *in vivo* situation. Like *in situ* hybridization, immunohistochemical studies can define the synthetic profile of cells *in vivo* and *in vitro*. Localization of antigenic determinants within cellular synthetic organelles can be achieved at the electron microscopic level.

This chapter will review the specific methodologies for immunohistochemistry, as well as general considerations on fixation, embedding, prestaining treatment, and antibody usage.

2. Theoretical considerations

In any immunohistochemical study, the investigator faces the problem of preserving tissue structure, while simultaneously preserving antigenicity of the molecules to be studied. Unfortunately, these goals are often somewhat contradictory, for instance when optimal fixation is used to achieve optimal tissue preservation, antigenicity is often decreased.

Tissue preparation has as its goal the stabilization of tissue components. This can be achieved by either physical or chemical means. The most frequently used physical method of tissue preparation is snap-freezing. This is often done using methylbutane at liquid N_2 temperature. The advantage inherent to snap-freezing is that it provides the least amount of chemical modification, thus typically maximizing antigenic preservation; of course, structural preservation is often less than optimal. Microwave irradiation has been used to fix tissues for light microscopy, but is not commonly used for electron microscopy. Chemical stabilization is achieved with a variety of fixatives; some of the most commonly used fixatives in immunohistochemistry are described below.

2.1 Tissue preparation

2.1.1 Fixation

The choice of fixative must be suited for the experiment at hand, that is the tissue to be used and the antigen(s) to be preserved. Although there are innumerable fixatives available for general histologic procedures, few allow simultaneous antigenic and structural preservation. The most commonly used fixatives are cross-linking agents, but only a few of them have proven generally useful. For example, glutaraldehyde has limited use in immunohistochemical studies of ECM components due to the formation of non-reducible aldehyde groups and strong cross-links. Formaldehyde, on the other hand, alone or combined with lysine and periodate, has proven to be dependable for immunohistochemistry. With any fixation, perfusion is preferable to immersion, and a brief initial perfusion with buffer is recommended. In the case of immersion fixation, formaldehyde penetrates tissues at approximately 1 mm per hour. Therefore, tissue blocks should be cut at no greater than 1–2 mm in thickness. With richly vascularized organs, such as fetal liver or placenta, multiple changes of fixative are recommended to prevent quenching by blood (1). The total time of fixation is important. For instance with formaldehyde, fixation for longer than three hours often results in a decreased reaction of antibodies with ECM components. For periodate–lysine–paraformaldehyde (PLP) the method of McClean and Nakane is referenced (2).

2.1.2 Post-fixation treatments

After fixation of tissues, the next goal is the removal of any free fixative. Therefore, washing tissues subsequent to fixation is a critical step. In our laboratories, where formaldehyde fixation is used predominantly, multiple washes with PBS and sucrose are used. Sucrose, not only quenches the fixative, but is also a cryoprotectant. For example, a common artefact in many sections is a lack of staining at the centre, while at the periphery there is intense reaction. This artefact is often related to fixation and may be due to over- or under-fixation of the central portions of the blocks. The use of thick blocks (> 2 mm), results in inadequate fixation of the block centre, autolysis, and decreased reactivity. On the other hand, if fixative has been delivered, but not adequately removed (insufficient washing), the centre of the block may be *over-fixed*, resulting in extensive denaturation. We recommend washing tissue blocks overnight with multiple changes, using at least 500 ml of the PBS–sucrose mixture, followed by a one hour wash with PBS–sucrose and glycerol. Not only is this an adequate cryoprotectant, but it also results in uniform staining throughout the section.

2.2 Embedding

Most tissues lack enough consistency to permit adequate sectioning; therefore, embedding is necessary to obtain thin sections. Embedding usually involves the replacement of tissue water by a liquid that eventually can be hardened. Therefore, freezing tissues may be considered as an embedding procedure since it results in the *replacement* of liquid water by solid ice resulting in sufficient hardening to allow for sectioning at 4–6 μm. Ultra-cryomicrotomy (3) makes possible the sectioning of frozen tissues for use in electron microscopy immunohistochemistry (100 nm). It should be noted that antibodies have a limited capacity to penetrate tissues, therefore sections thicker than 12 μm should not be used for immunohistochemistry (false-negatives). In this regard, obtaining sections thin enough to permit complete antibody penetration is difficult with a vibratome.

The most commonly used embedding medium in histology laboratories is paraffin. While typically providing superior sectioning and allowing blocks to be stored for long periods, antigenicity is often reduced or lost. Improved localization in paraffin-embedded tissues has been obtained using a paraffin with a low melting point (56°C) and rapid dehydration through graded alcohols and xylenes into paraffin. Nevertheless, enzyme pre-treatments are often needed to restore antigenicity. Although the exact mechanism by which enzymatic digestion restores some antigenicity is not clear, with the proper enzyme, concentration, buffer, temperature, and length of digestion, enzyme pre-treatments can greatly enhance antigen localization. A number of enzymes are available for these procedures and often must be titrated to the specific tissue and antigen. Specific methods used in our laboratory will be described below.

For electron microscopy two major approaches are in vogue: pre- and post-embedding methods (4, 5). As a general rule, pre-embedding provides maximum antigenicity, but less than optimal structural preservation. The reverse is true of post-embedding methods. For immunogold studies, post-embedding methods using Lowycril (6) as the embedding medium at low temperatures ($-25\,°C$) have been adopted by a number of laboratories. Specific methods are described below.

2.3 Pre-staining treatments

An obvious goal of any immunohistochemical procedure must be to enhance the specific signal, while minimizing the background noise. Several pre-staining procedures have been designed with this objective in mind.

2.3.1 Intrinsic peroxidase activity

Peroxidatic activities are present in a number of cells, including neutrophils and macrophages (myeloperoxidase and catalase). In these cells, incubation with chromogens and hydrogen peroxide will result in deposition of the reaction product, unrelated to antibody localization. Intrinsic peroxidase activity can be removed by oxidation with hydrogen peroxide or acid treatment. The concentration of acid or hydrogen peroxide should be titrated to minimize possible deleterious effects on structure. If an enzyme other than peroxidase is used as a marker, blocking of intrinsic peroxidase activity is, of course, unnecessary.

2.3.2 Aldehyde reduction

Free aldehyde radicals may react with immunoglobulins through non-immune mechanisms by binding ϵ-amino groups, which results in high background. There are several methods available to reduce free aldehyde groups. For example, amino acids such as lysine, containing ϵ-amino groups, bind and block free aldehyde radicals, whereas sodium borohydride reduces aldehyde groups and may restore some of the conformation lost during fixation.

2.3.3 Enzyme digestions

There are a number of enzymatic digestions available that can restore some of the antigenic conformation lost during fixation and embedding. This loss may result from extensive cross-linking, aggregation, heat denaturation, or a combination of these phenomena. It has been shown that controlled enzymatic digestion can often restore some of this antigenicity. Depending on the method of fixation, embedding, and antigen, a different enzyme treatment will be employed. For paraffin-embedded tissues we have typically used pepsin (0.3 mg/ml in 0.5 M acetic acid) for 50 min at $37\,°C$. Others have recommended the use of trypsin and pronase in a similar manner (7, 8). In frozen sections, glycosidases have been found to decrease non-specific staining and increase penetration and specific staining (9). In any event, enzyme digestion should be used at the lowest possible concentration and for the shortest possible time to maximize antigenic detection, while minimizing tissue destruction and subsequent antigen translocation.

2.3.4 Blocking of non-specific binding sites

To minimize any non-immune binding of immunoglobulins to tissues (binding to Fc sites, non-specific protein–protein interactions, etc.) tissues are pre-treated with non-immune serum from the same species as the secondary antibody.

2.4 Antibodies

Both monoclonal and polyclonal antibodies have a wide application in immunohistochemistry. Polyclonal antibodies often recognize multiple epi-topes on a single molecule; whereas, monoclonal antibodies recognize a single epitope. Polyclonal antibodies have as one advantage their ability to recognize multiple epitopes. In most cases, all epitopes would not be de-natured by embedding and fixation; therefore polyclonals will often react with tissues so processed. Monoclonal antibodies have exquisite specificity, pro-vide a consistent reactivity between batches, and the potential for unlimited supply. However, they tend to have lower affinities than polyclonals and face the risk of denaturation of their single epitope. IgM monoclonals should be avoided since they are large molecules and tend to aggregate, resulting in little tissue penetration and higher background.

Polyclonal antibodies may be used in a variety of forms including complete serum, purified IgG, or affinity-purified antibodies. IgGs are typically purified by salt precipitation (ammonium sulfate), followed by chromatography (usually DEAE). The monospecificity of the antibodies may be determined by radioimmunoassay, ELISA, and/or immunoblotting. Cross-absorption or affinity chromatography is useful for removing undesired cross-reactivities. Typically antibodies have a limited capability to penetrate tissues, but this can be enhanced using Fab fragments. *Checkerboard titration* is used to establish the working dilutions for each new antibody.

2.5 Markers

Markers are necessary to visualize the localization of the individual anti-bodies. Fluorescein was the first of these to be employed. Immunofluorescence is an effective localization system, but has a number of disadvantages, in-cluding its temporary nature and the need for a specialized microscope. Numerous other markers have been employed and provide advantages in the above-mentioned parameters over fluorescein.

2.5.1 Particulate markers

The particulate markers are electron dense and thus may be used in electron microscopy (*Figure 1*). Ferritin and colloidal gold are the most commonly used particulate markers. Ferritin is not visible by light microscopy, has a high molecular weight (limited tissue penetration), a lower electron density than

Peter S. Amenta and Antonio Martinez-Hernandez

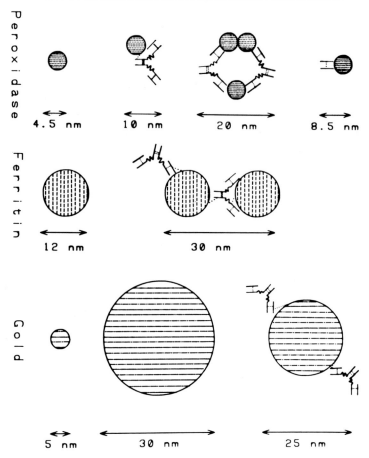

Figure 1. Diagram of markers frequently used in electron immunohistochemistry. Horse-radish peroxidase has a small radius (4.5 nm) and can be conjugated to IgG molecules by chemical means. The binding of antibodies to horseradish peroxidase with their antigen results in the formation of an immune complex, the Peroxidase–Anti-Peroxidase complex (PAP). The penetration problems created by the relative large size of PAP can be mini-mized using complexes formed by the Fab fragment and horseradish peroxidase.

Ferritin, has a large radius (12 nm), low electron density, and a proclivity for non-specific interactions with other proteins, making it the least desirable of the three markers.

Colloidal gold particles of different sizes are available. The commonly used diameters range from 1 to 40 nm. Antibodies labelled with particles of different diameters allow the simultaneous localization of more than one antigen. Gold particles have high electron density and they are the preferred marker for post-embedding localization.

Reproduced from reference 4, with permission.

gold, and a propensity for non-specific interactions. Colloidal gold has been used in immunohistochemistry to circumvent the limitations of ferritin. Of added benefit is the range in sizes of the colloidal gold particles, allowing for the simultaneous localization of more than one antigen. Silver enhancement

methods allow the use of colloidal gold in light microscopy immunohistochemistry.

2.5.2 Enzyme markers

The small size of the enzyme markers and consequent ability to penetrate
tissues, makes them particularly useful for immunohistochemistry. Enzyme
markers reacting with their substrate, in the presence of a suitable chromogen, results in an accumulation of reaction product. Increased concentrations
and incubation times may result in increased sensitivity, but also in excessive
background or diffusion of the reaction product. We use horseradish peroxidase with H_2O_2 and diaminobenzidine (DAB) hydrochloride. Alkaline
phosphatase is another enzyme marker often used.

2.6 Detection systems

A number of detection systems are available (*Figure 2*). The direct method
labels the primary antibody with the desired marker. In the indirect method a

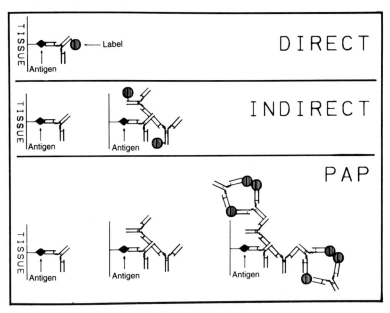

Figure 2. Diagram illustrating the principles of three commonly used detection methods.
The simplest system is the direct method, in which the primary antibody itself is labelled
with a suitable marker. In the indirect method, an antibody directed against the primary
antibody is labelled with the marker. In the PAP method, the unlabelled secondary antibody is followed by a peroxidase–anti-peroxidase complex, in which the antibody is from
the same species as the primary. Each subsequent antibody addition increases the
method's sensitivity. The direct method requires labelling each antibody individually, thus
the indirect and PAP methods are more desirable, since only the secondary, or the PAP are
labelled. Multiple variations on these basic principles exist.
Reproduced from reference 4, with permission.

secondary antibody directed against the primary antibody is the labelled reagent. A number of systems have been used to increase the sensitivity over that obtainable with the direct and indirect methods. The peroxidase–antiperoxidase technique provides excellent specificity and sensitivity. More recently avidin–biotin systems have been used. Streptavidin, a non-glycosylated, bacterial protein, has fewer non-specific interactions than the glycosylated egg-white avidin, resulting in lower backgrounds. There are multiple variations using more than three antibodies, each additional step tends to increase sensitivity and decrease the background (8).

2.7 Chromogens

Chromogens are compounds that yield a colorized derivative following a chemical reaction, typically an enzymatic one. A number of chromogens are available for immunohistochemistry. The most frequently used are DAB and amino-ethyl carbazole (AEC). The former provides a brown reaction product, is alcohol fast, and suitable for a wide range of counterstains and mounting media. DAB has been listed as a laboratory carcinogen, thus it should be used with caution. DAB, unlike AEC, can be used in electron microscopy, since osmium tetroxide forms an electron-dense complex with it. Amino-ethyl carbazole forms a red reaction product; however, AEC is alcohol soluble, thus must be mounted in an aqueous-based glycerol medium.

A number of amplification methods have been employed to intensify the reaction product, including osmium tetroxide, imidazole, and cobalt chloride. These products, while intensifying reaction product, also increase background staining.

2.8 Pre-embedding vs post-embedding techniques

In pre-embedding methods the antibodies are applied prior to tissue embedding; whereas, in post-embedding methods the antibodies are applied after embedding. As mentioned, post-embedding methods are often encumbered by decreased or lost antigenicity.

2.9 Interpretation of results

Immunohistochemistry depends on observer interpretation. Therefore, positive and negative controls are essential. Negative controls are those where the primary antibody is replaced by non-immune serum or in the case of monoclonal antibodies, tissue culture media or ascites fluid. An antibody with known reactivity should be used as a positive control. Internal controls within the test tissue (i.e., *those structures known to react with the test antibody*) are extremely useful in the evaluation.

310

3. Equipment and reagents, methods, and formulations

3.1 Equipment and reagents for *Protocols 1 to 15*

- Sodium phosphate, monobasic
- Sodium phosphate, dibasic
- Sodium chloride
- Paraformaldehyde powder (Polaron)
- Saturated sodium hydroxide solution
- Whatman grade 4 paper
- Liquid nitrogen
- Dewar flask
- Stainless-steel beaker
- Methylbutane
- Plastic wrap
- Coated slides
- Double-distilled water
- Staining dishes
- Baking dish
- Applicator sticks
- Paper towels
- Vortex mixer
- Glass slides
- Diamond pen
- Microcentrifuge
- Microcentrifuge tubes
- Plastic test tubes and caps
- Sodium borohydride
- Periodic acid (Kodak)
- Trizma acid
- Trizma base
- Diaminobenzidine hydrochloride
- Hydrogen peroxide
- 25% glutaraldehyde (for electron microscopy)
- Glycine (for electron microscopy)
- Osmium tetroxide (for electron microscopy)
- Micropipette
- Appropriate normal sera, link antibody, and tertiary antibodies
- OCT-embedding media (for frozen sections)
- Paraffin (Paraplast; for paraffin embedding)
- Resins for electron microscopy (see appropriate protocols below)
- A peristaltic pump (Harvard)
- Tubing
- Vascular catheter

Protocol 1. Preparation of formaldehyde (buffered 4% formaldehyde)

1. Suspend 8 g paraformaldehyde powder (Polaron) in 100 ml of double-distilled water in a beaker.

2. Heat the suspension on a magnetic stirrer to 55 °C.

3. Transfer the beaker to a non-heated stir plate and clear the solution with drops (8–10) of saturated sodium hydroxide.

4. When the solution reaches room temperature filter through two layers of Whatman grade 4 paper.

5. Mix with equal volumes of sodium phosphate buffer (pH 7.4) to reach a final concentration of 4% formaldehyde in either 0.1 or 0.2 M sodium phosphate buffer.

Protocol 2. Preparation of PBS solutions

Phosphate-buffered saline is made in aliquots of 10 litres at 0.01 M (pH 7.2).

1. Prepare stock solutions of 0.05 M mono- and dibasic sodium phosphate with 144 ml of dibasic and 56 ml of monobasic sodium phosphate added to 4 litres of distilled water, containing 85 g of sodium chloride.

2. Adjust the pH to 7.05 with sodium hydroxide (1 M) and bring to a final volume of 10 litres with distilled water.

3. Adjust the solution to pH 7.20 with either 1 M hydrochloric acid or sodium hydroxide.

4. Prepare the sucrose/PBS solution at a concentration of 4% (w/v) and keep at 4 °C.

5. Prepare the PBS with 4% (w/v) sucrose and 7% (v/v) glycerol solution and keep at 4 °C.

Protocol 3. Perfusion fixation

1. Depending on the organ to be studied, choose a blood vessel to effectively deliver the perfusate. For example, in the liver, we have perfused through the portal vein, while simultaneously allowing for drainage from the inferior vena cava. Other organs, such as kidney and heart, may be perfused through the aorta.

2. Do not allow perfusion pressures to exceed systolic pressures in order to prevent significant distortion of normal histologic relationships. Excessive perfusion pressure in the liver will result in marked dilatation of the sinusoids and aggregation of hepatocytes into nodules.

3. Perfuse for 1 min with phosphate-buffered saline, prior to delivery of the fixative, to remove blood.

4. Carry out vascular perfusion via a peristaltic pump for approximately 30 min at room temperature (to prevent vasoconstriction), use approximately 500 ml of fixative.

5. Subsequently, fix tissues (in the same fixative) on a rotator at 4°C. During fixation, rotate vials containing the tissues at a slow speed to provide a tumbling action, so that all sides of the blocks will be equally exposed. Depending on the antigen to be localized, the length of fixation required may be less than three hours. Keep tissues wet throughout all stages of the fixation. Drying artefacts at the periphery of the block can lead to high background.

Protocol 4. Immersion fixation

Human tissues or tissues which are difficult to perfuse, for instance fetal tissues, must be processed by immersion.

1. Cut tissues into the smallest blocks feasible.

2. Place blocks in a vial containing at least 10 times the tissue volume of fixative at 4°C.

3. During the first half-hour change the fixative every 10 min, and as the tissue hardens, cut progressively smaller tissue blocks to reach a final size of 5 × 5 × 2 mm.

4. Carry out the remaining fixation as described in *Protocol 1*.

Protocol 5. Freezing

1. Immerse a stainless steel beaker in liquid nitrogen. Place the methyl-butane in the stainless steel beaker, and when a frozen layer of methyl-butane forms at the base of the beaker, the solution is considered to be at liquid nitrogen temperature.

2. Immerse cork discs covered with a thin layer of Tissue-tek embedding medium in the methylbutane.

3. Place tissue blocks on the frozen embedding medium (upper layer partially melted to ensure tissue bonding). After a one hour wash in the PBS–sucrose–glycerol solution, cover the blocks with a drop of OCT and immerse in methylbutane. Avoid cracking the tissue block (large blocks, inadequate cryoprotection, etc.). Contact with methylbutane would result in dehydration of tissue.

Protocol 6. Preparation of coated slides

Numerous glues are available commercially, for instance Neoprin and polylysine. We use albumin–glutaraldehyde coated slides and have found them to provide the maximum degree of tissue adherence. Gelatin-(de-natured collagen) coated slides should be avoided when studying extra-cellular matrix. The albumin–glutaraldehyde coated slides are prepared as follows:

1. Place 1 egg white in 1 ml of concentrated ammonium hydroxide. Add 500 ml of distilled water and stir for 10 min.

2. Filter the solution through 6 layers of gauze followed by filtration through Whatman Grade 4 paper. Foaming of the albumin solution should be avoided.

3. Prepare staining dishes as follows: one staining dish containing 1% concentrated HCl in 70% ethanol and 2 dishes of absolute acetone.

4. After 1 min incubation in each solution allow the slides to air dry.

5. Immerse the slides in the albumin solution for 10 sec.

6. Immerse the slides in 2.5% glutaraldehyde in 0.1 M sodium phosphate buffer for 1 min.

7. Wash the slides in PBS 3× for 10 min each and dry at 60 °C for 2 h. The slides may be used after reaching room temperature.

Protocol 7. Frozen sectioning

1. Cut 5–8 μm sections on a cryostat with a retractable advance mechanism (to minimize tissue damage during the upstroke of the block).

2. Allow sections to adhere to coated slides for 20 min at room temperature.

3. Inscribe slides around the tissue section with a diamond pencil, to retain antibody solutions on the section.

i. Removal of intrinsic peroxidatic activity

Immerse tissue sections into periodic acid (Kodak) at concentrations of 0.01–0.05 M in PBS at room temperature for 15 min. The specific concentrations are listed under the appropriate staining protocols.

ii. Reduction of free aldehyde groups

Immerse tissue sections in a 0.05% solution of sodium borohydride for 30 or 60 min in PBS.

Protocol 8. Antibody application

1. Centrifuge all serum and antibody solutions in a microcentrifuge for 5 min at 4°C immediately prior to application. This lessens the accumulation of particles on stained tissue sections.

2. Remove the PBS on the slide, using a piece of gauze around the tissue section to avoid dilution of the antibody. On the other hand, retain enough PBS on the section, to prevent tissue drying.

3. Apply antibodies with a micropipette, with sufficient volume to flood the tissue section (30–80 μl).

Protocol 9. Construction of a moist chamber

1. Use any device capable of providing an air-tight humid environment for the incubation steps. We have used baking dishes with the bottoms covered by paper towels saturated with distilled water.

2. Use applicator sticks to keep the slides above the wet paper towels.

3. Cover the entire dish with plastic wrap to make an airtight chamber.

Protocol 10. Chromogen

CAUTION: DAB is a potential carcinogen and should be handled appropriately.

1. Place the slide rack in a solution of 50 mg of DAB in 150 ml Tris–HCl buffer (0.5 M, pH 7.6).
2. Judge the actual staining time by checking under the microscope for the formation of a brown precipitate, 7–15 min incubations are common.
3. Wash the slides 3 × in PBS.

---

4. Staining procedures

4.1 Light microscopy

Protocol 11. Unfixed tissues

1. Place tissue blocks in cryostat embedding medium and snap-freeze in methylbutane at liquid nitrogen temperature.
2. Collect cryostat sections, 3–5 μm thick, on albumin coated, glutaraldehyde-cross-linked slides.
3. Air dry sections for 30 min and encircle with a diamond-tip pen.
4. Fix sections in 100% acetone at 4 °C for 15 min.
5. Hydrate sections in PBS at 4 °C for 15 min.
6. Inhibit intrinsic peroxidase activity with 0.03 M periodic acid in PBS for 15 min at room temperature (RT).
7. Rinse 3 × in PBS for 5 min each at 4 °C.
8. React with normal serum of the same species as the secondary antibody (1/10 dilution) for 15 min at RT in a moist chamber.
9. Rinse 3 × in PBS for 5 min each at 4 °C.
10. React either with a suitable dilution of the primary antibody (as determined by 'checkerboard' titration) or with normal serum of the same species as the primary antibody (negative control) at similar dilutions for 1 h at RT in a moist chamber. Apply all antibodies and normal sera with enough volume to cover the tissue section (50–75 μl).
11. Rinse 3 × in PBS for 5 min each at 4 °C.
12. React with a suitable dilution of the secondary antibody (as determined by 'checkerboard' titration) for 1 h at RT in a moist chamber.
13. Rinse 3 × in PBS for 5 min each at 4 °C.

316

14. React with a suitable dilution of peroxidase–antiperoxidase (PAP) (as determined by 'checkerboard' titration) for 1 h at RT in a moist chamber.

15. Rinse 3 × in PBS for 5 min each at 4 °C.

16. React in a solution containing DAB 50 mg/150 ml and 20 μl of 5% H_2O_2 in 0.5 M Tris–buffer, pH 7.6, in the dark for 10 min.

17. Rinse 3 × in PBS for 5 min each at 4 °C.

18. Dehydrate in graded ethanols.

19. Mount with Permount.

Protocol 12. Frozen sections–fixed tissue

1. Perform fixation at RT. As described above perfusion is the preferred method of fixation; however, immersion fixation has to be accepted when the former is impractical (see *Protocols 3* and *4*).

2. Decant the fixative and wash the blocks at 4 °C overnight, under continuous, gentle agitation with multiple changes of PBS–4% sucrose. An electron microscopy tissue rotator is the recommended device for the agitation.

3. One hour prior to embedding, wash for 1 h the blocks in PBS containing 4% sucrose and 7% glycerol.

4. Collect cryostat sections, 3–5 μm thick, on slides (see *Protocols 6* and *7*).

5. Encircle sections with a diamond-tip pen and air dry for 30 min.

6. Hydrate sections in PBS at 4 °C for 15 min.

7. To inhibit intrinsic peroxidase activity, treat the sections with 0.03 M Periodic acid in PBS for 30 min at RT.

8. React sections with sodium borohydride (50 mg/ml) in PBS at 4 °C for 1 h to reduce free aldehyde groups.

9. Rinse the sections 3 × in PBS for 5 min each at 4 °C.

10. React the sections with normal serum of the same species as the secondary antibody (1/10 dilution) for 15 min at RT in a moist chamber.

11. Rinse 3 × in PBS for 5 min each at 4 °C.

12. React with either a suitable dilution of the primary antibody as determined by 'checkerboard' titration (experimental) or with the normal serum of the same species (negative control) at a similar dilution for 1 h at RT in a moist chamber. Apply all antibodies and normal sera with a micropipette with enough volume to cover the tissue section (50–75 μl).

Protocol 12. *Continued*

13. Rinse 3× in PBS for 5 min each at 4 °C.

14. React with a suitable dilution of the secondary antibody (as determined by 'checkerboard' titration) for 1 h at RT in a moist chamber.

15. Rinse 3× in PBS for 5 min each at 4 °C.

16. React with a suitable dilution of the peroxidase–antiperoxidase (PAP) (as determined by 'checkerboard' titration) for 1 h at RT in a moist chamber.

17. Rinse 3× in PBS for 5 min each at 4 °C.

18. React with DAB (50 mg/150 ml in 0.5 M Tris–buffer, pH 7.2) in the dark for 10 min

19. Add 20 μl of a 5% (v/v) solution of H_2O_2 in water to the staining dish, mix well and incubate the slides for 5–10 min.

20. Rinse 3× in PBS for 5 min each at 4 °C.

21. Dehydrate in graded ethanols and mount with Permount.

Protocol 13. Paraffin–embedded tissues

1. Perform fixation as described above (see *Protocols 3* and *4*).

2. Cut blocks with razor blades, place in scintillation vials, and fix for 3 h.

3. Dehydrate in graded ethanols, clear in xylene, and embed in paraffin (55–60 °C). Place sections (3 μm) on coated slides and keep at 45 °C overnight. Deparaffinize and rehydrate via xylene, graded ethanols, and PBS (2 min changes).

4. React sections with 0.5 M periodic acid in PBS for 20 min to inhibit intrinsic peroxidase activity.

5. Rinse 3× in PBS for 5 min each at 4 °C.

6. React sections with $NaBH_4$ (50 mg/ml) in PBS at 4 °C for 1 h to reduce free aldehyde groups.

7. Flood sections with 0.3 mg/ml pepsin (Sigma) in 0.5 M acetic acid, place in a moist chamber and incubate at 37 °C for 30 min

8. Perform the remainder of the staining procedure as described for fixed tissues (*Protocol 12*).

4.2 Electron microscopy

Protocol 14. Pre-embedding methods

Reagents

- Embedding resin:

Medcast	31.0 ml	28.34%
Araldite	24.0 ml	21.94%
DDSA	50.0 ml	45.70%
DMP-30	4.4 ml	4.02%

Method

1. Perform fixation and processing as described in *Protocols 3, 4*, and *12*.

2. Cut frozen sections for electron microscopy (6–8 μm) and collect on coated slides.

3. Air dry sections for 30 min and encircle with a diamond-tip pen.

4. Hydrate sections in PBS at 4 °C for 15 min.

5. To inhibit intrinsic peroxidase activity, treat the sections with 0.03 M periodic acid in PBS for 30 min at RT.

6. Wash the sections 3× in PBS for 10 min each at 4 °C.

7. To reduce free aldehyde groups, react the sections with 0.05% $NaBH_4$ in PBS at 4 °C for 1 h.

8. Wash the sections 3× in PBS for 10 min each at 4 °C.

9. React the sections with normal serum of the same species as the secondary antibody (1/10 dilution) for 30 min at RT in a moist chamber.

10. Wash the sections 3× in PBS for 10 min each at 4 °C.

11. React the sections with a suitable dilution of the primary antibody (as determined by 'checkerboard' titration) or with the normal serum of the same species, at a similar dilution, overnight at 4 °C in a moist chamber. Apply all antibodies and normal sera with a micropipette and enough volume to cover the tissue section (usually 50–75 μl).

12. Wash the sections 3× in PBS for 10 min each at 4 °C.

13. React the sections with a suitable dilution of the secondary antibody (as determined by 'checkerboard' titration) for 2 h at RT in a moist chamber.

14. Wash the sections 3× in PBS for 10 min each at 4 °C.

15. React the sections with a suitable dilution of peroxidase–antiperoxidase (PAP) (as determined by 'checkerboard' titration) for 2 h at RT in a moist chamber.

16. Wash the sections 3× in PBS for 10 min each at 4 °C.

Protocol 14. *Continued*

17. Fix the sections with 2.5% glutaraldehyde in 0.1 M sodium phosphate buffer, pH 7.4, for 30 min at RT.

18. Wash the sections 3× in PBS for 10 min each at 4 °C.

19. React the sections with 0.1 M glycine in PBS for 30 min at RT.

20. Wash the sections 3× in PBS for 10 min each at 4 °C.

21. React the sections with DAB (50 mg/150 ml, in 0.5 M Tris–buffer, pH 7.6) in the dark for 10 min.

22. Add hydrogen peroxide (20 μl of a 5% solution in water) to the staining dish, mix well and 'incubate' the slides for 5–10 min.

23. Wash the sections 3× in PBS for 10 min each at 4 °C.

24. React the sections with 1% osmium tetroxide in 0.1 M sodium phosphate buffer containing 0.026 M NaCl, pH 7.2, for 1 h at RT. Osmium tetroxide vapours are toxic, particularly to corneal tissues: care should be taken to work under a chemical hood. Break an ampoule of osmium tetroxide (1 g/vial) inside a tightly stoppered, brown-glass bottle and dissolve in 50 ml of distilled water (2%). Complete solubilization may take 24 h, but may be accelerated by placing the tightly stoppered bottle in a sonicator bath. Dilute this 2% aqueous OsO_4 solution 1:1 with 0.2 M sodium phosphate buffer containing 0.026 M NaCl (pH 7.2) immediately before use.

25. Wash the sections 3× in PBS for 10 min each at 4 °C.

26. Dehydrate the sections in graded ethanols followed by three 8 min propylene oxide dehydrations.

27. Infiltrate the section with resin, overnight in a 1:1 mixture propylene oxide resin (without DMP-30 catalyst), followed by a 1 h infiltration in a 2:1 resin:propylene oxide mixture (containing DMP-30), and a 1 h infiltration in resin.

28. Invert a capsule containing the plastic mixture (with DMP-30) on top of the section and allow the capsule to remain on the section overnight at RT.

29. Place the slides at 60 °C for 48 h to polymerize the resin.

30. Remove the blocks (containing the sections) from the slides by briefly heating them over a gas flame and snapping the block off quickly.

Protocol 15. Post-embedding methods

1. Fix tissues according to either *Protocol 3* or *4*.

2. Dehydrate the tissues in 50% ethanol at 4 °C for 15 min, followed by graded ethanols at −20 °C.

3. Perform infiltration at $-20\,^{\circ}\text{C}$ with equal parts of 100% ethanol: Lowicryl solution for 2 h, followed by 1:2 ethanol:Lowicryl solution for 2 h, and finally in 100% Lowicryl overnight.

4. After infiltration, place tissue blocks at the tip of capsules, fill the capsules with Lowicryl, and cross-link the resin with UV light (370 λ) at $-20\,^{\circ}\text{C}$ for 48 h. Continue the cross-linking at RT for an additional 48 h.

5. Mount ultrathin sections (60–80 nm) on Formvar-coated nickel grids (400 mesh).

6. Perform all incubations in a moist chamber at RT. Invert grids on a drop of 10% H_2O_2 for 10 min, rinse in 2 changes of PBS, place on a drop of PBS with 0.5% Tween-20 (PBS–T), then incubate on a drop of 0.01 M HCl for 10 min at RT.

7. To reduce free aldehyde groups, place grids on a drop of NaBH$_4$ (5 mg/15 ml PBS–T) for 1 h, rinse in PBS, then sequentially expose to NGS, primary antibody, secondary antibody (goat anti-rabbit), and PAP for 1 h, with intervening PBS rinses. Determine optimal antibody dilutions by serial dilution. Then react the sections with DAB–HCl (50 mg/100 ml in 0.1 M Tris-buffer, pH 7.62) with H_2O_2 for 20 min

7a. For immunogold, the initial steps are the same as described for the PAP staining. However, instead of the goat anti-rabbit GAR, rinse the grids with Tris–HCl, pH 8.2, and react with GAR-gold diluted with Tris-buffer at working dilution. Rinse the sections with Tris–HCl, distilled H_2O, dry, and lightly counter-stain with uranyl acetate and lead citrate.

5. Examples of immunolocalization of extracellular matrix components

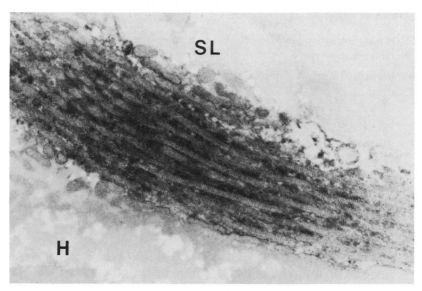

Figure 3. Localization of collagen type I in normal rat liver. Collagen fibrils in Disse space demonstrate the characteristic cross-banding. Pre-embedding staining; PAP method. H, hepatocyte; SL, sinusoidal lumen. From reference 10, with permission.

Figure 4. Laminin localization in rat uterus. 4a. Light microscopy. Laminin is demonstrated in the basement membranes of endometrial glands, blood vessels, and smooth muscle cells. Frozen section; PAP method. 4b. Electron microscopy. The basement membrane of an endometrial gland (BM), and a capillary (CBM) contain laminin. Pre-embedding staining; PAP method. CL, capillary lumen; GL, glandular lumen. From reference 11, with permission.

(a)

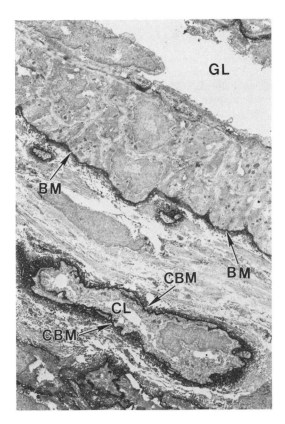

(b)

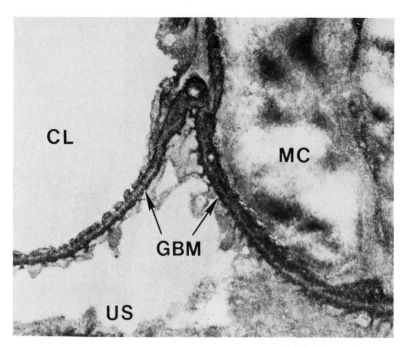

Figure 5. Electron microscopic localization of laminin in rat renal glomerulus. Laminin is preferentially distributed in both lamina rarae of the glomerular basement membrane (GBM). Pre-embedding staining; PAP method. CL, capillary lumen; MC, mesangial cell, US, urinary space. From reference 12, with permission.

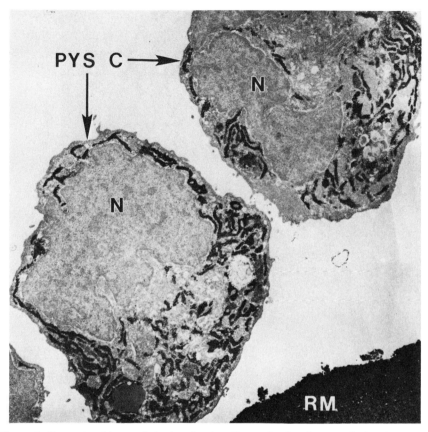

Figure 6. Example of intracellular localization. Rat parietal yolk sac cells (14-day gestation) reacted with antilaminin antibodies. Laminin is demonstrated within the cisternae of the rough endoplasmic reticulum and in Reichert's membrane (RM), the yolk sac basement membrane. Pre-embedding staining; PAP method. N, nucleus; PYSC, parietal yolk sac cell.

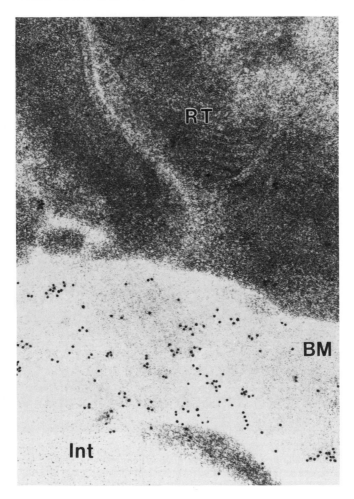

Figure 7. Collagen type IV localization in renal tubular basement membrane. Gold particles (15 nm) demonstrate antigenic sites within the basement membrane (BM). Post-embedding method (Lowicryl). Int, interstitium; RT, renal tubular cell.

References

1. Amenta, P. S., Gay, S., Vaheri, A., and Martinez-Hernandez, A. (1986). *Coll. Relat. Res.*, **6**, 125.
2. McLean, I. W. and Nakane, P. K. (1974). *J. Histochem. Cytochem.*, **22**, 1077.
3. Tokuyasu, T. K. (1984). In *Immunolabelling for electron microscopy* (ed. J. M. Polak and M. Varndell), pp. 71–82. Elsevier, Amsterdam.
4. Martinez-Hernandez, A. (1987). In *Methods in enzymology. Structural and contractile proteins: extracellular matrix* (ed. L. W. Cunningham), Vol. 145, pp. 78–103. Academic Press, New York.

5. Bendayan, M. and Zollinger, M. (1983). *J. Histochem. Cytochem.*, **31**, 101.
6. Carleman, E., Garavito, M., and Villiger, W. (1982). *J. Microsc. (London)*, **126**, 123.
7. Barsky, S. H., Rao, N. C., Restrepo, C., and Liotta, L. A. (1984). *Am. J. Clin. Pathol.*, **82**, 191.
8. Bullock, G. R. and Petrusz, P. (1982). *Techniques in immunocytochemistry*. Academic Press, New York.
9. Andrews, L. P., Clark, R. K., and Damjanov, I. (1985). *J. Histochem. Cytochem.*, **33**, 695.
10. Martinez-Hernandez, A. (1984). *Lab. Invest.*, **51**, 57.
11. Karkavelas, G. *et al.* (1989). *J. Ultras. Mol. Struc. Res.*, **100**, 137.
12. Martinez-Hernandez, A. (1984). *J. Histochem. Cytochem.*, **32**, 289.

Isolation and analysis of RNA from connective tissues and cells

LINDA J. SANDELL

1. Introduction

Extraction of intact functional RNA from connective tissues presents special challenges. Most extracellular matrix (ECM) molecules are very large and consequently their mRNAs are among the longest known. The difficulty in isolating these large mRNAs is compounded by the large amount of ECM surrounding the cells. In addition, the number of ECM mRNAs per cell may be quite low due to the low turnover rate of the protein. In some connective tissues, however, where the ratio of cells to matrix is high, RNA can be isolated from the tissue directly. In other cases where the matrix is resistant to denaturation or where the matrix is particularly abundant, better results are achieved when the matrix is removed from the cells before isolation of RNA. In this chapter methods will be described that have been used successfully to isolate RNA from cartilage and a variety of other connective tissues, such as bone, tendon, and smooth muscle, and from connective tissue cells in culture.

2. Isolation of cells

It is often advantageous to initially isolate cells from tissues such as cartilage, and even from cultures where a large amount of ECM has accumulated. Separation of mRNA from ECM proteoglycans, which are present at high concentration in cartilage, poses a significant problem because both classes of molecules are extremely large, have a high buoyant density, and are strongly electronegative: thus, they separate poorly from each other with most fractionation techniques. Contamination of the RNA with proteoglycans adversely affects the results with subsequent preparative or analytical procedures such as agarose electrophoresis or oligo (dT) affinity chromatography. The proteoglycan heparin has also been reported to interfere with polymerase chain reaction (PCR) amplification of DNA (1). A further advantage of isolating cells prior to RNA extraction is that the viable cell number can be determined by counting an aliquot stained with Trypan blue. In our experience, almost all problems involving low yield of poor quality RNA from connective tissues can be solved by isolation of the cells to remove contaminating matrix molecules.

Connective tissue cells such as smooth muscle cells or fibroblasts do not accumulate an extensive proteoglycan-rich extracellular matrix and can be trypsinized or mechanically scraped into the extraction buffer. The procedures described here work well for cultured chondrocytes, however, some adjustments may be needed for tissues and cell cultures containing high levels of mineral, elastin, hyaluronan, or other specific ECM molecules. If the cultures have accumulated large amounts of matrix either due to age or a high rate of secretion stimulated by ascorbic acid, it is very difficult to digest the matrix and dead cells may accumulate within the ECM.

Protocol 1. Enzymic isolation of cells

Reagents

- Hank's Basic Salt Solution (HBSS)
- Collagenase (Type 2 Worthington)
- Trypsin (Gibco/BRL) or pronase (Calbiochem)
- DMEM
- FBS
- Trypan blue

Method

1. Wash cells with 5–10 ml/100 mm dish Hank's Basic Salt Solution (HBSS, 3 ml/100 mm dish).

2. Digest matrix with 2 ml of a solution containing 0.4% collagenase and 0.25% trypsin at 37 °C for 45 min to 1.5 h. The end-point of digestion is the observation, under the light microscope, of single cells.

3. Alternatively, digest the cells with 0.75% pronase for 10 min, followed by 0.15% collagenase, both in DMEM.

4. Remove cells from the dishes, pool and wash 2 × with medium + 10% FBS, then 2 × with HBSS.

5. Count cells in a haemocytometer. Determine cell viability by exclusion of Trypan blue. Viability should be > 95%.

---

3. Preparation of RNA

Working with RNA has been described as a 'state of mind'. Somewhat different from a sterile technique, the objective is to protect the RNA from degradation by RNases. These enzymes are very stable, need no co-factors and are present throughout the cellular environment. The goal then, is to inactivate, remove, and not introduce RNase.

3.1 Preparation of RNase-free equipment and solutions

Experiments with RNA should be carried out in an RNase-free environment and solutions made with RNase-free water. When possible, all solutions including water, should be treated with the RNase inhibitor diethyl pyro-carbonate (DEPC 0.1%, Sigma). (**Note**: Tris solutions cannot be treated with DEPC as Tris inactivates DEPC.) Following DEPC treatment, allow solutions to stand overnight and then autoclave to remove excess DEPC and reaction products. Sterilize all solutions and bake glassware overnight at $> 180°C$; autoclaving alone will not inactivate many RNases. Plasticware is usually RNase-free if sterile, however, pipette tips should be autoclaved before use (fill tip box with gloved hands). It is a good idea to dedicate a special area of the laboratory and to designate equipment solely for the preparation of RNA, with all reagents marked as RNase-free. **Always** wear gloves to protect the RNA from RNases present on the skin and follow the principles of a sterile technique when contamination is suspected. We have found the following hints helpful in the successful preparation of high molecular weight mRNA.

(a) Once the preparation is begun and the RNases have been released from their cellular compartment, it is important to purify RNA in a quick and timely manner.

(b) Keep all solutions and RNAs cold, and whenever possible, centrifuge RNA at $-10°C$ (any of the EtOH or NaOAc precipitation steps will not freeze at this temperature).

(c) When making multiple preparations using a Dounce homogenizer, and rinse the homogenizer between uses with DEPC–H_2O solution.

(d) Equipment that will contact RNA should be RNase-free.

3.1.1 Isolation of total RNA from cells

The method used for the isolation of RNA from cells (RNA Method 1 (*Protocol 2*) or RNA Method 2 (*Protocol 3*)) depends on: (a) the number of cells available for extraction, and (b) the subsequent use of the RNA. Method 1 (*Protocol 2*) (originally modified from the method of Strohman (2)) is used for large numbers of cells ($> 10^7$) and yields very high quality RNA, suitable for any purpose and is good for up to 10 years (and probably longer). Method 2 (*Protocol 3*) by Chomczynski and Sacchi (3) is used when cell numbers are limited (generally $< 10^7$), yields RNA of sufficient purity for Northern blots; however, to maintain integrity of the RNA, it is stored in 0.5% SDS, making it inadequate for subsequent enzyme procedures. The advantages and disadvantages of each method are summarized in *Table 1*. Method 3 (*Protocol 4B*, which is a modification of *Protocol 4A*, is effective for RNA isolation from both cells and tissues and is now the protocol of choice in our laboratory) described in the next section is a particularly good technique for the isolation of RNA from tissue, but it can also be used for the isolation of RNA from cultured cells.

Table 1. Comparison of RNA isolation methods

	RNA Method 1 Guanidine hydrochloride	RNA Method 2 Guanidinium isothiocyanate
Quality of RNA	Very pure	Less pure
Number of cells	$> 10^7$	$< 10^7$
Time for preparation	2–3 days	2 h
Use of RNA	Any use; RNA can be stored indefinitely	Northern slot blots; PCR, cDNA[a]; RNA should be used quickly

[a] Leave out SDS in storage solution.

Protocol 2. Preparation of RNA using guanidine hydrochloride (RNA Method 1)

Equipment and reagents

- Gu–HCl solutions (Note: two different solutions are made, one with and one without EDTA)
  — Solution A: 8 M Gu–HCl, 20 mM NaOAc, 1 mM DTT
  — Solution B: Solution A + 20 mM EDTA
- DEPC
- NaOAc
- 20 mM EDTA, pH 7–7.5 (from 0.5 M EDTA stock solution)
- 10 mM Hepes, pH 7.2 (from 0.5 M stock solution)
- Chloroform/butanol (4:1, v/v)
- Dounce homogenizer
- Glacial HOAc
- Ethanol

A. *Preparation of solutions*

1. Make up Gu–HCl solutions A and B as follows:

   (a) Weigh out 304 g Gu–HCl. Add very little H_2O at first. Stir with slight warming in a beaker with 400 ml marked and add H_2O to just under 400 ml. Once in solution, remove and transfer to a graduated cylinder.

   (b) Add 1.8 ml of 4.5 M NaOAc, pH 5.2 and H_2O up to approximately 380 ml. Divide into 2 aliquots of 190 ml each.

   (c) Add 8 ml of 0.5 M EDTA to one aliquot and q.s. each to 200 ml with H_2O.

   (d) To each bottle, add 200 μl DEPC and mix vigorously. Let stand overnight with the tops ajar to release CO_2 and EtOH generated by DEPC breakdown. Autoclave and store at room temperature.

   (e) Add dithiothreitol (DTT) to a final concentration of 1 mM just before using.

2. Make up 4.5 M NaOAc (for 250 ml) as follows:

   (a) Weigh out 95 g NaOAc and add H_2O. After most NaOAc has gone into solution, bring up to 250 ml.

(b) Adjust pH to 5.2 with glacial acetic acid. (**Note**: solution will be almost saturated.) Treat with 250 μl of 0.1 % DEPC.

(c) Let stand overnight.

(d) Autoclave and store in freezer at −20 °C. This solution will tend to precipitate; if necessary, warm slowly, but make sure the concentration remains the same.

3. Make up 20 mM EDTA, pH 7–7.5 solution as follows:

(a) Prepare a 0.5 M EDTA stock solution (18.6 g EDTA in 100 ml H_2O).

(b) Treat the diluted solution with 0.1% DEPC and autoclave.

4. Make up 10 mM Hepes, pH 7.2 as follows:

(a) Prepare a 0.5 M stock, add 0.1% DEPC, treat and autoclave the diluted solution.

(b) Store in the refrigerator.

5. Make up chloroform/butanol solution as follows:

(a) Make a solution of 4:1 chloroform:butanol.

(b) Prepare fresh solution every week using chloroform less than 6-months-old. Do not heat or treat with DEPC.

B. *RNA preparation*

1. Disperse the cell pellet (*Protocol 1*) in 0.25 ml of Solution A per 10^6 cells.

2. Homogenize in a Dounce homogenizer (about 15 strokes with each of the A and B pestles) until solution is no longer viscous. The high viscosity is due to the presence of high molecular weight DNA which is sheared into small pieces by homogenization. Since small DNA fragments do not precipitate efficiently, the remaining DNA should remain soluble when precipitated with sodium acetate.

3. Remove cell debris by centrifugation at 5000 *g* at 4 °C for 15 min.

4. Transfer supernatant to a clean 15 ml or 30 ml glass Corex tube. Precipitate high molecular weight nucleic acids from the supernatant by adding 1/2 vol. of ice-cold 100% EtOH and incubating at −20 °C for 4 h. Centrifuge at 6500 *g* for 25 min at −10 °C.

5. Resuspend the pellet in approximately 1/2 vol. of the original volume in Solution B. Transfer to microcentrifuge tubes.

6. Precipitate with 1/2 vol. 100% EtOH at −20 °C for 4 h.

7. Centrifuge for 5 min in a microcentrifuge in the cold.

8. Repeat precipitation (steps 5–7) twice.

9. To remove residual Gu–HCl, quickly rinse precipitate with 100 μl of 10 mM Hepes, pH 7.2; centrifuge for 2 min and discard rinse. RNA should not go into solution at this step.

Protocol 2. *Continued*

10. Dissolve pellet in 500 μl 10 mM Hepes. If RNA is difficult to solubilize in aqueous layer, heat to approximately 65 °C and vortex. Repeat if necessary using 100 μl 10mM Hepes. Extract with an equal volume of chloroform:butanol (4:1). Centrifuge for 2 min in a microfuge at 4 °C, remove and save the aqueous layer (upper).

11. To the combined aqueous portion, add 1/10th volume of 3 M NaOAc, 2 volumes of 100% EtOH. Mix well and incubate at −20 °C for 4 h or −70 °C for 30 min to precipitate RNA. Centrifuge for 5 min in a micro-centrifuge. Repeat precipitation twice.

12. Wash final pellet twice with 66% EtOH/0.1 M NaOAc, once with 100% EtOH, resuspend in 200 μl of Hepes and store at −70 °C.

Protocol 3. Preparation of RNA using guanidinium isothiocyanate (RNA Method 2)

Equipment and reagents

- Denaturing Solution: 4 M guanidinium isothiocyanate, 25 mM Na citrate, 0.5% sarcosyl. (250 g guanidinium isothiocyanate +1 293 ml DEPC-treated water, 17.6 ml 0.75 M sodium citrate, pH 7.0, and 26.4 ml 10% sarcosyl). Add just before use: 7.2 μl β-mercaptoethanol and 4 μl Antifoam A/1 ml of solution.

- Chloroform:isoamyl alcohol (49:1)
- 2 M NaOAc

Note: Solutions for this method are available as a commercial product from Biotecx Laboratories, Inc.

- H_2O-saturated phenol
- Isopropanol

Method

1. Trypsinize or scrape the cells from the dish.
2. Collect the cells by centrifugation at 1000 *g* for 15 min at 4 °C.
3. Add 0.5 ml denaturing solution, mix well using a pipettor to disrupt the cells.
4. Add 0.05 ml 2 M NaOAc, pH 4.1, mix well.
5. Add 0.5 ml water-saturated phenol, mix well.
6. Add 0.1 ml chloroform:isoamyl alcohol, mix well.
7. Cool on ice for at least 15 min (may be up to 2 h).
8. Centrifuge 10 000 *g* for 20 min at 4 °C. Transfer the upper layer into a new tube.
9. Add one volume (0.5 ml) isopropanol. Store the sample at −20 °C for at least 1 h. (Can stop here and store sample overnight.)
10. Centrifuge 10 000 *g* for 20 min at 4 °C.
11. Dissolve the pellet into 0.4 ml of denaturing solution. Be careful to remove any solution from the side of the tube.

12. Add one volume (0.4 ml) of isopropanol. Can stop here. Store the sample at $-20\,°C$ for at least 1 h. Centrifuge as in step 10.

13. Wash the pellet twice with 75% ethanol and once with 100% ethanol. Dry the pellet briefly using a Speed-Vac and dissolve in 200 μl of 10 mM Hepes, pH 8.0.

14. Test the integrity of the RNA (heat at $95\,°C$ for 5 min and cool) on an agarose mini-gel stained with ethidium bromide (see *Protocol 5*). Background streaks or aberrant migration of the ribosomal RNA can be due to contamination with guanidinium isothiocyanate or proteoglycans. To reduce guanidinium contamination, make sure no solution is retained in the pellet during purification. Proteoglycans must be removed by more extensive enzyme digestion and removal of the ECM before lysis of the cells.

3.1.2 Extraction of total RNA from tissues

Several methods have been published for isolating RNA either from fetal epiphyseal cartilagenous tissues (1), from rat bone or xiphoid cartilage (4), or from cells that contain a high concentration of proteoglycans (5). Adams and colleagues (6) have recently reported a method employing ultracentrifugation in caesium trifluoroacetate (CsTFA) for isolating RNA from adult articular cartilage. CsTFA, a highly chaotropic salt capable of denaturing proteins while concomitantly maintaining the integrity of isolated nucleic acids, has been used to purify RNA from lysates containing a high concentration of high buoyant density carbohydrates (7). RNA isolated by this method is suitable for the preparation of cDNA libraries, polymerase chain reaction, poly A$^+$ RNA isolation, *in vitro* translation, and Northern blot analysis. The yield of total RNA from adult cartilage is approximately 50 μg/g wet wt. tissue. Another procedure, described here, was developed by Glisin *et al.* (8) and has been used by Nemeth and colleagues (5) for rat skeletal tissue. The procedure has been modified for isolation of RNA from adult bovine cartilage (T. Hering, personal communication), and recently, we have successfully used this method for the isolation of RNA from cultured chondrocytes.

Protocol 4A. Extraction of RNA from tissue (Method 3)

Equipment and reagents

- 6 M guanidinium–hydrochloride (Gu-HCl)
- 20% sarcosyl
- Antifoam A
- 5.6 μM caesium chloride (CsCl)
- DEPC-treated water
- 2 M potassium acetate, pH 5.0 (adjust pH with glacial HOAc)
- 3 M and 4 M NaOAc, pH 5.0
- Chloroform:isobutanol (4:1)
- Freeze mill (Spex mill or Bessman tissue pulverizer, Fisher)
- Tekmar tissue homogenizer
- Beckman SW 50.1 rotor and tubes

Linda J. Sandell

Protocol 4A. *Continued*

Day 1

1. Powder 500 mg tissue in a freeze mill; store in liquid nitrogen. For young tissue, homogenization in a Virtis or Polytron homogenizer may be adequate.

2. Transfer powdered tissue to a 15 ml cooled Corex glass tube and add 3.9 ml 6 M Gu–HCl and sarcosyl to 1%. Add 100 μl Antifoam A and homogenize with a Tekmar tissue homogenizer. Cool sample intermittently by dipping into liquid nitrogen.

3. After homogenization, centrifuge samples at 10 000 *g*, 2 h at 18 °C.

4. Discard pellet and save supernatant. Supernatant can be frozen at this step. Freeze quickly so that Gu–HCl remains in solution and store at −70 °C; thaw at 37 °C. Filter supernatant through a 0.5 μ filter. This filter step may not be necessary for cell lysates.

5. To the supernatant, add 90 μl of 2 M potassium acetate, pH 5.0 (adjusted with glacial acetic acid), vortex and layer over 1.2 ml of a 5.6 M CsCl cushion in a SW 50.1 ultracentrifuge tube (5.2 ml capacity). Place samples in a Beckman SW 50.1 ultracentrifuge rotor and centrifuge at 150 000 *g* for 18 h at 18 °C.

Day 2

6. Remove supernatant from the ultracentrifuge tube using a long Pasteur pipette. Hold the pipette tip at the air/liquid interface and move down the tube as the solution is removed. Leave approximately 100 μl over the RNA pellet. The pellet may be very difficult to see. Invert the tube to decant any remaining supernatant. Cut the bottom off the tube with a scalpel to avoid contamination with protein and DNA on the wall of the upper portion of the tube.

7. Dissolve RNA in 150 μl of DEPC-treated water (leave water on the pellet for 1 h to dissolve, then triturate with a micropipettor to aid dissolution). Wash tube with 50 μl H₂O.

8. Extract once with an equal volume of chloroform:isobutanol (4:1). Add another 100 μl of water to the chloroform:isobutanol, vortex, remove aqueous phase, and add to previous aqueous phase to ensure complete recovery of RNA.

9. Precipitate RNA with 4 vol. of 4 M NaOAc, pH 5.0, at −20 °C overnight.

Day 3

10. Recover RNA by centrifugation for 30 min in a microcentrifuge at 4 °C. Dissolve the pellet in 100 μl DEPC-treated water. Add 0.1 vol. 3 M NaOAc, pH 5.0, and 2.5 vol. 100% EtOH. Precipitate at −20 °C for at least 2 h. Recover RNA by centrifugation for 30 min at 4 °C.

11. Wash pellet with 500 μl of cold 70% EtOH, followed by 500 μl cold 100% EtOH.

12. Dissolve RNA in 10–20 μl of DEPC-treated water and determine the concentration at 260 nm. RNA can be stored frozen at −70°C or in 15 μg aliquots in 0.3 M NaOAc and two volumes of EtOH. The yield from 500 mg of fresh cartilage is approximately 10–20 μg, however, after explant culture, the yield can be 5–10 fold greater. The yield of RNA will be ~10 μg/10^6 cells.

Protocol 4B. Modification of *Protocol 4A* for extraction of RNA from cells (20)

Equipment and reagents

- 4 M Solution X (4 M Gu isothiocyanate; 5 mM EDTA; 50 mM Na citrate) for 100 ml:
  dd H$_2$O, 35 ml
  4 M Gu isothiocyanate, 47.2 g
  5 mM EDTA, 0.18 g
  50 mM Na citrate, 5.0 ml of 1 M, pH 7.0
  Stir on heat until dissolved, q.s. to 100 ml with dd H$_2$O
  Add 500 μl 2 mercaptoethanol
  Store at room temperature
- 20% Sarcosyl
- 5.6 M CsCl/25 mM EDTA Antifoam A
- 18 and 23 gauge needles and 5 ml syringe
- 3 M NaOAc
- Beckman SW 41 rotor and tubes

Method

1. Remove cells from plates or tissue by enzyme digestion (*Protocol 1*). Centrifuge at 2000 *g* for 10 min at 4°C in a 50 ml centrifuge tube. Wash pellet once with 50 ml PBS or HBSS. Pellet cells and remove supernatant completely.

2. To an iced 15 ml Falcon tube, add 2.4 ml of Solution X, 120 μl of 20% sarcosyl and 25 μl of Antifoam A. Vortex and pipette into the tube containing the pellet. Vortex and shear DNA by drawing cells 5 X and 10 X sequentially through 18 gauge and 23 gauge needles using a 5 cc syringe. Add solution X to a final volume of 8 ml and mix well.

3. Add 4 ml of 5.6 M CsCl solution to 12.4 polyallomer ultracentrifuge tube on ice. Layer the cell lysate solution directly on top of the caesium layer. Fill remaining portion of the tube with solution X. Centrifuge in a SW-41 rotor at 150 000 *g* for 18 h at 18°C.

4 Perform steps 6, 7, and 8 as in *Protocol 4A*. Precipitate RNA by adding 0.1 vol of 3 M NaOAc and 2 vol 100% EtOH. Vortex and store at −70°C overnight.

5. Perform steps 11 and 12 of *Protocol 4A* from tissue.

3.1.3 Isolation of mRNA

In *Protocols 2, 3* and *4*, total cellular RNA was isolated. These preparations include cytoplasmic RNAs (tRNA, rRNA, mRNA) as well as nuclear precursors. For most procedures this RNA is adequate. We prefer to work with total RNA because there is less likelihood of mRNA degradation during further purification procedures and most methods of analysis can be performed without additional purification. If cytoplasmic RNA (that is, without nuclear RNA precursors) is desired the reader is referred to detailed methods described in references (9) and (10). mRNA can be isolated by affinity chromotography between the poly (A) tail of the mRNA and oligo dT-cellulose or poly-U Sepharose purchased from most biological companies. Detailed protocols are available in Clemans, this series (1990) (11).

4. Analysis of RNA

4.1 Quantitation

The concentration of RNA can be measured by spectrophotometric determination. The ratio of absorbance at 260 nm (the absorbance of nucleic acids) to 280 nm (absorbance of protein) will provide an indication of the purity of the sample. Values of 1.8 and 2.0 indicate pure preparations of DNA and RNA, respectively. Potential contaminants such as phenol, caesium chloride and Gu–HCl, absorb broadly in the 240 nm range. An absorbance of 1.0 corresponds to approximately 50 µg/ml for double-stranded DNA, 40 µg/ml for single-stranded DNA and RNA, and approximately 20 µg/ml for single-stranded oligonucleotides.

4.2 Cell-free translation

For translation, RNA should be intact and free of contaminants such as heparin, DNA, SDS, and salts that could interfere with translation. RNA can be translated *in vitro* essentially by following the method of Pelham and Jackson (12) using a commercially available micrococcal nuclease-treated rabbit reticulocyte lysate translation system (BRL) incorporating [^{35}S] methionine at a concentration of 1.5 µCi/µl reaction volume. Typical translation incubations (40 µl) contain 13.3 µl nuclease-treated reticulocyte lysate, 82 mM KOAc, 1.16 mM $MgCl_2$, 50 µM of all amino acids except methionine, 64 µCi [^{35}S]methionine, and 200 µg/ml RNA (13, 14). Reaction mixtures are incubated for 90 min at 30°C. Heat denaturation is particularly important for large connective tissue mRNAs and those with a high C + G content such as collagen, fibronectin, or elastin mRNAs

Cell-free synthesized ECM proteins can be identified by separation on SDS-PAGE (collagens and large proteoglycans), immunoprecipitation, or Western blotting (see Hames, this series) (15). The collagenous nature of collagens is verified by digestion with collagenase (see Chapter 1). Purified bacterial collagenase (Advanced Biofactures) can be added directly to the translation mix after incubation or added to the immunoprecipitate to a final concentration of 250 U/ml collagenase, 10 mM NEM, 2.5 mM Tris base (pH 7.4), and 33 mM $Ca(OAc)_2$ (16).

4.3 Northern and slot-blot analysis of mRNA

Specific mRNAs are detected and quantitated by blot transfer analysis. RNA can be separated by electrophoresis through agarose gels and transferred to a solid substrate, called a Northern blot (17). Detailed protocols for the electrophoresis of nucleic acids are given in *Gel electrophoresis of nucleic acids* (18). Alternatively, RNA can be applied directly to the substrate without prior size fractionation (19). We prefer to use nitrocellulose as the substrate material because RNA is transferred efficiently and the blot can be monitored for transfer by staining with ethidium bromide. In addition, nitrocellulose exhibits very low background of non-specific binding of DNA probes and the blot can be re-used. Some investigators prefer to use a nylon substrate; however, in our experience, high background is often encountered minimizing the general usefulness of the blot. (Note: recently, we have used the nylon substrates hydrogen bond N or N^+ (Amersham) with success.) For both Northerns and slot-blots, the RNA must be fully denatured before applying to the nitrocellulose. Particular care must be taken to denature connective tissue mRNA molecules which have a high C + G content. For proper interpretation of the Northern blot hybridization, it is important to load on to the gel both positive and negative control RNAs. *Figure 2* is a Northern blot hybridized to α1(I) and α1(II) procollagen mRNAs showing specific hybridization for these very homologous probes.

Protocol 5. Formaldehyde denaturing agarose gels and Northern blots

Equipment and reagents

- 10 × MOPS—1 litre (final conc.): 40.86 g MOPS (200 mM); 6.8 g NaOAc (80 mM), 20 ml 0.5 M EDTA (10 mM), pH 7.0, adjust pH to 7.0 using NaOH. (DEPC-treat buffer (0.1%) overnight and autoclave.) Store solution cold (4 °C) and light tight (foil wrapped).
- Ethidium bromide
- 1 × SSPE: 0.15 M NaCl, 10 mM sodium phosphate, pH 7.4
- 1 mM EDTA
- 10 × gel-loading buffer: 50% glycerol, 1 mM EDTA (pH 8.0), 0.25% Bromophenol blue, 0.25% xylene cyanol FF (Bio-Rad). Treat with DEPC by adding 1 μl/ml, heat to 37 °C overnight and boil for 10 min to remove residual DEPC.
- 1–2% agarose gel in MOPS
- 37% formaldehyde
- 0.1 M Tris–HCl, pH 8.0

A. *Agarose gel electrophoresis*

1. Prepare 1 × MOPS using DEPC-treated water. Use for the gel and running buffer.

2. Prepare a 1.2% agarose gel (this concentration of agarose is optimal for separating large mRNAs) in 1 × MOPS. Dissolve in a microwave and cool to under 65 °C. Add 16.7 ml of 37% formaldehyde. Swirl until mixed well and pour gel. Gel should be poured on a level surface in a fume hood.

3. Prepare RNA samples (15 μl total volume; increase each component accordingly for 24 μl and 30 samples) as follows:

Protocol 5. *Continued*

- Formamide   5 μl
- Formaldehyde   2 μl
- 10 × MOPS   1.5 μl
- 10 × gel-loading buffer   1.5 μl
- RNA in H_2O 5 μl

After the addition of RNA, heat to 65 °C for 5 min, then place in an ice-water bath for 1 min. Centrifuge briefly in a microcentrifuge to make sure all the contents are at the bottom of the tube.

4. Load the gel. Running buffer is 1 × MOPS. Run the gel at 20 V overnight or 70 V for 3–4 h. **Note:** longer running time at lower voltage will result in tighter bands.

5. After the dye front has migrated three-quarters down the length of the gel, remove and stain the entire gel in 0.5 μg/ml ethidium bromide (EtBr) in 1 × MOPS running buffer for 20–30 min. Destain in running buffer for 15–25 min and photograph the gel. **Note:** Do not add EtBr to samples before running gel as the efficiency of transfer of the RNA to the nitrocellulose may be reduced.

B. *Northern transfer*

1. After photographing the gel, place the gel in a tray and alkali-denature with 50 mM NaOH, 10 mM NaCl. Cover the entire gel with the denaturing solution. Place on a platform rocker for 30–45 min.

2. Decant the denaturing solution, rinse twice with deionized water, and neutralize the gel with 0.1 M Tris–HCl, pH 7.4. Be sure that the entire gel is covered by neutralizing solution and place on the platform rocker for an additional 30–45 min.

3. Discard the neutralizing solution and cover the gel with 20 × SSPE with rocking for 30–45 min.

4. The gel is now ready for transfer to nitrocellulose as shown in *Figure 1*.

Protocol 6. Preparation of RNA for slot-blots

Equipment and reagents

- Dot-blot or slot-blot apparatus
- Nitrocellulose or nylon membrane (Amersham)
- DEPC-treated water
- 10 × SSPE (*Protocol 5*)
- Formaldehyde
- Stratelinker light (Stratagene) (optional)
- 0.45 μm Millipore filter

Method

1. RNA slot-blots are prepared by a modified method of Thomas (19), using a 'dot-blot' or 'slot-blot' apparatus.

2. Prepare the nitrocellulose by soaking in DEPC-H_2O for 10 min, then in 10 × SSPE for 10 min. Filter 10 × SSPE using a 0.45 μm filter to eliminate clogging of the nitrocellulose pores.
3. Denature total cellular RNA in 10 × SSPE and 5% formaldehyde by heating at 60 °C for 15 min then chilling on ice for 15 min.
4. Apply denatured RNA to nitrocellulose in a final volume of 50 μl in the denaturing buffer. After 15 min, apply a gentle vacuum to the nitrocellulose, and wash wells twice with 200 μl of 10 × SSPE. Remove nitrocellulose from the apparatus, air-dry and bake under vacuum for 2 h at 80 °C. In place of baking, nylon membranes can also be treated with a 'Stratalinker' light for 1 min to cross-link RNA to the substrate.
5. Samples are usually loaded in three concentrations: 1 ×, 2 ×, and 4 × RNA and in duplicate. This procedure allows quantitation of specific mRNAs by scanning densitometry of exposures which exhibit linearity of signal.

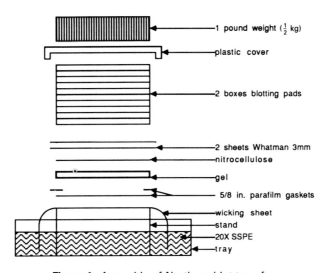

Figure 1. Assembly of Northern blot transfer.

4.4 Hybridization of RNA to DNA probes

mRNA molecules are detected by hybridization to genomic DNA, cDNA, cRNA, or oligonucleotide probes. Probes for many human, mouse, chicken, and bovine connective tissue mRNAs are available from a variety of sources including investigators, certain commercial sources, and the DNA probe bank (ATCC/NIH Repository of Human and Mouse DNA Probes and Libraries). The catalogue from the ATCC/NIH Repository contains valuable information and is available from the American Type Culture Collection,

Linda J. Sandell

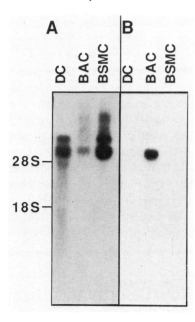

Figure 2. Northern blot of RNAs isolated from 'dedifferentiated' human articular chondrocytes (DC), bovine articular cartilage chondrocytes (BAC), and bovine smooth muscle cells (BSMC). Panel A is probed with a human cDNA insert from the plasmid HF677 (35); hybridized blots were washed at 54°C in 0.1 × SSPE. In α1(I) mRNAs, the difference in band size of the two bands in the DC sample is the stop site of transcription: these two bands can be used to distinguish the α1(I) mRNA pattern from the α1(II) mRNA pattern. The larger bands observed in the BSMC and BAC lanes are nuclear precursors. In panel A, the α1(I) probe shows hybridization to α1(I) mRNA in the chondrocyte RNA: the pattern of mRNA bands indicates that this is not cross-hybridization to α1(II) mRNA. Panel B was probed with a 3.8 kb genomic fragment from the 3' end of the human COL1A2 gene (36). Hybridized blots were washed at 65°C to eliminate cross-hybridization with the α1(I) mRNA. Blots were exposed overnight.

Rockville, MD. The bank can provide probes for many of the collagens, fibronectin, metalloproteinases, thrombospondins, laminins, and elastin. Some proteoglycan probes such as rat serglycin, human decorin, and human versican are available from Telios Pharmaceuticals. References to many connective tissue probes can be found in the book, *Extracellular matrix genes* (20) and from the Gen Bank repository of DNA sequence. Oligonucleotides can be synthesized from published sequences, but care must be taken to ensure specific sequences that do not cross-hybridize with other closely related genes or ribosomal RNAs. Each oligonucleotide must be tested for specific hybridization by Northern blot analysis. *Table 2* lists the sequence of some oligonucleotides we have used successfully to detect ECM mRNAs on Northern blots.

There are advantages and disadvantages that distinguish the use of cDNAs

Table 2. Oligonucleotides for detection of ECM mRNAs

Source Oligo name	Length/GC%		Sequence	(5'-3')/Reference
Human				
Collagens				
α1(I)	24	46	TGA–TTG–GTG–GGA–TGT–CTT–CGT–CTT	d'Allesio *et al.* (1988) (23)
α2(I) Exon 1/2	27	44	CTT–GTA–AAG–ATT–GGC–ATG–TTG–CTA–GGC	deWet *et al.* (1987) (24)
α1(IIA) Exon 1/2	24	67	TGC–CAG–CCT–CCT–GGA–CAT–CCT–GGC	Ryan *et al.* (1990) (25)
α1(IIB) (1/3)	24	67	CTC–CTG–GTT–GCC–GGA–CAT–CCT–GGC	Ryan *et al.* (1990) (25)
α1(II) (exon 8)	21	48	GCC–TTC–TGA–TCA–AAT–CCT–CCA	Sandell *et al.* (1991) (26)
Type VI	24	50	GTT–CAG–GTA–TTG–TGT–GGT–CTC–CAG	Saitta *et al.* (1990) (27)
Proteoglycans				
Aggrecan	25	52	ACA–GCG–ATA–TGA–CGT–GTC–CCT–CTG–T	Doege *et al.* (1991) (28)
Biglycan	25	52	CGT–TCA–TCA–TGA–ATG–GCC–CAT–GGT–C	Fisher *et al.* (1989) (29)
Decorin	24	54	TCC–AGC–CCA–GGA–AAC–TTG–TGC–AAG	Krusius and Ruoslahti (1986) (30)
Versican	24	50	CTG–TTC–TGA–TGC–TGC–TAT–CGT–GGA	Zimmermann and Ruoslahti (1989) (31)
Fibromodulin	24	63	CTG–CCT–CTC–CAG–GTG–CCT–CAG–CTT	Oldberg *et al.* (1989) (32)
Bovine				
α1(IIB) collagen	24	46	GTC–ATC–TCC–ATA–GCT–GAA–GTG–GAA	Sangiorgi *et al.* (1985) (33)
Aggrecan	22	55	TTC–AGG–CCG–ATC–CAC–TGG–TAG–T	Oldberg *et al.* (1987) (34)

and oligonucleotides as probes. cDNAs are longer and can incorporate more isotope. Consequently, cDNAs detect fewer copies of mRNA. There is a greater likelihood also of successfully using cDNA probes across species as the per cent difference in DNA sequence will be low. However, the preparation of cDNAs requires facilities for the bacteriological methods required for growing plasmids in bacteria. The advantages of oligonucleotides are that they can essentially be used as shelf reagents once synthesized, and they are very specific. Due to their size and single-stranded nature, however, they cannot be labelled to a high specific activity, therefore the signal may be lower. Oligonucleotides will be less useful than cDNA probes for cross-species hybridization unless the sequence for each species is known to be homologous. *Figure 3* shows the use of oligonucleotides and cloned cDNAs on the same filter. *Figure 4* shows hybridization to two proteoglycan core protein oligonucleotide probes.

For a reference to standardize the loading of RNA on the gel, a variety of

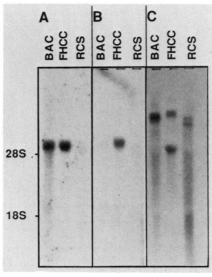

Figure 3. Hybridization to oligonucleotide and cDNA probes. Northern blot of RNAs isolated from bovine articular chondrocytes (BAC), fetal human costal cartilage chondrocytes (FHCC), and rat chondrosarcoma cells (RCS). The same filter was re-used after removing the probe: the filter was washed twice with 90°C RNase-free water for 15 min each wash. Removal of labelled probe was checked by exposure to X-ray film. Oligonucleotides are easily removed while cDNA probes require more washing. Minimum washing is recommended for extending the life of the filter. Panel A was probed with the type IIB oligonucleotide (*Table 1*). Note hybridization to BAC and FHCC: positive signal was observed in the RCS sample upon longer exposures. Panel B was probed with the type IIA oligonucleotide; Panel C was probed with a rat aggrecan cDNA probe provided by Dr Kurt Doege. Note the difference in mRNA sizes between the three species. The lower band in the FHCC cells is probably degraded mRNA; and the smear in the RCS sample is degraded RNA. Blots were exposed for 23 h.

probes can be used as long as their concentration is not altered by the experimental conditions. A probe for actin mRNA is often used, however, it is not usually appropriate for chondrocytes which exhibit changes in actin concentration and isoform upon dedifferentiation. The housekeeping genes, glyceraldehyde-6-phosphate dehydrogenase and ribosomal elongation factor 1α (21), are useful. A probe specific for 28S or 18S ribosomal RNA can also be used (22), however, because mRNA is present in an approximately two orders of magnitude lower concentration, we generally prefer to use a mRNA control. An example of using the elongation factor 1α probe is shown in *Figure 5*.

Many connective tissue proteins are now known to be encoded by closely related groups of genes called gene families. Due to sequence similarity and often a relatively high C + G content, care must be taken to detect and eliminate cross-hybridization to related genes. This is accomplished by the judicial selection of positive and negative control RNAs and stringent hybridization conditions. Probes can be used 'across species', however, it must be shown that these heterologous probes hybridize to specific mRNAs on a Northern blot. Often, more than one band on the Northern blot will hybridize to the probe. These multiple mRNA bands can be due to the presence of

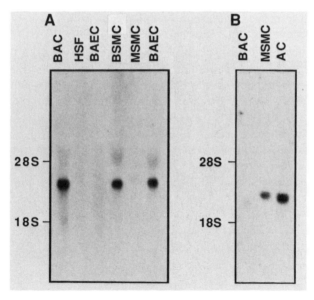

Figure 4. Northern blot of an oligonucleotide hybridized to RNAs isolated from bovine articular chondrocytes (BAC), human skin fibroblasts (HSF), bovine aortic endothelial cells (BAEC), bovine smooth muscle cells (BSMC), monkey smooth muscle cells (MSMC), and articular cartilage (AC). Panel A was hybridized to an oligonucleotide encoding fibromodulin; Panel B was hybridized to an oligonucleotide encoding biglycan. Panel A was exposed for 5 days, panel B was exposed overnight.

nuclear precursors, differences in coding sequence generated by alternative splicing of the pre-mRNA, or to additional mRNAs generated by downstream transcription stop sites. The number and pattern of multiple mRNA bands is often, although not necessarily, different between species.

For radiolabelling cloned DNA probes, we prefer the method of nick-translation, although cRNA 'riboprobes' and probes labelled by the random primer method can also be used (see references 10 and 9 for further methodologies).

Protocol 7. Hybridization to cloned DNA probes

Equipment and reagents

- [α-^{32}P]dCTP or commercial labelling kit (e.g. Nick-translation kit, BRL)
- Solution A: 50% formamide, 5 × SSPE, 5 × Denhardt's solution
- Solution B: Solution A, 10% dextran sulfate (Sigma), 100 ng/ml labelled cDNA
- Solution C: 2 × SSPE, 0.1% SDS
- Solution D: 0.1 × SSPE, 0.1% SDS
- Kodak XAR X-omat film

Method

1. Cloned DNAs are labelled with [α-^{32}P]dCTP by nick translation (10). Commercial kits are appropriate for labelling (for example, BRL Nick-translation kit). The cloned insert cDNA is generally isolated from a plasmid to reduce background hybridization; some plasmids do not require isolation of the insert cDNA.

2. Prehybridize filters in Solution A for 2 h to overnight at 42 °C.

3. Transfer to hybridization Solution B. Denature the probe by heating to 100 °C for 5 min and cooling on ice before adding to the hybridization mix. If already mixed, heat probe longer to ensure denaturation. Hybridize at 42 °C for 18–24 h.

4. Wash filters in 2 changes, 15 min each, of Solution C followed by 2 changes with Solution D at 54 °C, air-dry on Whatman paper to remove excess moisture and expose to Kodak XAR X-omat film. Monitor background radioactivity with a Geiger counter. Higher temperatures may be necessary to eliminate background or cross-hybridization to closely related mRNAs.

Protocol 8. Oligonucleotides as DNA probes

Equipment and reagents

- Acrylamide (40%, deionized)
- Urea
- 10 × TBE, pH 8.3
- Ammonium persulfate (25%)
- TEMED
- Forward reaction buffer: 300 mM Tris–HCl, pH 7.8, 75 mM 2-mercaptoethanol, 50 mM MgCl$_2$, 1.6 μM ATP
- [γ-^{32}P]ATP (5000 Ci/mM)
- T4 polynucleotide kinase (5 units/μl)
- TE buffer, pH 8.0

- Gel-loading buffer: 0.1% xylene cyanol FF (Bio-Rad), 0.1% Bromophenol blue, 10 mM EDTA (Sigma), pH 8.0, in formamide
- Hand-held UV lamp (260/280 nm)
- Sephadex G-50
- USB SurePur kit
- 5′-Terminus labelling kit (BRL)

- 25% formamide/5 × SSPE
- 10% dextran sulfate
- 2 × SSPE/0.1% SDS
- Phenol:chloroform
- Salmon sperm DNA, sheared

A. *PAGE purification of oligonucleotides*

1. Dilute DNA in sterile water to approximately 4 μg/ml, and determine the concentration at 260 nm.

2. Assemble the gel apparatus and cast one or two gels, depending on the number of oligonucleotides you want to purify. The gel concentration is 20% acrylamide/8 M urea. For two gels mix the following:
   - Acrylamide (40%, deionized)   50 ml
   - Urea   48 g
   - 10 × TBE, pH 8.3   5 ml

   Mix until urea dissolves, then add:
   - Ammonium persulfate (25%)   200 μl
   - TEMED   70 μl

3. To prepare 100 μg of each oligonucleotide sample (in 25 μl), add 5 μl gel-loading buffer.

4. Load 25 μg per well and run the gel until the first dye front reaches the bottom of the gel slab.

5. Disassemble the apparatus, and visualize oligonucleotide bands with a hand-held UV 260/280 lamp. The bands will be visible under the short wave (260 nm) light. Slice out the bands with a razor blade, and place the gel pieces in a 3 ml syringe.

6. Using the plunger, crush the pieces into microcentrifuge tubes, and add enough sterile water to cover the acrylamide pieces. Seal the tubes with Parafilm, and incubate in a 60°C water bath overnight to elute the oligonucleotide from the gel.

7. Remove the tubes and pipette out the liquid, being careful not to take up too many acrylamide pieces. Add more water to the gel pieces and elute again overnight if yield is low from the first elution.

8. Determine the volume of the solution, and ethanol precipitate the DNA (add 1/10 volume sodium acetate and 2 volumes absolute ethanol to each tube; store at −20°C for at least 2 h). Centrifuge tubes in a microcentrifuge for 20 min at 19000 g at 4°C and pour off supernatant. Dry pellet, and dissolve in 100 μl of sterile water.

9. Pour a Sephadex G-50 gravity column in a Pasteur pipette. Apply the DNA solution to the column, and collect ten 100 μl fractions using sterile water. Read the fractions on UV at 260/280 nm and pool the fractions contain-

Protocol 8. *Continued*

ing the oligonucleotide. Ethanol precipitate, and dilute the oligonucleotides to the appropriate concentration (we use 1.5 pmol/ml).

B. *Purification of oligonucleotides by USB SurePur kit*

Recently, this reagent kit has become available for the purification of
oligonucleotides by a chromatographic method. We have found this kit
easy and efficient to use.

C. *Radiolabelling oligonucleotides*

1. A 5'-terminus labelling kit can be purchased from BRL. Mix the following in a microcentrifuge tube:
   - oligo DNA (1.5 pmol/ml)   10 μl
   - Sterile water   4 μl
   - 5 × Forward reaction buffer   5 ≅l
   - [γ-^{32}P]ATP (5000 Ci/mmol)   5 μl
   - T4 polynucleotide kinase (5 units/μl) 1 μl

2. Incubate at 37 °C in a water bath for 30 min.

3. Add 25 μl TE buffer to the reaction tube to stop the reaction. Add 50 μl
phenol:chloroform (1:1, v/v), vortex, and centrifuge in microcentrifuge
for 2 min.

4. Separate aqueous layer, and apply to a Sephadex G-50 gravity column.
Collect 150 μl fractions, and count fractions. Pool fractions with labelled
oligonucleotide.

D. *Hybridization*

1. Prehybridize filter in a solution of 25% formamide, 5 × SSPE, 0.2 mg/
ml salmon sperm DNA, 1 × Denhards, 1% SDS.

2. Add labelled oligonucleotide to hybridization solution (same as prehybridization solution, except with 10% dextran sulfate).

3. Calculate the melting temperature of the oligonucleotide:mRNA hybrid
using the following formula:

$$T_m = 16.6 \log[Na^+] + 0.41(\% \, GC) + 81.5 - 675/L - 0.65(Pf)$$

where: $[Na^+]$ = molar concentration of sodium ion (for hybridization
mix it is 0.825 M), %GC = whole number percentage of G + C in oligo
sequence; L = length of oligonucleotide, and Pf = whole number
percentage of formamide (here it is 25).

(a) For hybridization temperature, take T_m, 20 °C.

(b) For wash temperature, recalculate the T_m for slightly different conditions: the sodium ion concentration is 0.325 M (2 × SSPE), and
there is no formamide in the wash solution so the last term of the
equation drops out. The wash temperature is now the new T_m, 15 °C.

4. Pour off prehybridization solution and add hybridization solution containing labelled oligonucleotide. Hybridize filter overnight at computed hybridization temperature.

5. Wash filter with 2 × SSPE/0.1% SDS for 10 min at computed wash temperature. Check background with a Geiger counter, and repeat wash if high. Decrease concentration of SSPE in wash solution if higher stringency is necessary for subsequent washes.

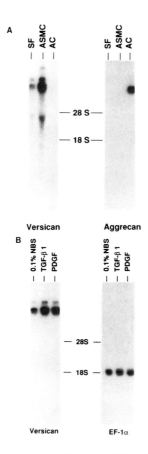

Figure 5. Northern transfer analysis using the control hybridization probe EFl-α. Panel A, total cellular RNA (10 μg/lane) from cultured human skin fibroblasts (SF), monkey aortic smooth muscle cells (ASMC), and bovine articular chondrocytes (AC) probed first with a nick-translated cDNA for versican and then with a nick-translated cDNA for aggrecan. Panel B, total cellular RNA (5 μg/ml) from unstimulated (lane 1), or TGF-β1 (lane 2), or PDGF (lane 3) stimulated ASMCs in culture, probed first with a cDNA for versican and then with a cDNA for elongation factor 1-α (EF 1-α) serving as a reference mRNA. Reproduced from reference 37, with permission.

4.4.1 Additional considerations

Formaldehyde denaturation, extended prehybridization, and the calculation of the hybridization temperature are important for obtaining reproducible, quantitative results with low background. For initial hybridizations used to empirically establish hybridization conditions, the filter can be exposed after washing in 2 × SSPE (if the background is low as determined by a Geiger counter). Hybridization can be quantitated by liquid scintillation counting of the individual slots or bands or scanning the autoradiograph with a densitometer, or by phosphoimage analysis. Linear hybridization conditions should be established and multiple exposures can be used to stay within the linear absorbance range of the densitometer. Background on the autoradiograph can be a large problem. It can be avoided by making sure everything is free of dust and salt and gloves are worn at all times. If background cannot be eliminated, the prudent solution is to remake all components of the hybridization mix.

Acknowledgements

The assistance of Margo Weiss in the preparation of this chapter is greatly appreciated. Special thanks to past and present laboratory members: Dr Tom Hering, Dr Maureen Ryan, Dr Oliver Zamparo, Andrew Nalin, and James Sugai for contributing to the methods and reviewing the manuscript.

References

1. Beutler, E., Gelvart, T., and Kuhl, W. (1990). *Biotechniques*, **9**, 166.
2. Strohman, R., Moss, P., Micou-Eastwood, J., Spector, D., Przybyla, A., and Patterson, B. (1977). *Cell*, **10**, 265.
3. Chomczynski, P. and Sacchi, N. (1987). *Anal. Biochem.*, **162**, 156.
4. Leboy, P. S., Shapiro, I. M., Uschmann, B. D., Oshima, O., and Lin, D. (1988). *J. Biol. Chem.*, **263**, 8515.
5. Nemeth, G. G., Heydemann, A., and Bolander, M. E. (1989). *Anal. Biochem.*, **183**, 301.
6. Adams, M. E., Huang, D. Q., Yao, L. Y., and Sandell, L. J. (1992). *Anal. Biochem.*, **202**, 89.
7. Zarlenga, D. S. and Gamble, H. R. (1987). *Anal. Biochem.*, **162**, 569.
8. Glisin, V., Crkkvenjakov, R., and Byus, C. (1974). *Biochemistry*, **13**, 1633.
9. Ausubel, F. M., Brent, R., Kingston, R. E., Moore, D. D., Seidman, J. G., Smith, J. A., and Struhl, K. (1989). *Current protocols in molecular biology*. Greene Publishing Associates and Wiley-Interscience, New York.
10. Sambrook, J., Fritsch, E. F., and Maniatis, T. (1989). *Molecular cloning: a laboratory manual* (2nd edn). Cold Spring Harbor Laboratory Press, Cold Spring Harbor, New York.
11. Clemens, M. (1990). In *Transcription and translation: a practical approach* (ed. S. J. Higgins), pp. 211–30. IRL Press, Oxford.

12. Pelham, H. R. B. and Jackson, R. J. (1976). *Eur. J. Biochem.*, **67**, 247.
13. Sandell, L. J., Sawhney, R. S., Yeo, T.-K., Poole, A. R., Rosenberg, L. C., Kresse, H., and Wight, T. N. (1988). *Eur. J. Cell. Biol.*, **46**, 253.
14. Hering, T. M. and Sandell, L. J. (1990). *J. Biol. Chem.*, **265**, 2375.
15. Hames, B. D. (1990). In *Gel electrophoresis of proteins* (ed. B. D. Hames and D. Rickwood) (2nd edn), p. 1. Oxford University Press, Oxford.
16. Sandell, L. J. and Veis, A. (1980). *Biochem. Biophys. Res. Comm.*, **92**, 554.
17. Alwine, J. C., Kemp, D. J., Parker, B. A., Reiser, J., Renart, J., Stark, G. R., and Wahl, G. M. (1979). In *Methods in enzymology* (ed. Wu, R.), Vol. 68, p. 220.
18. Rickwood, D. and Hames, B. D. (1990). *Gel electrophoresis of nucleic acids: a practical approach* (ed. D. Rickwood and B. D. Hames) 2nd edn. Oxford University Press, Oxford.
19. Thomas, P. (1980). *Proc. Natl. Acad. Sci. USA*, **77**, 5201.
20. Sandell, L. J. and Boyd, C. A. (1990). In *Extracellular matrix genes* (ed. Boyd and Sandell), pp. 1–56. Academic Press, Orlando.
21. Järveläinen, H. T., Kinsella, M. G., Wight, T. N., and Sandell, L. J. (1991). *J. Biol. Chem.*, **266**, 23274.
22. Sandell, L. J. and Daniel, J. C. (1988). *Connect. Tissue Res.*, **17**, 11.
23. d'Allesio, M., Bernard, M., Benson-Chanda, V., Chu, M.-L., Weil, D., and Ramirez, F. (1988). *Gene*, **67**, 105.
24. de Wet, W., Bernard, M., Benson-Chanda, V., Chu, M-L., Dickson, L., Weil, D., and Ramirez, F. (1987). *J. Biol. Chem.*, **262**, 16032.
25. Ryan, M. C., Sieraski, M., and Sandell, L. J. (1990). *Genomics*, **8**, 41.
26. Sandell, L. J., Morris, N., Robbins, J. R., and Goldring, M. R. (1991). *J. Cell Biol.*, **114**, 1307.
27. Saitta, B., Stokes, D. G., Vissing, H., Timpl, R., and Chu, M-L. (1990). *J. Biol. Chem.*, **265**, 6473.
28. Doege, K. J., Saski, M., Kimura, T., and Yamada, Y. (1991). *J. Biol. Chem.*, **266**, 894.
29. Fisher, L. W., Termine, J. D., and Young, M. F. (1989). *J. Biol. Chem.*, **264**, 4571.
30. Krusius, T. and Ruoslahti, E. (1986). *Proc. Natl. Acad. Sci. USA*, **83**, 7683.
31. Zimmermann, D. R. and Ruoslahti, E. (1989). *EMBO J.*, **8**, 2975.
32. Oldberg, A., Antonsson, P., Lindblom, K., and Heinegard, D. (1989). *EMBO J.*, **8**, 2601.
33. Sangiorgi, F. O., Benson-Chanda, V., deWet, W. J., Sobel, M. E., and Ramirez, F. (1985). *Nucleic Acids Res.*, **13**, 2815.
34. Oldberg, A., Antonsson, P., and Heinegard, D. (1987). *Biochem. J.*, **243**, 255.
35. Chu, M.-L., Weil, D., de-Wet, W., Bernard, M., Sippola, M., and Ramirez, F. (1985). *J. Biol. Chem.*, **260**, 4357.
36. Ryan, M. C. and Sandell, L. J. (1990). *J. Biol. Chem.*, **265**, 10334.
37. Schönherr, E., Järveläinen, H. T., Sandell, L. J., and Wight, T. N. (1991). *J. Biol. Chem.*, **266**, 17640.

16

Generation of cosmid DNA libraries and screening for extracellular matrix genes

DAVID TOMAN and BENOIT DE CROMBRUGGHE

1. Introduction

The application of molecular biological techniques to study the structure, function, and transcriptional regulation of extracellular matrix genes has provided a wealth of knowledge at the DNA level. A nine base pair repeat encoding Gly–X–Y at the amino acid level has been implicated as the ancestor of the collagen molecule. This repeat has been assembled into a 54 base pair exon which has been duplicated and combined with other functional groups to generate at least part of the family of collagen genes. Individual members of the collagen family show tissue specificity and provide different structural roles in many multicellular organisms. Analyses of the role of collagens in human connective tissue diseases have shown a wide spectrum of phenotypic effects which are currently studied in order to more fully understand their functional role in an organism. Mouse transgenic models are probing structure/function relationships with other components of the extracellular matrix. Identification and characterization of specific transcription factors and *cis*-acting elements are unravelling the regulatory pathways which determine the tissue- and developmental-regulation of expression of these genes.

The size and complexity of the collagen genes has posed roadblocks in the elucidation of their molecular components. Efforts directed toward cloning of several of these genes have provided a basic understanding of these genes and have laid a foundation for further studies. *Table 1* lists the size and complexity of the known collagen genes within the collagen gene family of extracellular matrix proteins. Many experimental approaches used in the study of extracellular matrix genes require the entire gene to be contiguous on one fragment of DNA. Since several of the genes for the collagen molecules are large, it is useful to have a method which can isolate and identify fragments of genomic DNA which contain the entire gene locus.

Two methods have been used in the isolation of genomic fragments

Table 1. Collagen genes: sizes, exon numbers, and length of coding regions

Type	Gene size[a] (kb)	No. exons	Coding region (nt)	Reference
α1(I)	18	51	4392	(1, 2)
α2(I)	38	52	4098; 4092[b]	(3–5)
α1(II)	30; 29	54	4257, 4464[c]	(6–9)
α1(III)	44; 38	51	4401	(10–12 and D. Toman, personal communication)
α1(IV)	>100[d]	52	5055; 5007	(13–16)
α2(IV)	> 60	> 19	5136; 5121	(14, 17, 18)
α3(IV)	ND	ND	>1413	(19)
α4(IV)	ND	ND	>1359	(20)
α5(IV)	130	51	5055	(21)
α1(V)	ND	ND	5514	(22)
α2(V)	> 8	> 6	4410	(23, 24)
α3(V)	ND	ND	ND	
α1(VI)	> 36; 21	34	3084; 3057	(25–28)
α2(VI)	> 35; 26	28	2733–3045	(28–30)
α3(VI)	> 26	> 10	9411	(31–33)
α1(VII)	> 26	>119	>8400	(D. Greenspan, personal communication; A. Christiano, personal communication)
α1(VIII)	> 53	4	2232	(34–37)
α2(VIII)	ND	> 1	>1907	(38)
α1(IX)	100	> 19	2063, 2793; 2037, 2760	(37, 39)
α2(IX)	10	32	2031	(37)
α3(IX)	ND	ND	2025	(40)
α1(X)	7; 5	3	2040; 2025; 2046	(37, 41–43)
α1(XI)	ND	ND	>3807	(44)
α2(XI)	ND	> 6	>2883	(45)
α3(XI) same as α1(II)				
α1(XII)	> 12	> 7	9372	(37, 46)
α1(XIII)	130	> 39	1842	(47–49)
α1(XIV)	ND	ND	>588	(50)
α1(XV)	> 50	ND	>2127	(51)
α1(XVI)	ND	ND	4809	(52)

[a] The size and number of exons listed may be from a different species than the listing for the coding region.
[b] Semi-colon separates values obtained from two or more species.
[c] Comma separates two sizes due to use of alternate exons which are mutually exclusive.
[d] Denotes either a minimum size of the gene or minimum exon number, i.e. the gene has not been fully characterized.
[e] α1(II), α3(VI), α2(XI), and α1(XIII) mRNAs are subject to alternate splicing.
ND = not listed in table.

containing some or all of the gene locus for the collagen genes. Cloning of fragments of extracellular genes within lambda phage vectors has been used extensively in the field of collagen research. A less-often used approach has been the use of cosmid clones to isolate larger fragments of these genes. The major advantage in the cloning of extracellular matrix genes using cosmids is that approximately twice the size of a linear DNA fragment can be isolated within one clone. Analysis of far-upstream and downstream regions of an extracellular matrix gene as well as the cloning of a complete functional gene with its regulatory regions are more easily performed with cosmid clones. However, it is generally acknowledged that the mechanics of isolating a clone from a cosmid library are slightly more complicated than from a lambda phage library. The purpose of this chapter is to provide a complete series of protocols which can be used to successfully construct and screen a cosmid library. Many of the steps are similar to those used in the construction of lambda phage libraries. We have used these methods to clone several of the fibrillar collagen genes.

The most significant advantage today of using lambda phage libraries is the availability of many of these libraries from commercial vendors at a reasonable cost. Currently, the commercial availability of cosmid libraries is limited. Other types of genomic library construction, such as P1 and YAC systems, are beyond the scope of this chapter.

1.1 Outline of basic method

To clone into cosmids, genomic DNA is cleaved in a semi-random fashion by partial restriction enzyme digestion with isolation of the 35–45 kb size range (see *Figure 1* for an illustration of the overall strategy of cosmid DNA cloning). This fractionated DNA is ligated to a linearized cosmid vector under conditions which favour the formation of concatameric DNA, that is a linear stretch of DNA containing multiple copies of both the vector and the genomic DNA in tandem. Addition of lambda packaging extract to the ligated DNA results in the cleavage of the concatamers at the cos site (cohesive end site) of the cosmid vector and incorporation of linear fragments of DNA containing the genomic DNA insert into the mature lambda bacteriophage particle. The lambda particle is used only as a vehicle to introduce the recombinant DNA into bacteria. Upon infection of *Escherichia coli*, the DNA within the packaged phage is injected into the bacterial cell where it circularizes via the cos sites. The cosmid then replicates as a plasmid within the *E. coli* host and expresses its selectable marker. By plating a library of bacteria containing cosmid clones on filters followed by denaturation and fixation of the DNA to the filters, and subsequent hybridization using a specific DNA probe, a clone can be isolated which contains a gene with its flanking regions and regulatory sequences.

1.2 Verification of clone

It is important to be able to verify the identity of an isolated genomic sequence using an independent method. This is especially important when isolating one member from a family of related genes such as the collagen genes. There are several methods that can be used in the determination of the identity of a clone. One method is by comparison of the maps generated by

David Toman and Benoit de Crombrugghe

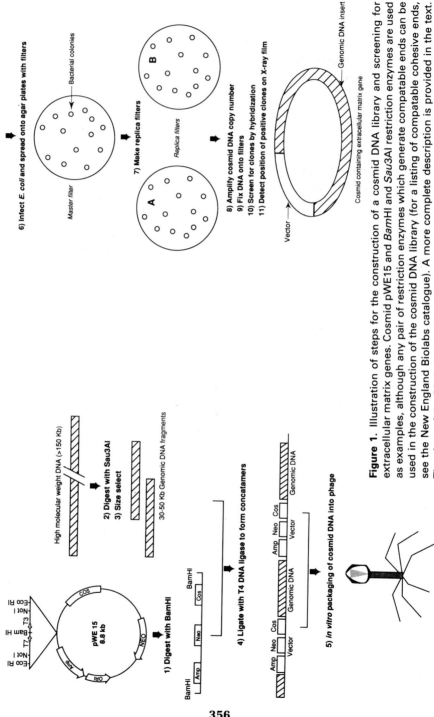

Figure 1. Illustration of steps for the construction of a cosmid DNA library and screening for extracellular matrix genes. Cosmid pWE15 and *Bam*HI and *Sau*3AI restriction enzymes are used as examples, although any pair of restriction enzymes which generate compatable ends can be used in the construction of the cosmid DNA library (for a listing of compatable cohesive ends, see the New England Biolabs catalogue). A more complete description is provided in the text. The sizes of the DNAs in the illustration are not drawn to scale.

356

restriction enzyme digestion between the purified cosmid DNA and total genomic DNA by Southern analysis (52). In this method, the radioactively-labelled DNA probe is hybridized to genomic DNA which has been digested by various restriction enzymes and transferred on to a nylon membrane. By performing a Southern analysis using several restriction enzymes, a diagnostic restriction enzyme pattern of the gene can be developed. The isolated cosmid clones are digested with the same restriction enzymes and the pattern generated by the digest is compared to the genomic restriction enzyme pattern. Another method is the use of an independent DNA probe specific to the gene. This probe should not have been used in the screening of the library. A third, and the most definitive method of verifying the identity of a gene, is by DNA sequencing and comparison of the resulting sequence with available published DNA sequence of the gene from either the identical or a similar species. The general procedure for a DNA sequencing project is to generate templates for sequencing by subcloning fragments of the cosmid clone into plasmid vectors. The subcloning is usually performed in a series of steps:

(a) mapping of the cosmid clone using restriction enzymes which have a hexa-, hepta-, or octanucleotide recognition sequence;

(b) subcloning of larger fragments of the cosmid clone;

(c) often further restriction enzyme mapping of these subclones using a different subset of restriction enzymes; and

(d) generating smaller subclones suitable as DNA sequencing templates.

Since DNA sequencing of a cosmid clone requires several steps, it is common to use one of the first two approaches initially to verify the identity of the gene and then follow with DNA sequencing of the exons of the gene.

2. Preparation of cosmid vector

The total size of DNA, both cosmid vector and genomic DNA insert, which can be packaged into a lambda particle ranges from 38 kb to 52 kb. Usually, the size of the cosmid vector is approximately 5–8 kb, so the amount of genomic DNA that can be isolated in one clone ranges from a minimum of 30 kb to a maximum of 47 kb. This size range is approximately twice the size of DNA which can be isolated from lambda phage libraries (which usually is within the size range of 9 to 23 kb). Inclusion of additional features into the cosmid vector increases the size of the vector and thereby decreases the size distribution of genomic DNA inserted into this vector by the same amount. Several common cosmid vectors and their properties are described in *Table 2*. Descriptions of each of the components of cosmid vectors are discussed below.

The cosmid vector contains a minimum of four features which make it useful for genomic DNA cloning and isolation. First, the cosmid vector, like

Table 2. Features of several cosmid vectors

Cosmid vector	Size (kb)	Unique site	Flanking sites	Bacterial selection	Mammalian selection	Additional features	References
pJB8	5.4	*BamHI*	*EcoRI*	amp[f]	—	—	54
pWE15[a,b,c]	8.8	*BamHI*	*NotI*	amp	neo[j]	T3, T7 promoters	55
pWE16	8.8	*BamHI*	*NotI*	amp	meth[k]	T3, T7 promoters	55
Lorist 6	5.2	*BamHI, ScaI, NotI, HindIII*	—	kan[g]	—	Sp6, T7 promoters	56
SuperCos[a]	7.6	*BamHI*	*NotI, EcoRI*	amp	neo	T3, T7 promoters, 2 cos sites	57, 58
pHC79[d,e]	6.4	*EcoRI, HindIII*	—	amp, tet[h]	—	—	59
pSV13	7.7	*PstI*	—	amp, chl[i]	MPA[l]	—	59
pTCF	7.7	*BamHI*	*SalI*	amp	neo	—	60
pGNC	8.3	*BamHI*	*SalI*	amp	HAT[m]	—	60
pNN1	7.7	*BamHI*	*SalI*	amp	MPA	—	60
pHSG274	3.4	*BamHI*	*EcoRI, SalI*	kan	neo	—	61
Cos202	10	*BglII*	—	amp	hyg[n]	EBNA-1 gene[o]	62

[a–e] Available commercially from [a]Stratagene Cloning Systems, [b]Pharmacia LKB Biotechnology, [c]Clontech Laboratories, Inc., [d]Boehringer–Mannheim Biochemicals, or [e]Life Technologies, Inc. (Gibco BRL).
[f–i] Resistance to the antibiotics [f]ampicillin, [g]kanamycin, [h]tetracycline, and [i]chloramphenicol.
[j–n] Growth in media containing [j]neomycin or G418 sulfate, [k]methotrexate, [l]adenine, xanthine, and mycophenolic acid (MPA), [m]hypoxanthine, aminopterin, and thymidine (HAT media), or [n]hygromycin.
[o] EBNA-1 gene directs episomal replication in mammalian cells.

any plasmid vector, contains both an origin of replication and a selectable marker—usually a drug resistance marker such as beta-lactamase for ampicillin resistance. For this reason, a cosmid vector will behave as any plasmid vector and can be introduced into competent *E. coli* by standard transformation or electroporation procedures. Second, the cosmid vector can be isolated using standard plasmid purification techniques (53). Third, the cosmid vector also contains lambda cos DNA sequences which are required for *in vitro* packaging of recombinant DNA molecules into the bacteriophage lambda particles. Finally, the cosmid vector contains a unique restriction enzyme recognition site into which the isolated genomic DNA fragments can be inserted.

Cosmid vectors may have additional features engineered into them. One valuable feature is for the cosmid vector to contain flanking restriction enzyme recognition sites which surround the unique restriction enzyme site used for insertion of the genomic DNA. The restriction enzyme sites for the rare cutters such as *Not*I or *Sal*I often are used. By a standard restriction enzyme digestion using a restriction enzyme which recognizes the flanking sites, the genomic DNA can be removed from the cosmid vector, usually intact, and purified. This feature is useful for linearizing the genomic DNA and removing the cosmid vector prior to injection of the DNA into mouse oocytes for the generation of transgenic mice or before electroporation into mammalian cells in culture.

For structural and functional analysis of complex genomes, isolation of several overlapping cosmid clones may be necessary to determine a physical map for large regions of genomic DNA. Generally, a technique described as 'chromosome walking' is used to sequentially isolate neighbouring overlapping cosmid clones either uni- or bi-directionally (55). In this method, a probe, corresponding to one of the terminal fragments of a previously isolated cosmid clone, is used to hybridize to the cosmid library to identify an adjoining cosmid clone. A feature which may be added to a cosmid vector in order to simplify this approach is the inclusion of transcriptional promoters flanking both sides of the insertion site for the genomic DNA. An RNA probe can be generated in a transcription reaction using an RNA polymerase which recognizes one of these flanking promoters. Common promoter motifs contained within cosmid vectors are the recognition sites for the RNA polymerases of both T3 and T7 phage (55). Transcription reaction kits using either T3 or T7 polymerases are available commercially. This same approach may be used for chromosome jumping, in which larger regions of genomic DNA have to be covered (62).

Another feature which may be included in some cosmid vectors is eukaryotic selectable markers. Transfection of eukaryotic cells with DNA has proven to be a valuable tool in the study of gene expression. Large regions of eukaryotic DNA representing an entire gene, complete with upstream and internal regulatory regions, can be transfected into cultured animal cells. Isolation of genomic DNA clones from cosmid libraries constructed with cosmid vectors containing selectable markers, such as neomycin and hygromycin driven by eukaryotic promoters, can be used for these studies without further manipulation.

Additional features which may be included into cosmid vectors are:

(a) incorporation of two cos sites instead of one (57);

(b) inclusion of a eukaryotic origin of replication (62); and

(c) introduction of a stuffer fragment to specify a smaller size distribution (64).

2.1 Digestion and phosphatase treatment of cosmid vector

Protocol 1. Digestion and phosphatase treatment of cosmid DNA

Equipment and reagents

- 0.7% agarose gel
- Preparative comb (or see step 1)
- pJB8 or pWE15 cosmid vector (or vector of choice)
- *Bam*HI
- Calf intestinal phosphatase (CIP) (Boehringer-Mannheim cat. no. 108138)
- 10 × Dye buffer
- DNA size marker

- *sec*-Butanol
- Phenol:chloroform (1:1 v/v)
- NaOAc, pH 5.2
- Ethanol
- $T_{10}E_1$ buffer: 10 mM Tris–HCl, 1 mM EDTA, pH 8.0
- *E. coli*
- Ampicillin-containing agar plate

Method

1. Cast a 0.7% agarose gel using a preparative comb. If a preparative comb is unavailable, one can be made by taping several wells of a standard comb together. A 1.5 mm preparative well of 3.5 cm in length for a 1 cm thick gel will hold approximately 350 μl of sample. A well of the dimensions listed will efficiently fractionate up to 10 μg of DNA; otherwise, overloading of the gel will occur.

2. Digest 10 μg of the pJB8 or pWE15 cosmid vector (or vector of choice) with 5-fold excess of *Bam*HI for 1 h at 37 °C. Generally, three individual digests are performed. One of these digests will be used to check the efficiency of step 3.

3. Add 0.2 U of calf intestinal phosphatase (CIP) directly to all but one of the digestions and incubate for an additional 30 min.

4. Add 10 × dye buffer to each sample and load the samples directly on to the agarose gel. Include a DNA size marker on the gel. Run the gel until the cleaved cosmid DNA migrates 10 cm into the gel. View the gel by placing it on a long wavelength (312 nm) UV box. Locate the cleaved cosmid band in each lane and excise it with a razor blade.

---

5. Purify the DNA by electroelution into a dialysis bag. Separately purify the cosmid vector which was not treated with CIP. If the eluted DNA is in a large volume, concentrate the sample by sequentially adding equal volumes of *sec*-butanol, shake the sample, and discard the upper phase until the volume is reduced to approximately 400 μl. Extract the DNA once with an equal volume of phenol/chloroform. Add 1/10 volume of 3 M sodium acetate, pH 5.2 to the aqueous phase and 2 volumes of ethanol to precipitate the DNA. Alternatively, purify the cosmid vector using one of the commercial kits available (Prep-A-Gene from Bio-Rad Laboratories or Qiaex from Qiagen, Inc.).

6. Resuspend the DNA in $T_{10}E_1$ buffer at a concentration of 1 μg/μl. Store at − 20 °C until use. The efficiency of removal of the terminal phosphate groups on the cosmid vector is a critical factor in the success of construction of a cosmid library. Preparations of CIP can vary in quality among manufacturers.

7. Ligate 1 ng of the cosmid vector using the procedure in Section 6.1. Do this for both the CIP-treated and untreated cosmid vector.

8. Transform competent *E. coli* using standard procedures and spread on an ampicillin-containing agar plate.

9. Count the number of colonies on each plate. The plate with the CIP-treated cosmid vector should have 50–100-fold fewer colonies.

3. Extraction of high molecular weight DNA

It is imperative that the size of the genomic DNA isolated for the construction of a library must be large. As a general rule for the construction of genomic libraries, the purified DNA should be at least four times the target size of the partially digested DNA used in the ligation reaction with the vector. For construction of a successful cosmid library, the isolated genomic DNA has to be of a size greater than 150–200 kb as measured on an agarose gel. Using purified DNA of this size for partial restriction enzyme digestion reaction will generate correctly-sized genomic DNA fragments with both ends cleaved by the restriction enzyme. Using undersized DNA as the starting material decreases the size of the library by increasing the chances of ligating a fragment of genomic DNA which contains an end which was not cleaved by the restriction enzyme. A sheared end is not capable of having a cosmid vector ligated to it. This genomic DNA fragment will terminate the concatamer and thereby reduce the amount of DNA available for packaging into the bacteriophage particle. If the isolated DNA does not meet this size requirement, reisolation of the DNA is highly recommended.

The starting material for DNA isolation can be either tissue, blood, or a primary culture of cells. There are several advantages and disadvantages to each source of genomic DNA. Purification of high molecular weight DNA from tissue will yield the greatest amount of DNA, but the likelihood of shearing the DNA is greater than the other methods. Several common tissues used for DNA isolation are the liver, spleen, and kidney. A common method for isolating human DNA is from blood samples. The amount of DNA isolated from blood will be lower than the other two methods described. Isolation of high molecular weight DNA from tissue culture has the advantages of low shear force during isolation with moderately high yield. The drawback for using tissue culture is the time and work involved in starting a primary culture. Often, this extra work of starting a primary culture is rewarded. For established cell lines, selective pressure for the removal of mutations within non-essential genes is lost. Do not use an established line of cells unless it is known that function of the gene sought is essential for the survival of the cell. This criterion often is not fulfilled for extracellular matrix genes. If the genomic DNA will be used ultimately to generate transgenic mice in which a specific gene is inactivated by homologous recombination, then the source of the DNA should be from an isogenic mouse or ES cell line to avoid intra-species effects (65).

The general procedure for the isolation of high molecular weight DNA is by digestion of cells and cellular protein with Proteinase K in the presence of SDS and high concentrations of EDTA to inhibit endogenous DNases. Pro-teinase K is preferentially used because of its ability to rapidly inactivate endogenous nucleases in the presence of SDS. The proteolytically-digested protein is removed by several phenol/chloroform extractions and dialysis of the DNA. The recommended amount of genomic DNA that needs to be isolated should be at least 750 μg after this step. The size of the isolated genomic DNA is determined on a 0.3% agarose gel or on a pulse-field gel which is set to fractionate DNA in the range of 50 to 300 kb. This DNA extraction procedure does not remove RNA, but this does not affect sub-sequent steps. The RNA is removed ultimately by sucrose gradient fraction-ation. If removal of RNA from the genomic DNA is desired, then 20 μg/ml of DNase-free RNase can be added during the Proteinase K digestion step, and the digestion temperature lowered to 37°C.

3.1 DNA extraction procedure

The initial steps of isolating genomic DNA differ depending on the starting material. These initial steps are described in *Protocol 2* for cells in culture, in *Protocol 3* for tissue, and *Protocol 4* for blood. The procedure then continues in *Protocol 5*.

3.1.1. DNA extraction from cell culture

A confluent culture of cells will yield approximately 25–100 μg of genomic DNA per 100 cm^2. The cells should be cultured in several large, i.e. 600 cm^2 tissue culture plates. The number of plates needed are dependent on the type of cells grown and will have to be empirically determined. It is much easier to manipulate culture plates than regular culture flasks for isolating undegraded DNA. It is recommended that at least one culture flask of cells continue to be propagated as a reserve until the amount and average size of the DNA has been verified.

Protocol 2. Isolation of DNA from cells in culture

Equipment and reagents

- Tissue culture plates (600 cm², Nunc cat. no. 166508)
- PBS
- Lysis buffer: 25 mM Tris–HCl, 20 mM EDTA, pH 8.0, 1% SDS, 100 μg/ml Proteinase K

Method

1. Grow cells to confluence using standard cell culture techniques.

2. Wash cells twice with PBS. Remove as much liquid as possible after each wash. Tip the culture plate to a 30 degree angle to allow for drainage and then remove the remaining liquid with aspiration or with a pipette.

3. Add 1 ml of lysis buffer per 100 cm² of tissue culture dish surface area. Swirl the solution gently around the dish. As the cells lyse the viscosity of the solution increases.

4. Scrape the cell lysate into a 50 ml conical tube. Alternatively, use the cell lysate from one plate as the lysis buffer for a second and third plate if a smaller volume of more concentrated DNA is desired.

5. Place the sample in a 50°C water bath overnight.

6. Proceed to *Protocol 5*.

3.1.2 Extraction of DNA from tissue

As a general rule, a tissue low in connective tissue is recommended for the isolation of genomic DNA. Liver, spleen, or kidney are commonly used. Spleen gives good yields of DNA. If liver is used, the animal should be starved for 24 h prior to sacrifice in order to deplete glycogen levels.

Protocol 3. Isolation of DNA from tissue

1. Cut the tissue into small pieces and drop into liquid nitrogen. Frozen tissue can be stored in a − 70°C freezer until needed.

2. Add resuspension buffer (resuspension buffer is lysis buffer without the SDS—see *Protocol 2* for formulation) to a 50 ml conical tube. A general rule is 20 ml of resuspension buffer per gram wet weight of tissue.

Protocol 3. *Continued*

3. Add liquid nitrogen into the mortar bowl and then add the frozen pieces of tissue. Use the pestle to quickly grind the tissue into a powder.

4. Add the frozen powder to the resuspension buffer by either tapping the contents of the mortar into the conical tube or by scraping. Cap and invert the tube to uniformly mix the tissue into the buffer.

5. Add SDS to the resuspension buffer to a final concentration of 1%. Gently invert the tube several times. The solution should be viscous.

6. Incubate the sample in a 50 °C water bath overnight.

7. Proceed to *Protocol 5*.

3.1.3 Extraction of DNA from blood

The amount of DNA which can be isolated from human blood will vary considerably between individuals but should be approximately 250 µg per 15 ml of blood.

Protocol 4. Isolation of DNA from blood

Equipment and reagents

- T$_{10}$E$_{10}$ buffer: 10 mM Tris–HCl, 10 mM EDTA, pH 8.0
- Resuspension buffer (see *Protocol 3*)
- SDS
- EDTA or citrate blood-collecting tubes

Method

1. Collect blood in tubes containing either EDTA or citrate—**not** heparin. Use the blood immediately for DNA isolation or quickly freeze it for future use.

2. Transfer the blood to a 50 ml conical tube. Add 35–40 ml of cold T$_{10}$E$_{10}$ to the blood and place on ice for 5 min. The red blood cells will lyse under these conditions.

3. Centrifuge the diluted blood for 10 min at 800 *g* in a table-top centrifuge located in a refrigerated room.

4. Discard the supernatant and loose pellet. The loose pellet is red blood cell ghosts.

5. Resuspend the pellet in 10 ml of cold T$_{10}$E$_{10}$ and then fill to 40 ml. Remove any coagulant with forceps when the pellet is resuspended in buffer.

6. Re-centrifuge the sample for 10 min at 800 *g* in the cold and

discard the supernatant. The pellet should show little colour at this point.

7. Resuspend the pellet thoroughly in 15 ml of resuspension buffer (see *Protocol 3*).

8. Add SDS to a final volume of 1%. The solution will become viscous as the cells lyse.

9. Place the sample in a 50 °C water bath overnight.

10. Proceed to *Protocol 5*.

3.1.4 Continuation of DNA extraction procedure from *Protocols 2, 3,* and *4*

After overnight incubation, the solution should be transparent. Continue the genomic DNA isolation procedure by following the steps described below.

Protocol 5. Continuation of DNA extraction procedure

Equipment and reagents

- Phenol:chloroform (1:1 (v/v))
- Chloroform
- $T_{10} E_1$ buffer: 10 mM Tris–HCl, 1 mM EDTA, pH 8.0
- Fluorimeter (Model TKO 100, Hoeffer)

Method

1. Add an equal volume of phenol/chloroform to the sample. Gently shake the sample on a vertically-rotating platform for 15 min. Be careful not to shear the DNA.

2. Centrifuge the sample at 4000 *g* for 20 min at room temperature.

3. Carefully remove the tubes from the centrifuge. Do not disturb the white interphase present between the layers. Remove the upper layer which contains the DNA and avoid the interphase if possible. This is best accomplished by pouring the DNA from one tube to another and leaving approximately 10% of the aqueous layer behind. Often, the upper solution is viscous and some contamination with the interphase is unavoidable. Repeat steps 1–3 until no protein is visible at the interphase.

4. Add an equal volume of chloroform to the sample. Shake and centrifuge the sample as described in steps 1–3.

5. Dialyse the DNA overnight against two changes, $T_{10}E_1$ with a dilution factor of 1:100 for each change. The dialysis should be performed at room temperature for the first two hours to avoid precipitating the SDS from solution. After this period the dialysis should take place at 4 °C.

Protocol 5. *Continued*

6. Carefully pour the DNA from the dialysis bag into a 50 ml conical tube.

7. Determine the concentration of the DNA by measuring its concentration in a spectrophotometer at A_{260} and its purity using the A_{260}/A_{280} ratio. If the lysis solution did not contain RNase, then the amount of DNA is estimated to be a third of the total A_{260}-absorbable material. The DNA concentration should be greater than 25 μg/ml. The exact amount of DNA can be determined by fluorimetry.

8. For long-term storage, 1:20 vol. of chloroform should be added to the DNA to inhibit microbial growth.

3.2 Agarose gel electrophoresis

The gel recipe below fractionates DNA up to 50 kb efficiently. DNA above this size is not fractionated as efficiently, but it is still sufficient to determine the relative size of isolated DNA. The gel itself is fragile, so care has to be exercised. Alternatively, a pulse-field gel electrophoresis can be used.

Protocol 6. Isolation of genomic DNA by agarose gel electrophoresis

Equipment and reagents

- 0.3% agarose gel
- 50 × TAE running buffer stock: 2 M Tris–HCl, 0.5 M sodium acetate, 0.05 M EDTA, 1.44 M acetic acid, pH 8.0
- Loading dye (20% Ficoll 400; 0.25% bromophenol blue, 0.25% xylene cyanol)
- Large-bore micropipette tips
- DNA size standards (Gibco–BRL, cat. no. 5618SA and 5627SA)
- Ethidium bromide (10 mg/ml in H_2O)

Method

1. Make a 0.3% agarose gel in 1 × TAE containing 1 μg/ml EtBR. Pour the gel in the cold room and allow it to solidify. Assemble the gel apparatus and add 1 × TAE as the running buffer.

2. Load the samples in a total of 50 μl including the loading dye. Use large-bore tips (available from several scientific supply companies, or make them by cutting a pipette tip using a razor blade) when handling genomic DNA. Add DNA size standards which cover a size range up to 50 kb and beyond.

3. Electrophorese in the cold room at 1 V/cm for 16 h. Recirculation of the running buffer may be needed.

4. Take a photograph of the gel.

4. Partial digestion of genomic DNA

The most practical way to generate near-random cleavage of genomic DNA is by using a restriction enzyme which recognizes a frequently occurring 4-base sequence within the eukaryotic DNA, such as *Sau*3AI or its isoschizomer *Mbo*I. Because there is a high number of *Sau*3AI target sites within genomic DNA, partial digests will generate overlapping fragments of genomic DNA. A truly random cleavage of genomic DNA can be accomplished by shearing using mechanical force, that is by sonication. However, this method destroys the ends of the DNA, and repair of these ends takes several additional steps.

The amount of restriction enzyme needed to digest the genomic DNA to an average size class of 35–50 kb has to be empirically determined for each sample of genomic DNA. This can be accomplished by adding increasing dilutions of the restriction enzyme to the DNA for 1 h and then analysing each digest on an agarose gel with DNA size standards. Once the optimal conditions are determined, the reaction is scaled-up proportionately for the preparative digest.

4.1 Method

The example in *Protocols 7* and *8* and shown in *Figure 2* will use the restriction enzyme *Sau*3AI and the high-salt buffer recommended and supplied by the

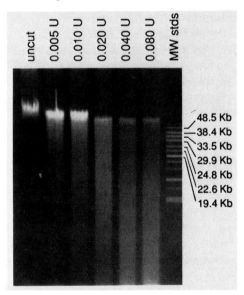

Figure 2. Determination of the concentration of *Sau*3AI necessary for the partial digestion of mouse DNA to give a distribution of DNA fragments enriched for 30–50 kb fragments. 1.5 μg of mouse genomic DNA was digested with various dilutions of *Sau*3AI for 1 h at 37°C. After digestion, the DNA was fractionated on a 0.3% agarose gel and visualized by ethidium bromide staining. For this experiment, 0.01–0.02 U of *Sau* 3AI was optimal.

David Toman and Benoit de Crombrugghe

manufacturer as a 10 × 'H' buffer stock. *Sau*3AI-cleaved DNA can be ligated to a *Bam*HI-cut cosmid vector. Other enzymes and their buffers can be used. This procedure avoids the use of dispensing 1 μl aliquots as most pipettes are not very accurate in this range and this inaccuracy will cause errors in the scale-up to the preparative digest. Wide-bore tips should be used in dispensing the genomic DNA.

Protocol 7. Analytical digests of genomic DNA with *Sau*3AI

Equipment and reagents
- 10 × 'H' buffer: 0.5 M Tris–HCl, pH 7.5, 0.1 M MgCl₂, 1 M NaCl, 10 mM DTT
- *Sau*3AI restriction enzyme
- 0.5 M EDTA, pH 8.0
- 0.3% agarose gel
- High molecular weight markers
- Wide-bore pipette tips

Method

1. Add 34 μl of genomic DNA into each of 6 different microcentrifuge tubes using wide-bore tips to avoid shearing the DNA.
2. Add 4 μl of 10 × 'H' buffer and stir to mix. Do not pipette up and down as this may shear the DNA. Place the tubes in a 37 °C water bath for at least 5 min.
3. Make several dilutions of *Sau*3AI. First, make 1000 μl of 1 × 'H' buffer by diluting 100 μl of the 10 × 'H' buffer stock with 900 μl of water. Add 400 μl of 1 × 'H' buffer to the first microcentrifuge tube and 25 μl to each of the 5 subsequent tubes. Then add 2 μl of *Sau*3AI from a 10 U/μl stock vial to the first tube (containing 400 μl of 1 × 'H' buffer) and stir gently. Remove 25 μl from the first tube and add to the second tube for a 1:2 dilution. Repeat this serial dilution for the remainder of the tubes.
4. Dispense 2 μl of each of the diluted *Sau*3AI into each DNA-containing tube in the water bath. The final amounts of *Sau*3AI will span from 0.05 to 0.0015 U/sample. Incubate at 37 °C for 1 h.
5. Add 2 μl of 0.5 M EDTA to each tube to stop the reaction.
6. Analyse the digests by electrophoresis through a 0.3% agarose gel with the high molecular weight markers as standards (add 4 μl of the 10 × 'H' buffer and 2 μl of 0.5 M EDTA to the markers to normalize the buffer composition). If the DNA is too dilute to visualize on the agarose gel, then digest a large volume of DNA, then concentrate it by ethanol precipitation, and resuspend the DNA in 40 μl.
7. Note the enzyme concentration which gives the highest amount of DNA in the 35–45 kb range. An easy method to compare different digestion patterns is to take two sheets of paper and place them across the gel, one above the 45 kb marker and the other below the 35 kb marker, and simply compare the intensities of DNA between the sheets.

4.1.1 Preparative digest of genomic DNA

Digestion of 500 µg of genomic DNA is recommended for efficient recovery of the 35–50 kb size range of digested DNA. The DNA concentration should be identical to those used for the analytical digests which gave the size distribution of 35–50 kb. If the amount of genomic DNA is not limiting, three digestions with differing amounts of restriction enzyme can be performed. The digestions correspond to (a) the amount of restriction enzyme as empirically determined by the analytical digest, (b) twice as much enzyme needed, and (c) one-half as much enzyme needed. This approach will correct for any variation in scale-up of the analytical procedure.

Protocol 8. Preparative digests of genomic DNA with *Sau*3AI

Reagents

- 10 × 'H' buffer (see *Protocol* 7)
- *Sau*3AI restriction enzyme
- 0.5 M EDTA, pH 8.0
- 0.3% agarose gel
- Phenol:chloroform (1:1 v/v)
- 3 M NaOAc, pH 5.2
- Ethanol
- $T_{10}E_1$ buffer (see *Protocol 1*)

Method

1. Aliquot 500 µg of genomic DNA into a 50 ml conical tube. Avoid shearing the DNA by using a wide-bore pipette. Add the appropriate amount of 10 × 'H' buffer so the final concentration of 'H' buffer is 1 ×. Place the tube into a 37 °C water bath for 10 min.

2. Add the empirically-determined amount of restriction enzyme *Sau*3AI to the genomic DNA. Incubate for 1 h at 37 °C.

3. Stop the reaction by adding an appropriate amount of 0.5 M EDTA. Run a 40 µl aliquot on a 0.3% agarose gel to ensure that the proper size distribution has been obtained.

4. Extract the DNA with phenol/chloroform gently as not to shear the DNA.

5. Add 1/10 volume of 3 M NaOAc and 2.5 volumes of ethanol to precipitate the DNA. Place the tube at 0–4 °C overnight.

6. Collect the DNA by centrifugation at 4000 *g* for 15 min at 4 °C. Redissolve the DNA in 500 µl $T_{10}E_1$. DNA of this size dissolves slowly.

---

5. Fractionation of genomic DNA

After preparative digestion of the genomic DNA is performed, the partially-digested DNA is size-fractionated by sedimentation through a sucrose density

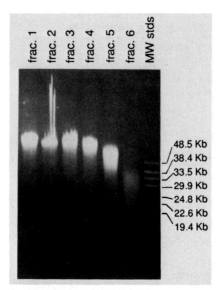

Figure 3. Sucrose gradient fractionation of mouse genomic DNA which was partially digested with *Sau*3AI. Two millilitre gradient fractions were collected after centrifugation in an SW 41 rotor followed by precipitation of the DNA with ethanol and resuspension in 40 μl of TE. 0.4 μg of DNA from each gradient fraction was loaded on to a 0.3% agarose gel and visualized with ethidium bromide staining. Fraction 1 corresponds to the bottom of the gradient. Fraction 5 was enriched for the 30–50 kb size class and was used in the construction of a cosmid library.

gradient. Fractions enriched for the 35–45 kb size are determined on an agarose gel as shown in *Figure* 3, pooled, and concentrated by precipitation. The procedure described herein uses 500 μg of DNA which is centrifuged using a 38 ml 5–20% linear sucrose gradient in a SW 28 rotor (Beckman Instruments, Inc.). This procedure will give an ample amount of size-fractionated DNA for the synthesis of several cosmid libraries. Alternatively, the procedure can be scaled-down to use 100 μg of genomic DNA and centrifugation in an SW 41 rotor with a 13 ml sucrose density gradient. Alternative methods for size-fractionating partially-cleaved DNA are the use of NaCl density gradients and fractionation of the DNA using preparative agarose gel electrophoresis (53).

5.1 Gradient centrifugation

The digested DNA is enriched for the 35–50 kb size range by centrifugation through a 5–20% sucrose gradient. The sucrose gradient is made by using any of several types of gradient makers available commercially and following the manufacturer's instructions.

Protocol 9. Selection of DNA fragments by centrifugation through sucrose gradients

Equipment and reagents

- 5% and 20% sucrose solutions in 100 mM NaCl, 10 mM Tris–HCl, 0–5 mM EDTA, pH 7.4
- Gradient maker
- Bromophenol blue
- SW 28 Beckman rotor
- SW 41 Beckman rotor
- 3 M NaOAc, pH 5.2
- Ethanol
- $T_{10}E_1$ buffer (see *Protocol 1*)
- 0.3% agarose gel

Method

1. Make two 38 ml 5–20% sucrose gradients in polyallomer tubes by placing 19 ml of 5% sucrose solution in one chamber of the gradient maker and 19 ml of 20% sucrose solution in the other chamber. Follow the manufacturer's instructions to pour the gradient.

2. Heat 500 μl of digested DNA to 65 °C for 5 min, cool to room temperature, and carefully overlay the DNA on to the top of one of the gradient tubes. To visualize this process more efficiently, add a touch of Bromophenol blue to the DNA prior to loading on to the gradient. The second tube can be used to fractionate a standard such as 20 μg of lambda DNA or used just as a balance tube.

3. Centrifuge in a SW 28 Beckman rotor at 62000 *g* for 16 h at 20 °C.

4. Fractionate the gradient into 3 ml aliquots and precipitate the DNA with 1/10 volume of 3 M NaOAc and 2 volumes of ethanol. Place the DNA at 4 °C overnight.

5. Pellet the DNA by centrifugation in a SW 41 rotor at 210000 *g* for 1 h at 4 °C. The pellet may be invisible at the bottom of the tube.

6. Resuspend the pellets in 100 μl of $T_{10}E_1$. Re-precipitate the fractions in a microcentrifuge tube. Wash the pellets with 70% ethanol.

7. Resuspend the DNA in 40 μl of $T_{10}E_1$. Determine the concentration of the DNA using a spectrophotometer.

8. Load 0.4 μg of DNA from each fraction on to a 0.3% agarose gel. Choose the fractions which have a 35–50 kb size distribution.

---

6. Ligation and packaging

6.1 Ligation

The ligation reaction contains the vector DNA at a molar concentration of 10-times the molar concentration of the genomic DNA. Since the cosmid vector cannot self-ligate due to removal of its terminal 5′ phosphate, this ratio enhances the probability of generating a ligation product in which a genomic DNA molecule will be ligated at both ends by a vector molecule. The cos cleavage enzyme contained in the packaging reaction will cleave at the cos sites and liberate a linear molecule which is packagable by the extract.

High concentrations of DNA are used in the ligation reaction to drive the formation of concatameric molecules which are efficiently packaged using the packaging extract. The concentration of DNA in the ligation reaction must be greater than 200 ng/μl in order to favour the formation of these concatamers and not circular DNA molecules containing only one cos site.

Protocol 10. Ligation

Equipment and reagents

- T$_4$ DNA ligase
- 10 × ligation buffer: (60 mM Tris–HCl, 50 mM MgCl$_2$, 10 mM DTT, 10 mM ATP, pH 7.5)
- 0.3% agarose gel

Method

1. Prepare ligation mix as follows: 1.5 μg size-fractionated DNA, 1.5 μg dephosphorylated vector DNA, 1.5 μl 10 × ligation buffer, water q.s. 14 μl.

2. Remove 1 μl from the ligation mix and save for later analysis.

3. Add 1 μl of T$_4$ DNA ligase at a concentration of 1 Weiss unit/μl.

4. Incubate the ligation reaction at 14 °C for 16 h or 4 °C for 24–48 h.

5. Remove 1 μl from the ligation reaction and analyse this and the 1 μl sample taken in step 1 on a 0.3% agarose gel. If the ligation was successful, the genomic DNA and the cosmid vector DNA should be converted into a higher molecular weight.

6.2 Packaging

Packaging extracts are available from several scientific companies (Promega Corp. and Stratagene Cloning Systems) at a reasonable cost. A specialty packaging extract is available from Stratagene which preferentially packages larger cosmid DNA. We recommend the purchase of the packaging extract from one of these commercial sources. This will alleviate the need for an individual laboratory to prepare packaging extracts which is a time-consuming task and requires some experience. Detailed methodology for preparing packaging extracts is described in reference 53.

Follow the instructions which accompany the packaging extract. These instructions will state the amount and maximum volume of DNA to be added to each extract. Generally, several different packagings can be done from one ligation reaction. One key factor for a successful packaging reaction is to avoid the introduction of air bubbles when mixing the components of the extract—this will dramatically decrease the efficiency of packaging of the cosmid DNA. After the packaging reaction is completed, the sample is diluted in SM buffer or equivalent (composition of SM is described in the packaging extract product literature). A small amount (50 μl) of chloroform is added to the diluted sample to prevent growth of bacteria. The packaged phage are stable in the refrigerator for 1 month without loss of titre. The titre of the packaged cosmid will decrease 10-fold over a 1-year-period in the refrigerator. Long-term storage is accomplished by adding an equal volume of 15% DMSO in SM buffer, or equivalent, and storing in a − 70°C freezer.

7. Plating and screening

A key feature in the ability to generate a high-titre cosmid library is the host bacterial cell which is used. Both a quantitative and qualitative difference can exist between different bacterial cells as host. Three factors have to be considered:

(a) the absence of restriction systems within the host bacteria which may affect the general representation of genomic DNA sequences;

(b) the efficiency of infection; and

(c) the uniformity of growth rate upon infection with foreign cosmid DNA which affects the colony size on an agar plate.

Several bacterial hosts have been used for cosmid cloning. During the last decade, several restriction systems were first described in *E. coli* which could affect representation of particular sequences (66–68). These sequences, many which were described as unclonable, usually had DNA which was either highly methylated, contained repetitive sequences, or contained non-standard secondary or tertiary sequences, such as cruciform or alternating purine–pyrimidine tracts (Z-DNA). Several restriction systems were described in *E. coli* which affected the ability of genomic DNA to be cloned. *Rec*A is part of a repair system which will cause recombination at repetitive DNA sequences contained within the genomic DNA. The *mcr* restriction systems cleave at sites within genomic DNA which are methylated. The *hsd* system cleaves DNA that lacks methylation at the *Eco*K sites within eukaryotic DNA. Several strains of *E. coli* have been developed recently which lack the *mcr* and *hsd* DNA restriction systems as well as *Rec*A.

The efficiency of infection is dependent on the *E. coli* host strain. Several strains were tested in our laboratory with identical amounts of packaged phage, and the number of antibiotic-resistant colonies which grew on an agar plate varied as much as 100-fold. In addition, experiments showed that several of the host strains exhibited a variation in colony size when plated on to a single agar plate. Faster-growing colonies can effectively mask slower-growing colonies either by physical contact or by nutrient depletion and thereby introduce a bias into the representation of individual clones within the cosmid library. We have used the *E. coli* strain NM554 in our library constructions because of the lack of restriction systems described above, the greater efficiency of infection, and the greater uniformity of colony size when plated on agar plates. Other strains of *E. coli* may be used if they meet these criteria.

It is commonplace to amplify a DNA library (cDNA or genomic) prior to screening the library. The purpose of amplification is to be able to make one library and store multiple copies of it for future use or to distribute to other laboratories. A danger exists that the representation of clones within the amplified library is reduced. This loss in representation occurs by three mechanisms:

(a) unequal growth of bacterial colonies during amplification (especially problematic if amplification is done in liquid culture as compared to growth on an agar plate);

(b) possible rearrangements and loss of genomic DNA in the inserts (strongly dependent on the host bacterial strain); and

(c) introduction or creation (by the second possibility) of faster-growing bacterial colonies which 'take over' the library.

The best library is an unamplified one. Care must be exercised if obtaining the cosmid DNA library from another source, especially if the library was amplified.

There is an alternative to amplification of a library. The procedures in Section 5 for isolating genomic DNA cleaved to the 35–45 kb range allow for the construction of several libraries. 'In-house' construction provides the necessary 'raw ingredients' to make more cosmid libraries and freeze them as packaged phage as decribed in Section 6. Alternatively, the fractionated genomic DNA and the vector can be stored in $T_{10}E_1$ at 4°C. If the library needs to be amplified, the original library should be divided into 20–40 aliquots and each aliquot should be amplified individually on agar plates.

7.1 Analytical infection

An analytical infection is performed to quantify the number of individual members of the cosmid library—called the titre of the library. Several dilutions of the cosmid library are added to log-phase bacteria. The number of colonies are quantified and multiplied by the dilution factor to give the total members of the cosmid library. Using commercial packaging extracts, the control lambda DNA should give a titre of 10^8 p.f.u./µg of DNA (the instructions for determining titre are contained within the product information). The number of individual members of the cosmid library should be greater than 10^5 per µg of input DNA and 10^4 per ml. To check the quality of the cosmid library and to verify the average size of the insert DNA, several of the colonies produced by the analytical infection should be picked, grown in medium, and the cosmid DNA isolated by standard plasmid mini-prep procedures (53).

7.2 Titration

An analytical library is prepared in order to determine the optimal conditions for infections and to measure the titre of the packaged DNA.

Protocol 11. Titration

Equipment and reagents

- Luria broth
- 10 ml MgSO$_4$ from 100 × stock solution, sterile-filtered
- 0.2% maltose from 100 × stock solution, sterile-filtered
- E. Coli NM554

- SM buffer (50 mM Tris–HCl, pH 7.5, 100 mM NaCl, 10 mM MgSO$_4$, 0.01% (w/v) gelatin)
- Ampicillin
- Plasmid mini-prep procedure (53)
- Restriction enzyme
- Agarose gel

Method

1. Supplement 50 ml of Luria broth with 0.2% maltose and 10 mM MgSO$_4$ from sterile-filtered 100 × stocks. Place medium into a 250 ml Erlenmeyer flask.

2. Inoculate medium with a single colony of NM554 bacterial host. Shake at 200 r.p.m. overnight in a 30 °C incubator. Check the concentration of the overnight culture in a spectrophotometer to verify that the absorbance at 600 nm (A_{600}) is less than 1.0.

3. Centrifuge cells in a sterile conical tube for 10 min at 2000 g at 4 °C.

4. Gently resuspend the pellet in 15 ml of cold 10 mM MgSO$_4$.

5. Dilute the cells to 0.5 A_{600}/ml with 10 mM MgSO$_4$. Store the cells on ice. These cells should be used immediately.

6. Take 5 μl of the packaged DNA and make two or more serial dilutions with SM buffer. The dilutions should range from 10- to 100-fold.

7. Add 25 μl of each dilution of packaged DNA to 100 μl of NM554 cells. Incubate the cells and packaged DNA at room temperature for 30 min without shaking.

8. Add 900 μl of Luria broth to each tube and incubate at 37 °C for 1 h.

9. Remove 200 μl of solution and spread on to 150 mm plates of Luria agar supplemented with 25 μg/ml ampicillin.

10. Incubate plates at 37 °C for 18 h or until the colony size is large enough to count. The expected titre is 5 × 10^5 to 1 × 10^6 transformants per μg of genomic DNA.

Protocol 11. *Continued*

11. Purify the cosmid DNA from 10 colonies using a plasmid mini-prep purification scheme based on alkaline-SDS (53). Digest the DNA with a restriction enzyme and run the digestion product on an agarose gel. Verify that all samples have a genomic DNA insert and determine the DNA insert size for each colony.

7.3 Large-scale plating

The number of colonies required for a complete library can be calculated using the equation:

$$N = \ln(1 - P)/\ln(1 - f), \qquad [1]$$

where N is the number of colonies necessary to have a probability P (usually 95 to 99%) of containing a particular DNA sequence represented in the cosmid library (69). The term f is the proportion of the genome contained within an average clone. This figure is based on the average insert size determined by mini-prep analysis (as determined in step 11 of *Protocol 11*) compared to the total complexity of the genome. For a library constructed from mammalian DNA with an average insert size of 40 000 bp, f would be calculated as 4×10^5 bp/3×10^9 bp. Thus, with an average insert size of 40 000 bp, a cosmid library needs to contain 500 000 individual colonies in order to have a probability of 99% that it contains a DNA sequence represented only once within the genome.

The standard approach for plating a cosmid library is to first infect the host *E. coli* with the packaged cosmid DNA, and then evenly distribute the infected bacteria on to a filter which overlays an agar plate containing an antibiotic—usually ampicillin (refer to *Figure 1*). After overnight growth of the cosmid-containing bacteria on the filters, copies are made from each master filter. The master filter is regrown and then stored at 4°C. The replica filters are placed on ampicillin-containing agar plates to grow. The cosmid DNA within each bacterial cell is amplified by transferring the filter on to an agar plate containing chloramphenicol and incubating these plates overnight. This amplification will selectively increase the copy number of the cosmid within the bacterial host. The bacteria are lysed *in situ* and the DNA is fixed on to the underlying nylon membrane. Standard hybridization conditions are used with a radioactively-labelled DNA as probe. Clones which have DNA sequences corresponding to the probe are visualized after overnight exposure of filters to X-ray film. A timetable of the steps needed for one round of plating and screening of the cosmid library is shown in *Table 3*.

7.3.1 Infection and plating of the library

Aim for a library size of 500 000 colonies total with 25 000 colonies on each of 20 filters. Generally, the infection of the packaged cosmid DNA for the

Table 3. A recommended timetable for plating of the cosmid DNA library and screening by hybridization. The steps listed below are a general outline of the procedures described in the text.

Day 0 (Sun)	Start an overnight culture of plating bacteria.
Day 1 (Mon)	Infect bacteria with packaged cosmid DNA library. Spread infected bacteria on to plates and incubate overnight.
Day 2 (Tue)	Make 2 replica filters per master filter. Grow bacteria on replica filters until colonies are visible. Transfer replica filters to chloramphenicol-containing plates. Incubate plates overnight.
Day 3 (Wed)	Fix cosmid DNA on to replica filters.
Day 4 (Thurs)	Make radioactively-labelled probe and hybridize replica filters.
Day 5 (Fri)	Wash filters after hybridization. Expose filters to X-ray film.
Day 6 (Sat)	Develop X-ray films.

library is done in one tube. Alternatively, 20 separate infection reactions can be performed simultaneously, although there may be some variation in the number of colonies per plate. Fresh 150 mm plates of Luria agar with 25 mg/ml ampicillin should be made the previous day and stored until needed at 4°C. The ampicillin used in the preparation of the agar plates should be fresh.

Protocol 12. Infection and plating of the library

Equipment and reagents

- 137 mm nylon disc (Micron Separations Inc, cat. no. HY 13750, or Fisher Scientific)
- 150 mm TC plates (Fisher Scientific, cat. no. 08–757–14)
- Glass beads (Fisher Scientific, cat. no. K13500–4)
- Packaged cosmid library
- Bacterial host (see *Protocol 11*)
- Luria broth (LB) containing 0.2% maltose and 10 mM MgSO₄
- LB–ampicillin plate

Method

1. Prepare a fresh culture of baterial host as described in the titration section.

2. Warm the plates to room temperature. Number, and then layer a 137 mm nylon disc on to the top of each 150 mm plate. Place five glass beads on to the filter on each plate.

3. Next, prepare the infection of the host bacteria. A predetermined amount of the packaged cosmid library (which will generate 500 000 colonies) is added to 2 ml of bacterial host. Incubate the cells and packaged cosmid DNA together for 30 min at room temperature without shaking.

377

Protocol 12. *Continued*

4. Add 18 ml of Luria broth (LB) prewarmed at 37 °C and place in an incubator for 1 h at 37 °C. Mix gently to evenly distribute the bacteria after the 1 h incubation.

5. Add 1 ml of infection mix (made in the last two steps) to each plate containing a nylon filter (and beads). Shake the plate to distribute the bacteria evenly—this is important! It may take up to 10 min for the infection mix to be absorbed by the filter depending on the dampness of the plates. To verify the titre of this infection, remove 10 μl of the infection mix, dilute to 200 μl with LB, and plate on to a LB–amp (ampicillin) plate with five glass beads but without a filter. Shake the plate with the beads to distribute the bacteria. These plates contain the unamplified library and are called the master plates.

6. Place the plates in a 37 °C room until the colonies are about 0.2 mm in diameter. For NM554 cells, this will be approximately 17 h. Do not stack the plates during this incubation—stacked plates will warm at different rates and the colony sizes will not be uniform.

7.3.2 Replica plating

Two replica filters will be made using the master filter as the template. The master filter will be stored until later use. The clones identified by hybridization of the replica filters will be picked from the master plates. These master plates generally will store for 3 weeks at 4 °C. Therefore, multiple screenings of the library require all hybridizations to be done within a 3-week-period while the bacteria on the master plates are viable.

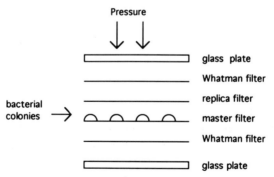

Figure 4. Cross-section of replica plating. Each replica filter is made by pressing the master filter to a new filter in a sandwich-type arrangement. A portion of the bacteria of each of the 20 000–30 000 colonies on the master filter are transferred to the replica filter. The replica filter will have a mirror-image of the master filter. Orientation marks are introduced when the replica and master filters are in contact with each other. Colonies identified on the replica filter after hybridization are located on the master filter by using the orientation marks to align the replica filter with the master filter.

Protocol 13. Replica plating

Equipment and reagents

As *Protocol 12*
- Sharpie pen
- Whatman No. 1 filter, 150 mm
- Glass plates
- 18-gauge needle
- Chloramphenicol

Method

1. Cool the bacteria in the cold room for 1 h to temporarily stop their growth.

2. Label two sets of nylon filters with the numbers of the master filters and the added designation of A or B. Use a Sharpie pen for this labelling.

3. Place each filter on an LB–amp plate with the number side face down.

4. Make two replicas of each master filter. These steps must be done quickly to prevent the master filter from drying (*Figure 4*):

   (a) Place a sterile 150 mm Whatman No. 1 filter, or equivalent, on to a glass plate, followed by the wetted blank replica filter 'A' (from step 2) number side up, the master filter colony side down, and another Whatman No. 1 filter, and a second glass plate. Press to squeeze the filters together.

   (b) Remove the two nylon filters. They should be stuck to each other. Punch three orientation marks through both filters using an 18-gauge needle. The three needle marks should be in an asymmetric pattern for later orientation.

   (c) Peel the filters apart, using filter forceps, and place the 'A' filter back on its plate with the colony side up.

   (d) Place the 'B' filter on top of the master filter and repeat the 'sand-wiching' technique.

   (e) Mark the 'B' filter with an 18-gauge needle by using the needle holes in the master filter as a guide.

   (f) Place the 'B' filter and the master filter, respectively, back on to their home plates—colony side up.

5. Place the plates in the 37 °C incubator until the colony size is 0.2 mm. The bacteria on the master plates will grow to that size in about 2 hours. The plates should be wrapped in Parafilm and stored at 4 °C. The colonies on the replica plates will take about 5 h to reach 0.2 mm in size.

6. Transfer the filters on each replica plate to a new plate of LB agar containing 150 µg/ml chloramphenicol with the colony side facing up. Make sure there are no air bubbles present between the filter and the surface of the agar plate.

7. Place in the 37 °C incubator for 20 h.

7.3.3 Processing the filters

In this procedure, the colonies will be lysed and the cosmid DNA fixed on to the nylon filters. The filters will go through a series of overlayings on to Whatman paper which has been soaked with different solutions. The Whatman paper should be saturated, but not covered, with each solution. The nylon filter should not be submerged into any of these solutions. Penetration of each solution will be by capillary action through the bottom of the nylon filter. With good planning, six filters can be processed simultaneously using large plastic containers to hold each of the solutions.

Protocol 14. Processing the filters

Equipment and reagents

- Replica filter from *Protocol 13*
- Whatman 3MM paper
- 10% SDS
- 0.5 M NaOH
- 1 M Tris–HCl, pH 7.4
- 0.5 M Tris–HCl, pH 7.4, 1.5 M NaCl
- $20 \times$ SSPE stock solution: 0.2 M Na_2HPO_4, 2.8 M NaCl, 0.02 M EDTA, pH 7.4 (used at $2 \times$ SSPE)
- Proteinase K
- Plastic containers (PGC scientifics, Inc., cat. no. 82-0213)
- Nylon filters

Method

1. Remove the replica filter from a plate and place on a sheet of dry Whatman 3MM paper or equivalent for 1 min.
2. Transfer the filter to another sheet of Whatman 3MM paper soaked in 10% SDS for 3 min.
3. Move the filter to paper permeated in 0.5 M NaOH for 12 min.
4. Shift the filter to paper soaked in 1 M Tris–HCl, pH 7.4 for 4 min.
5. Move the filter to paper soaked in 0.5 M Tris–HCl, pH 7.4, 1.5 M NaCl for 4 min.
6. Layer filter on to a dry sheet of paper for 4 min.
7. Submerge the filter into 1 litre of $2 \times$ SSPE at room temperature for temporary storage until the remainder of the filters are processed to this step.
8. Transfer the filters one at a time into a 500 ml solution of $2 \times$ SSPE containing 100 mg Proteinase K. Submerge each filter into the solution. Incubate the filters for 30 min at room temperature.
9. Air-dry the filters on Whatman paper. Make sure they are completely dry.
10. Generate a sandwich of alternating nylon and Whatman filters (paper towels can be used in place of Whatman filters for baking). Place this stack of filters in a vacuum oven at 80 °C from 2 h to overnight.
11. Store the baked filters at room temperature until ready for hybridization.

---

8. Hybridization

The hybridization reaction is customized, based on the type(s) of DNA probes and temperature and salt conditions which will be used for the screening of a cosmid library. There are several types of DNA probes which can be used for the screening of a cosmid library. Most commonly, a cDNA has been cloned in the laboratory. This cDNA has the advantage of having 100% homology to the exon sequences of the genomic DNA used in the construction of the cosmid library. Screening of the cosmid library using this probe can be performed under conditions of high stringency and, therefore, high specificity. Another possibility is to obtain a cDNA from a closely related species. In this case, the stringency of hybridization must be reduced. Reduction of stringency is expressed as a 1.5°C decrease in the hybridization temperature per 1% difference in homology between the probe and target sequence (70). Often, the genomic DNA sequence is not known, so the conditions of hybridization are determined empirically using the cross-species cDNA probe to detect the gene in a Southern analysis prior to screening of the cosmid library. The Southern hybridization should be done with the same nylon membrane and the same prehybridization and hybridization buffers as for the screening of the cosmid library. The most stringent hybridization conditions which generate a clear signal should be used in the screening of the cosmid library.

Another method to generate a cDNA is by employing polymerase chain reaction (PCR) (71). In this approach, the cDNA sequence of the gene should be available from at least two different species. A comparison of the DNA sequence between different species should identify regions of conservation. Knowledge of a conserved region will be beneficial both in the design of primers used in the generation of a PCR cDNA product and also in the hybridization of the cosmid library. The PCR primers are designed from conserved regions of the cDNA, and several reactions are performed using different annealing temperatures for each PCR. Since this approach relies on the specificity of the two primers to cross-hybridize to the cDNA made from another species, occasionally several different sets of primers may have to be designed and tested in order to obtain a PCR product. The cDNA generated by PCR must be analysed by DNA sequencing to verify the expected product.

Conditions for hybridization using a genomic DNA fragment as the probe are determined by whether it is from the same species as the genomic DNA used in the construction of the cosmid library. If the genomic DNA probe is from the same species, then the entire probe has 100% homology and high-stringency conditions can be employed in the hybridization reaction. If the genomic DNA is from a similar species, the stringency of hybridization is reduced. In this case, the intronic sequences are expected to be less conserved than the exon sequences, so this reduces the specificity of the probe. Again, the conditions of hybridization using the cross-species genomic DNA fragment as a probe must be determined empirically by Southern analysis.

The success of isolation of a cosmid clone containing a full-length copy of the gene is dependent on the location of the probe or probes. Ideally, probes corresponding to unique sequences in both the 5′ and 3′ ends of the gene are used sequentially in the screening of a cosmid library. The filters are stripped of the first probe prior to hybridization with the second probe. Clones which are identified by both probes are purified. If one probe will be used, it is recommended that it corresponds to the centre of the gene. In this manner, if a clone corresponding to a full-length gene is not identified, then two overlapping cosmid clones can be isolated.

The length of the DNA probe should be between 700–1500 bp. Methods to radioactively label the DNA probe include both nick-translation or random-priming techniques (53). Both labelling systems can be purchased as a kit from a variety of commercial vendors. The radioactively-labelled probes should be purified from unincorporated nucleotides and quantified in a scintillation counter. The specific activity of the probe should be greater than 1×10^8 per μg of probe DNA. Fresh radioactive nucleotide should be used in the labelling reaction. The amount of probe used in the hybridization reaction should be $0.5–1.0 \times 10^6$ c.p.m./ml of hybridization buffer in order to generate a signal which can be observed with an overnight exposure of the filters to X-ray film. Use of the random-priming method to label probe DNA will give a higher specific activity, but usually several reactions will have to be performed to generate the amount of labelled probe necessary to screen the library. The specific activity of the nick-translated probes can be increased by using two different radioactively-labelled nucleotides ([^{32}P]dCTP and [^{32}P]-dATP) in the labelling reaction if necessary. The quality of the labelling reaction should be checked on a polyacrylamide gel to verify the size distribution of the labelled probe. Additional methods to label the probe are by the use of PCR (72) and by generating RNA probes using RNA transcription reactions (55). The use of oligomers to screen genomic libraries is not recommended.

The conditions of hybridization are determined empirically by Southern analysis prior to the screening of the library. As a general rule, medium- to high-stringency hybridizations are defined as 65°C as the hybridization and wash temperatures, with the salt concentration of the most-stringent washes varying from $1 \times$ SSPE down to $0.2 \times$ SSPE (the lower the salt, the higher the stringency). Low-stringency hybridizations are performed at lower hybridization temperatures—generally from 55–63°C. Hybridization conditions can be modified by the use of formamide or dextran sulfate in the hybridization buffer, but it usually does not confer any advantage to the conditions described here.

Once a clone has been identified by hybridization, the radioactive signal can be removed from the filters in preparation for hybridization with a second probe. The filters can be rehybridized and 'stripped' several times. Therefore, the cosmid library can be screened with several probes to the gene and/or to different genes. Usually the limiting factor for multiple screenings of a cosmid library is the viability of the master plate.

8.1 Hybridization

The following steps are for a high-stringency hybridization using a homologous cDNA as a probe. The hybridization is most conveniently done in Nalgene containers which can accommodate up to 1 litre of solution. The filters are transferred between solutions individually and not in bulk. Although this method is more time-consuming, it prevents the filters from sticking to each other and avoids the creation of non-specific background on the filters. A shaking water bath set at 65°C is needed for the following steps.

Protocol 15. Hybridization

Equipment and reagents

- Nalgene containers (> 1 litre capacity) (Fisher Scientific, cat. no. 03–484E)
- 65°C shaking water bath
- SSPE (see *Protocol 14*)
- 2 × SSPE, 0.5% SDS
- 0.5 × SSPE, 0.5% SDS
- 0.2 × SSPE, 0.5% SDS
- Prehybridization buffer: 3 × SSPE, 10 × Denhardt's solution, 0.1% Carnation dried milk (w/v), 200 μg/ml denatured herring sperm DNA, 0.5% SDS

- [^{32}P]dCTP, Sp. Act. 3000 Ci/mmol
- Whatman No. 1 paper
- Hybridization buffer: 3 × SSPE, 2 × Denhardt's solution, 100 μg/ml denatured herring sperm DNA, 0.5% SDS
- 1 M NaOH
- 1 M HCl
- Saran wrap
- Kodak XAR5 film
- Plexiglass shield and ^{32}P protection

Method

1. Soak filters in a solution of 0.5 × SSPE, 0.5% SDS for 30 min at 65°C to remove bacterial debris.

2. Move filters individually into 500 ml of prehybridization buffer. Incubate the filters in prehybridization buffer with shaking for a minimum of 4 h at 65°C to block sites on filters. Make sure the filters do not 'climb up' the side of the container.

3. Make the DNA probe by either nick-translation or random-prime methods using ^{32}P-labelled dCTP as the radioactive nucleotide. Often several labelling reactions have to be performed to generate the amount of radioactively-labelled DNA for this hybridization. The labelling reaction should be done on the day of use with the freshest isotope available.

4. Transfer the filters individually to 500 ml of hybridization buffer at 65°C. Place a sheet of 125 mm diameter Whatman No. 1 paper or

---

Protocol 15. *Continued*

equivalent between each nylon filter when adding them to the hybridization buffer. The Whatman paper will allow the probe easy access to
the nylon filter and also will result in a lower background.

5. Denature the DNA probe using alkali. Add an equal volume of 1 M
   NaOH to the probe and incubate at room temperature for 10 min. Add
   a volume of 1 M HCl (equal to the volume of NaOH added to the
   probe) to the hybridization mix.

6. Add the probe to the hybridization mix to yield $0.5-1.0 \times 10^6$ c.p.m./
   ml. Incubate the solution in a 65 °C water bath with gentle shaking for
   16 h.

7. Wash the filters quickly at least three times in 2 × SSPE, 0.5% SDS,
   and a fourth time for 15 min at 65 °C with shaking until the c.p.m. of
   the wash solution are low. The c.p.m. of the wash solution can be
   monitored by a Geiger counter. The purpose of the three quick
   washes is to remove the labelled probe in solution. This is best
   accomplished by having a water bath which can hold two of the
   Nalgene containers side by side. Prewarm the solution in the second
   Nalgene container, turn off the shaker, and then individually move the
   filters from one container to the other. Resume shaking at 65 °C after
   transfer. The transfer of 40 filters takes a few minutes so protection
   from the radiation is necessary. It is recommended that this transfer
   be done behind a Plexiglass shield. The Whatman paper should be
   discarded into the radioactive waste container. Dispose of the hybridization solution and the first two washes in appropriate containers for
   radioactive waste disposal.

8. Wash the filters twice stringently with 0.2 × SSPE, 0.5% SDS at 65 °C
   for 15 min each time. The c.p.m. of the discarded solution should be
   very low.

9. The filters are transferred temporarily into 2 × SSPE at room temperature.

10. The filters are moved on to a suitable backing (used X-ray film
    or equivalent, and wrapped in Saran wrap) and placed into film cas-
    settes. The duplicate filters should be adjacent to each other and in
    the same orientation in the cassette. This will save time by allowing
    for quick identification and verification of positive signals on the next
    day. Six filters, i.e. 3 sets of duplicates, can fit into a 14″ × 17″ (35 ×
    43 cm) cassette. The filters are exposed to Kodak XAR5 film over-
    night at −70 °C in the presence of an intensifying screen. The filters
    must not dry if they are to be stripped and used for additional screen-
    ings.

8.2 Stripping filters

Protocol 16. Stripping filters

Reagents

- 0.1 × SSPE (*Protocol 14*), 100 mM Tris–HCl, pH 7.4; 0.5% SDS
- 5 × SSPE
- 0.4 M NaOH

Method

1. Add 0.4 M NaOH to the filters and incubate in a 42 °C water bath for 30 min with gentle shaking.

2. Transfer the filters to a solution of 0.1 × SSPE, 100 mM Tris–HCl, pH 7.4, 0.5% SDS for 30 min at 42 °C.

3. Move filters to a solution of 5 × SSPE and store at 4 °C until further use.

4. Remove some of the filters (especially the ones which had positive signals) and re-expose them overnight to X-ray film to verify the removal of radioactive signals.

---

9. Colony isolation

A region which contains the colonies which have been identified by hybridization of the replica filters is selected from the master filter.

9.1 Colony isolation

Protocol 17. Colony isolation

Equipment and reagents

- Saran wrap
- X-ray film
- Cloning ring (Bellco Glass, cat. no. 2090–00808)
- Luria broth + ampicillin (*Protocol 11*)

Method

1. The filter itself should have a low background after an overnight exposure. Mark the location of the orientation marks on the X-ray film using a Sharpie or other marking pen.

2. Look at the duplicate filters and locate the positive signals on both filters. Ensure that the positive signal is in the same exact position on both filters. A positive signal (as shown in *Figure 5A*) should have a soft 'shoulder', usually circular in appearance, and it often has a tail (similar

---

Protocol 17. *Continued*

to a comet in shape). Occasionally, a filter will have one or more pin-point spots. Usually these small, intense spots are not positives. Generally, more than one positive signal on a filter is rare. If the probe was labelled to a high-specific activity, then exposures longer than over-night are not required and are often misleading.

3. Isolate the region surrounding the positive signal from the master plate.

   (a) Lay the filter of the master plate with the colony side up on to a small piece of Saran wrap and place over the X-ray film on a light box. Move the filter until the orientation marks on the filter are aligned with those on the film.

   (b) Place a cloning ring, with some autoclaved vacuum grease at the bottom, over the positive signal. Work quickly so as not to let the bacteria dry on the filter.

   (c) Add 200 µl of Luria broth with ampicillin into the cloning ring and pipette up-and-down several times to disperse the colonies on the plate.

   (d) Transfer the medium containing the bacteria to a microcentrifuge tube.

4. Use the bacteria immediately for the second round of purification or add glycerol to 15% and freeze at −70 °C.

9.2 Colony purification

The next step is to replate the bacteria on to a new filter and rescreen. Generally, two additional screenings are performed. A plate should have several positive signals at the second screening (see *Figure 5B*). Usually, one of the positive colonies will not be in close proximity with other colonies on the plate. A third screening is done to verify that this colony is pure, and not two colonies which grew on top of each other (see *Figure 5C*).

For the second screening, approximately 500–1000 bacterial colonies should be plated on to each filter. Two different approaches can be utilized to obtain this number of bacteria per filter. One approach is to first determine the number of bacteria in each microcentrifuge tube by plating several dilutions and counting the colonies which appear. Then a precise number of bacteria can be plated. Alternatively, several dilutions of the bacteria can be plated on to filters and the one which most closely approximates the required number chosen to process further. This latter procedure is quicker and is described below.

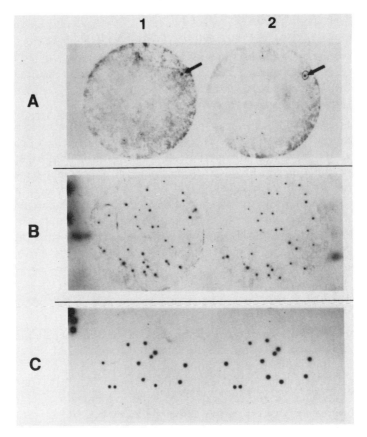

Figure 5. Autoradiography of filters hybridized with a ^{32}P-labelled cDNA probe for mouse type III collagen. The filters were exposed to Kodak XAR-5 X-ray film at − 70 °C. (A) Overnight exposure of replica filters from a primary screen of a mouse cosmid library (approximately 30 000 colonies per filter). The arrows show the position of the positive signal on each replica filter. (B) Overnight exposure of replica filters from a secondary screening. An isolate identified and selected in the first screening was replated at a lower density (approximately 1000 colonies per filter). (C) Final screening of a candidate clone. A single colony from the second screening which was distinctly separate from neighbouring bacterial colonies was selected and replated. Hybridization of the filter identified each colony on the plate as expected for a pure isolate.

Protocol 18. Colony purification

1. Plate 300 µl of several dilutions of bacteria on to new filters. The dilutions normally taken are in the range of 5000- to 50 000-fold.

2. Grow the bacteria overnight at 37 °C until they are visible to the eye. The bacteria should be allowed to grow to 0.5–1 mm in size during the second screening.

Protocol 18. *Continued*

3. Choose a plate which has approximately 500–1000 bacteria.

4. Treat the plate as described for the first screening.

9.3 Large-scale growth of purified colony

The alkaline-SDS procedure as described for plasmids is used with some modifications for cosmids (53). The salient points are described below. First, the copy number within a bacterial cell is much lower than for a standard plasmid—sometimes as low as 10 copies per cell. Generally, a much larger culture (or multiple flasks of cultures) is used as starting material for cosmid purification. In addition, the concentration of ampicillin added to the bacterial culture is reduced, due to the lower copy number of cosmids. Second, the culture time appears to have an effect on the yield. Overgrowth of the culture decreases the yield. It is recommended that the bacteria be harvested after 12–16 h of growth for optimal yield.

10. Summary

As the field of molecular biology continues to mature, the techniques that are inherent to this field become more refined and standardized. As these techniques become more routine, they can be applied to answer increasingly sophisticated scientific questions. In this chapter we have detailed the methodology used in our laboratory to make cosmid libraries and screen these libraries for several of the extracellular matrix genes—although the techniques described herein can be used to identify and isolate any gene. Many of the steps in this chapter are similar to the procedures utilized for screening a phage genomic library. Therefore, the experienced 'cloner' who is familiar with screening of phage genomic libraries should be able to make an easy transition to screening of cosmid DNA libraries. For the investigator who is venturing into this area for the first time, we hope to have clearly outlined the techniques required to perform each of the steps in the construction and screening of a cosmid DNA library. For each section we have discussed the rationale behind each step and have indicated potential pitfalls.

The study of the extracellular matrix has burgeoned in the last few years. The role of the extracellular matrix in normal and pathological conditions has been, and continues to be, demonstrated in both man and in other organisms. The isolation of each additional extracellular matrix gene will permit the further study of the regulation of the gene and the role of its protein product in a eukaryotic organism. Introduction of mutations within the structural gene and expression in a cell culture system or transgenic animal will elucidate important regions in the structure/function of the extracellular matrix protein. We hope the information in this chapter will contribute to the expansion of questions asked within the field.

Acknowledgement

Part of the work described in this paper was supported by NIH grant AR40335.

References

1. Chu, M.-L., de Wet, W., Bernard, M., Ding, J.-F., Morabito, M., Myers, J., Williams, C., and Ramirez, F. (1984). *Nature*, **310**, 337.
2. Tromp, G., Kuivaniemi, H., Stacey, A., Shikata, H., Baldwin, C. T., Jaenisch, R., and Prockop, D. J. (1988). *Biochem. J.*, **253**, 919.
3. de Wet, W., Bernard, M., Benson-Chanda, V., Chu, M.-L., Dickson, L., Weil, D., and Ramirez, F. (1987). *J. Biol. Chem.*, **262**, 16032.
4. Kuivaniemi, H., Tromp, G., Chu, M.-L., and Prockop, D. J. (1988). *Biochem. J.*, **252**, 633.
5. Boedtker, H., Finer, M., and Aho, S. (1985). *Ann. NY Acad. Sci.*, **460**, 85.
6. Sangiorgi, F., Benson-Chanda, V., de Wet, W., Sobel, M. E., Tsipouras, P., and Ramirez, F. (1985). *Nucl. Acids Res.*, **13**, 2207.
7. Baldwin, C. T., Reginato, A. M., Smith, C., Jimenez, S. A., and Prockop, D. J. (1989). *Biochem. J.*, **262**, 521.
8. Ryan, M. C. and Sandell, L. J. (1990). *J. Biol. Chem.*, **265**, 10334.
9. Metsaranta, M., Toman, D., de Crombrugghe, B., and Vuorio, E. (1991). *J. Biol. Chem.*, **266**, 16862.
10. Benson-Chanda, V., Su, M.-W., Weil, D., Chu, M.-L., and Ramirez, F. (1989). *Gene*, **78**, 255.
11. Chu, M.-L., Weil, D., de Wet, W., Bernard, M., Sippola, M., and Ramirez, F. (1985). *J. Biol. Chem.*, **260**, 4357.
12. Ala-Kokko, L., Kontusaari, S., Baldwin, C. T., Kuivaniemi, H., and Prockop, D. J. (1989). *Biochem. J.*, **260**, 509.
13. Brazel, C., Oberbaumer, I., Dieringer, H., Babel, W., Glanville, R. W., Deutzmann, R., and Kuhn, K. (1987). *Eur. J. Biochem.*, **168**, 529.
14. Tryggvason, K., Soininen, R., Hostikka, S. L., Ganguly, A., Huotari, M., and Prockop, D. J. (1990). *Ann. NY Acad. Sci.*, **580**, 97.
15. Wood, L., Theriault, N., and Vogeli, G. (1988). *FEBS Letters*, **227**, 5.
16. Muthukumaran, G., Blumberg, B., and Kurkinen, M. (1989). *J. Biol. Chem.*, **264**, 6310.
17. Hostikka, S. L. and Tryggvason, K. (1988). *J. Biol. Chem.*, **263**, 19488.
18. Saus, J., Quinones, S., MacKrell, A., Blumberg, B., Muthukumaran, G., Pihlajaniemi, T., and Kurkinen, M. (1989). *J. Biol. Chem.*, **264**, 6318.
19. Morrison, K. E., Germino, G. G., and Reeders, S. T. (1991). *J. Biol. Chem.*, **266**, 34.
20. Mariyama, M., Kalluri, R., Hudson, B. G., and Reeders, S. T. (1992). *J. Biol. Chem.*, **267**, 1253.

21. Zhou, J., Leinonen, A., and Tryggvason, K. (1994). *J. Biol. Chem.*, **269**, 6608.
22. Takahara, K., Sato, Y., Okazawa, K., Okamoto, N., Noda, A., Yaoi, Y., and Kato, I. (1991). *J. Biol. Chem.*, **266**, 13124.
23. Weil, D., Bernard, M., Gargano, S., and Ramirez, F. (1987). *Nucl. Acids Res.*, **15**, 181.
24. Woodbury, D., Benson-Chanda, V., and Ramirez, F. (1989). *J. Biol. Chem.*, **264**, 2735.
25. Walchli, C., Koller, E., Treub, J., and Treub, B. (1992). *Eur. J. Biochem.*, **205**, 583.
26. Bonaldo, P., Russo, V., Bucciotti, F., Bressan, G. M., and Colombatti, A. (1989). *J. Biol. Chem.*, **264**, 5575.
27. Saitta, B., Wang, Y.-M., Renkart, L., Zhang, R.-Z., Pan, T.-C., Timpl, R., and Chu, M.-L. (1991). *Genomics*, **11**, 145.
28. Chu, M.-L., Pan, T.-C., Conway, D., Saitta, B., Stokes, D., Kuo, H.-J., Glanville, R. W., Timpl, R., Mann, K., and Deutzmann, R. (1990). *Ann. NY Acad. Sci.*, **580**, 55.
29. Hayman, A. R., Koppel, J., and Trueb, B. (1991). *Eur. J. Biochem.*, **197**, 177.
30. Koller, E., Winterhalter, K. H., and Trueb, B. (1989). *EMBO J.*, **8**, 1073.
31. Stokes, D. G., Saitta, B., Timpl, R., and Chu, M.-L. (1991). *J. Biol. Chem.*, **266**, 8626.
32. Doliana, R., Bonaldo, P., and Colombatti, A. (1990). *J. Cell Biol.*, **111**, 2197.
33. Bonaldo, P., Russo, V., Bucciotti, F., Doliana, R., and Colombatti, A. (1990). *Biochemistry*, **29**, 1245.
34. Yamaguchi, N., Benya, P. D., van der Rest, M., and Ninomiya, Y. (1989). *J. Biol. Chem.*, **264**, 16022.
35. Yamaguchi, N., Mayne, R., and Ninomiya, Y. (1991). *J. Biol. Chem.*, **266**, 4508.
36. Muragaki, Y., Mattei, M.-G., Yamaguchi, N., Olsen, B. R., and Ninomiya, Y. (1991). *Eur. J. Biochem.*, **197**, 615.
37. Ninomiya, Y., Castagnola, P., Gerecke, D., Gordon, M. K., Jacenko, O., Lu Valle, P., McCarthy, M., Muragaki, Y., Nishimura, I., Oh, S., Rosenblum, N., Sato, N., Sugrue, S., Taylor, R., Vasios, G., Yamaguchi, N., and Olsen, B. R. (1990). In *Extracellular matrix genes*, (ed. L. J. Sandell and C. D. Boyd), p. 79. Academic Press, San Diego.
38. Muragaki, Y., Jacenko, O., Apte, S., Mattei, M.-G., Ninomiya, Y., and Olsen, B. R. (1991). *J. Biol. Chem.*, **266**, 7721.
39. Muragaki, Y., Kimura, T., Ninomiya, Y., and Olsen, B. R. (1990). *Eur. J. Biochem.*, **192**, 703.
40. Brewton, R. G., Ouspenskaia, M. V., van der Rest, M., and Mayne, R. (1992). *Eur. J. Biochem.*, **205**, 443.
41. Lu Valle, P., Ninomiya, Y., Rosenblum, N. D., and Olsen, B. R. (1988). *J. Biol. Chem.*, **263**, 18378.
42. Elima, K., Eerola, I., Rosati, R., Metsaranta, M., Garofalo, S., Perala, M., de Crombrugghe, B., and Vuorio, E. (1993). *Biochem. J.*, **289**, 247.
43. Thomas, T., Kwan, A. P. L., Grant, M. E., and Boot-Handford, R. P. (1991). *Biochem. J.*, **273**, 141.
44. Bernard, M., Yoshioka, H., Rodrigues, E., van der Rest, M., Kimura, T., Ninomiya, Y., Olsen, B. R., and Ramirez, F. (1988). *J. Biol. Chem.*, **263**, 17159.
45. Kimura, T., Cheah, K. S. E., Chan, S. D. H., Lui, V. C. H., Mattei, M.-G., van

der Rest, M., Ono, K., Solomon, E., Ninomiya, Y., and Olsen, B. R. (1989). *J. Biol. Chem.*, **264**, 13910.
46. Yamagata, M., Yamada, K. M., Yamada, S. S., Shinomura, T., Tanaka, H., Nishida, Y., Obara, M., and Kimata, K. (1991). *J. Cell Biol.*, **115**, 209.
47. Tikka, L., Elomaa, O., Pihlajaniemi, T., and Tryggvason, K. (1991). *J. Biol. Chem.*, **266**, 17713.
48. Pihlajaniemi, T. and Tamminen, M. (1990). *J. Biol. Chem.*, **265**, 16922.
49. Juvonen, M., Sandberg, M., and Pihlajaniemi, T. (1992). *J. Biol. Chem.*, **267**, 24700.
50. Gordon, M. K., Castagnola, P., Dublet, B., Linsenmayer, T. F., Van Der Rest, M., Mayne, R., and Olsen, B. R. (1991). *Eur. J. Biochem.*, **201**, 333.
51. Myers, J. C., Kivirikko, S., Gordon, M. K., Dion, A. S., and Pihlajaniemi, T. (1992). *Proc. Natl. Acad. Sci. USA*, **89**, 10.
52. Pan, T. C., Zhang, R. Z., Mattei, M. G., Timpl, R., and Chu, M. L. (1992). *Proc. Natl Acad. Sci. USA*, **89**, 6565.
53. Sambrook, J., Fritsch, E. F., and Maniatis, T. (1989). In *Molecular cloning. A laboratory manual*. Cold Spring Harbor Laboratory Press, New York.
54. Ish-Horowicz, D. and Burke, J. F. (1981). *Nucl. Acids Res.*, **9**, 2989.
55. Evans, G. A. and Wahl, G. M. (1987). In *Methods in enzymology* (ed. S. L. Berger and A. R. Kimmel), Vol. 152, p. 604. Academic Press, San Diego.
56. Gibson, T. J., Rosenthal, A., and Waterston, R. H. (1987). *Gene*, **53**, 283.
57. Bates, P. F. and Swift, R. A. (1983). *Gene*, **26**, 137.
58. Catalog, Stratagene Cloning Systems.
59. McCormick, M., Gottesman, M. E., Gaitanaris, G. A., and Howard, B.-H. (1987). In *Methods in enzymology*, Vol. 151 (ed. M. M. Gottesman), p. 397. Academic Press, San Diego.
60. Grosveld, F. G., Lund, T., Murray, E. J., Mellor, A. L., Dahl, H. H. M., and Flavell, R. A. (1982). *Nucl. Acids Res.*, **10**, 6715.
61. Brady, G., Jantzen, H. M., Bernard, H. U., Brown, R., Schutz, G., and Hashimoto-Gotah, T. (1984). *Gene*, **27**, 223.
62. Kioussis, D., Wilson, F., Daniels, C., Leveton, C., Taverne, J., and Playfair, J. H. L. (1987). *EMBO J.*, **6**, 355.
63. Poustka, A., Pohl, T. M., Barlow, D. P., Frischauf, A.-M., and Lehrach, H. (1987). *Nature*, **325**, 353.
64. Ala-Kokko, L. and Prockop, D. J. (1990). *Matrix*, **10**, 279.
65. te Riele, H., Maandag, E. R., and Berns, A. (1992). *Proc. Natl. Acad. Sci. USA*, **89**, 5128.
66. Clark, A. J. and Low, K. B. In *The recombination of genetic material* (ed. K. B. Low). Academic Press, New York.
67. Blumenthal, R. M. (1989). *Focus*, **11**, 41.
68. Raleigh, E. A., Murray, N. E., Revel, H., Blumenthal, R. M., Westaway, D., Reith, A. D., Rigby, P. W. J., Elhai, J., and Hanahan, D. (1988). *Nucl. Acids Res.*, **16**, 1563.
69. Clarke, L. and Carbon, J. (1976). *Cell*, **9**, 91.
70. Baldino, F., Jr., Chesselet, M.-F., and Lewis, M. E. (1989). In *Methods in enzymology* (ed. P. M. Conn), Vol. 168, p. 761. Academic Press, San Diego.
71. Bloch, W. (1990). *Biochemistry*, **30**, 2735.
72. Schowalter, D. B. and Sommer, S. S. (1989). *Anal. Biochem.*, **177**, 90.

Addresses of suppliers

Aldrich Chemical Company
Aldrich Chemical Company Ltd, The Old Brickyard, New Road, Gillingham, Dorset, SP5 4BR, UK.
Aldrich Chemical Company Inc., 1001 West Saint Paul Avenue, Milwaukee, WI 53233, USA.
American Diagnostica, 222 Railroad Avenue, PO Box 1165, Greenwich, CT 06836-1165, USA.
Amersham Corporation
Amersham International PLC, Lincoln Place, Green End, Aylesbury, Bucks, HP20 2TP, UK.
Amersham North America, 2636 Clearbrook Drive, Arlington Heights, IL 60005, USA.
Amicon Corporation
Amicon Ltd, Upper Mill, Stonehouse, Gloucester, GL10 2BJ, UK.
Amicon Division, 24 Cherry Hill Drive, Danvers, MA 01923, USA.
American Type Culture Collection (ATCC), 12301 Parklawn Drive, Rockville, MD 20852, USA.
Bachem, 3700 Market Street, Philadelphia, PA 19104, USA.
Beckman Instruments
Beckman Instruments UK Ltd., Progress Road, Sands Industrial Estate, High Wycombe, Bucks, HP12 4JL, UK.
Beckman Instruments Inc., 1050 Page Mill Road, Palo Alto, CA 94304, USA.
Becton Dickinson and Co.
Becton Dickinson UK Ltd., Between Towns Road, Cowley, Oxford OX4 3LY, UK.
Becton Dickinson Labware, Two Oak Park, Bedford, MA 01730, USA.
Bellco Glass, Inc.
c/o Scientific Laboratories Supplies, Unit 27, Nottingham South and Wilford Industrial Estate, Ruddington Lane, Wilford, Nottingham, NG11 7EP, UK.
Bellco Glass, Inc., 340 Edrudo Road, PO Box B, Vineland, NJ 08360-0017, USA.
Bio-Rad Laboratories
Bio-Rad Laboratories Ltd., Maylands Avenue, Hemel Hempstead, Herts HP2 7TD, UK.
Bio-Rad Laboratories, 2000 Alfred Nobel Drive, Hercules, CA 94547, USA.

Bio101

c/o Stratech Scientific Ltd, 61–63 Dudley Street, Luton, Beds, LU2 0HP, UK.

Bio Lab Inc., 1070 Joshua Way, Vista, CA 92083, USA.

Biotecx Laboratories, Inc., 6023 South Loop East, Houston, TX 77033, USA.

Boehringher-Mannheim Biochemicals

Boehringer-Mannheim UK (Diagnostics/Biochemicals) Ltd., Bell Lane, Lewes, East Sussex, BN7 1LG, UK.

Boehringer-Mannheim Corporation, Biochemical Products, 9115 Hague Road, PO Box 50816, Indianapolis, IN 46250, USA.

Boehringer-Mannheim GmbH Biochemica, PO Box 31 01 20, D-6800 Mannheim, Germany.

Branson Sonic Power Co.

c/o Lucas Dawes Ultrasonics Ltd, Concord Road, Western Avenue, London W3 0SD, UK.

Branson Ultrasonic Corporation, 41 Eagle Road, Danbury, CT 06810, USA.

Brinkman Instruments

c/o Chemlab Instruments Ltd., Hornminster House, 129 Upminster Road, Hornchurch, Essex, UK.

Brinkman Instrument Co., Cantiague Road, PO Box 1019, Westbury, NY 11590-0207, USA.

BRL, *see* Life Technologies, Inc.

Calbiochem Corporation

Calbiochem-Novabiochem (UK) Ltd., 3 Heathcoat Building, Highfields Science Park, University Boulevard, Nottingham NG7 2QJ, UK.

Calbiochem, PO Box 12087, La Jolla, CA 92039-2087, USA.

Canberra Packard, 800 Research Parkway, Meriden, CT 06450, USA.

Cappel, One Technology Court, Malvun, PA 19355, USA.

Clontech Laboratories, Inc.

c/o Cambridge Bioscience Ltd., 25 Signet Court, Stourbridge Common Business Centre, Swans Road, Cambridge CB5 8LA, UK.

Clontech Laboratories Inc., 4030 Fabian Way, Palo Alto, CA 94303, USA.

Collaborative Biomedical Products, *see* Becton Dickinson Labware.

Collaborative Research, *see* Becton Dickinson Labware.

Collagen Corp., 2500 Faber Place, Palo Alto, CA 94303, USA.

Corning/Costar

Corning/Costar (UK) Ltd., Victoria House, 28–38 Desborough Street, High Wycombe, Bucks HP11 2NF, UK.

Corning/Costar, 1 Alewife Center, Cambridge, MA 02140, USA.

Costar, *see* Corning/Costar.

Du Pont (Du Pont/NEN/Sorvall)

Du Pont UK Ltd, Wedgwood Way, Stevenage, Herts SG1 4QN, UK.

Du Pont Co., 549–3 Albany Street, Boston, MA 02118, USA.

Dynatech Labs, 14340 Sullyfield Circle, Chantilly, VA 22021, USA.

Elastin Products Co., PO Box 568, Owensville, MO 65066, USA.

Enzyme System Products, PO Box 2033, Livermore, CA 94550, USA.

Eppendorf, *see* Brinkman.

Falcon, *see* Becton Dickinson Labware.

Fisher Scientific, 711 Forbes Avenue, Pittsburgh, PA 15219-4785, USA.

Gen Bank, Natl. Center for Biotechnology Information, Natl. Library of Medicine, Bldg. 38A, Rm. 8N-803, Bethesda, MD 20894, USA.

Gibco, *see* Life Technologies, Inc.

Hamilton Co.

c/o Phase Separations Sales, Deeside Industrial Park, Deeside, Clwyd, CH5 2NU, UK.

Hamilton Co., 4970 Energy Way, Reno, NV 89502-4178, USA.

Harvard Apparatus, 22 Pleasant Street, South Natick, MA 01760, USA.

Heyltex, 10655 Richmond Ave., Suite 170, Houston, TX 77042, USA.

Hoeffer Scientific Instruments

Hoeffer UK Ltd., Newcastle, Staffs ST5 0TW, UK.

Hoeffer Scientific Instruments, 654 Minnesota St., Box 77387, San Francisco, CA 94107-0387, USA.

Hyclone Laboratories, Inc., 1725 South Hyclone Road, Logan, UT 84321, USA.

IBF, 7151 Columbia Gateway Dr., Columbia, MD 21046, USA.

IBI

IBI Ltd., 36 Clifton Road, Cambridge CB1 4ZR, UK.

IBI (International Biotechnologies Inc.), PO Box 9558, New Haven, CT 06535, USA.

ICN Biochemicals

Flow (ICN Flow), Eagle House, Peregrine Business Park, Gomm Road, High Wycombe, HP13 7DL, UK.

Flow (ICN Biomedical Inc.), 3300 Highland Ave., Costa Mesa, CA 92626, USA.

Kirkegaard and Perry Labs, Inc., 2 Cessna Court, Gaithersburg, MD 20879, USA.

Kodak (Eastman Kodak Co.)

c/o Phase Separations Sales, Deeside Industrial Park, Deeside, Clwyd CH5 2NU, UK.

Kodak (Eastman Kodak Co.), PO Box 9558, New Haven, CT 06535, USA.

Life Technologies (Gibco-BRL)

Life Technologies Ltd., Trident House, Renfrew Road, Paisley PA3 4EF, UK.

Life Technologies, Inc., PO Box 68, Grand Island, NY 14072-0068, USA.

LKB, *see* Pharmacia-LKB Biotechnology.

Micron Separations, Inc., 135 Flanders Rd., PO Box 1046, Westborough, MA 01581, USA.

Millipore

(Millipore UK Ltd.), The Boulevard, Blackmoor Lane, Watford, Herts WD1 8YW, UK.

Milligen (Millipore Intertech), 80 Ashby Road, Bedford, MA 01730, USA.

Nalgene

c/o FSA Laboratory Supplies, Bishop Meadow Road, Loughborough, Leics, LE11 0RG, UK.

Nalgene Co., 75 Panorama Creek Drive, PO Box 20365, Rochester, NY 21460-0365, USA.

NEN, *see* Du Pont/NEN/Sorvall.

New England Biolabs, 32 Tozer Road, Beverly, MA 01915-5599, USA.

NUNC, Inc.

see Life Technologies (for UK distribution).

NUNC, Inc., 2000 North Aurora Road, Naperville, IL 60563, USA.

PCR Research Chemicals, Inc., PO Box 1778, Gainesville, FL 32602, USA.

Pel-Freez, PO Box 68, Rogers, AR 72756, USA.

Peninsula Labs, Inc., 611 Taylor Way, Belmont, CA 94002, USA.

PGC Scientifics, PO Box 7277, Gaithersburg, MD 20898-7277, USA.

Pharmacia-LKB Biotechnology

Pharmacia Biosystems Ltd. (Biotechnology Division), Davy Avenue, Knowlhill, Milton Keynes, MK5 8PH, UK.

Pharmacia LKB Biotechnology Inc., PO Box 1327, Piscataway, NJ 08855-9836, USA.

Pierce

c/o Life Science Labs Ltd., Sedgewick Road, Luton, Beds LU4 9DT, UK.

Pierce, PO Box 117, Rockford, IL 61105, USA.

Pierce Europe BV, PO Box 1512, 3260 BA Oud-Beijerland, The Netherlands.

Polysciences, 7800 Merrimac Avenue, Niles, IL 60714, USA.

Popper and Sons, 300 Denton Ave., New Hyde Park, NY 11040, USA.

Promega

Promega Ltd, Delta House, Enterprise Road, Chilworth Research Centre, Southampton SO1 7NS, UK.

Promega Corp., 2800 Woods Hollow Road, Madison, WI 53711-5399, USA.

Qiagen, Inc.

c/o Hybaid Ltd., 111–113 Waldegrave Road, Teddington, Middx. TW11 8LL, UK.

Qiagen, Inc., 9600 De Soto Avenue, Chatsworth, CA 91311, USA.

RPI, 410 Business Center Drive, Mount Prospect, IL 60056-2190, USA.

Savant Instruments, 221 Park Ave., Hicksville, NY 11801, USA.

Schleicher and Schuell

c/o Anderman & Co. Ltd., 145 London Road, Kingston-upon-Thames, Surrey KT2 6NH, UK.

Schleicher and Schuell, PO Box 2012, Keene, NH 03431, USA.

Schleicher and Schuell, Postfach 4, D-3354 Dassell, Germany.

Seikagaku
c/ICN Flow, Eagle House, Peregrine Business Park, Gomm Road, High Wycombe, HP13 7DL, UK.
Seikagaku American, Inc., 30 West Gude Drive, Suite 260, Rockville, MD 20850, USA.
Seikagaku Kogyo Co. Ltd, Tokyo Yakugyo Bldg., 1-5, Nihonbashi-Honcho, 2-Chrome, Chuo-Ku, Tokyo, 103 Japan.
Separation Group, 17424 Mohave Street, Hesperia, CA 92345, USA.
Serva
c/o Universal Biologicals Ltd., 12–14 St. Ann's Crescent, London SW18 2LS, UK.
Serva, PO Box 1531, Paramus, NJ 07653-9900, USA.
Serva Feinbiochemica GmbH & Co KG, PO Box 10 52 60, Carl Benzstrasse 7, D-6900 Heidelberg 1, Germany.
Sherwood Medical, St. Louis, MO 63103, USA.
Shimadzu, 7102 Riverwood Drive, Columbia, MD 21046, USA.
Sigma
Sigma Chemical Co. Ltd., Fancy Road, Poole, Dorset, BH17 7NH, UK.
Sigma Inc., PO Box 14508, St. Louis, MO 63178, USA.
Sorvall, *see* Du Pont/NEN/Sorvall.
Spex Ind., Inc., 3880 Park Avenue, Edison, NJ 08820, USA.
Stratagene Cloning Systems
Stratagene Ltd., Unit 140, Cambridge Innovation Centre, Milton Road, Cambridge CB4 4FG, UK.
Stratagene Inc., 11011 North Torrey Pines Road, La Jolla, CA 92037, USA.
Supelco, Supelco Park, Bellefonte, PA 16823-0048, USA.
Telios Pharmaceuticals, 4757 Nexus Center Drive, San Diego, CA 92121, USA.
UBI, 199 Saranac Avenue, Lake Placid, NY 12946, USA.
United States Biochemical (USB) Corporation
c/o Cambridge Bioscience, 25 Signet Court, Stourbridge Common Business Centre, Swans Road, Cambridge CB5 8LA, UK.
USB, PO Box 22400, Cleveland, OH 44122, USA.
Vector Labs, 30 Ingold Road, Burlingame, CA 94010, USA.
Waters Associates
See Millipore (for UK distribution).
Waters (Millipore Intertech), 397 Williams Street, Marlborough, MA 01752-9162, USA.
Whatman Ltd
Whatman Scientific Ltd., Whatman House, St Leonards Road, Maidstone, Kent ME16 0LS, UK.
Whatman Ltd., PO Box 1359, Hillsboro, OR 97123-9981, USA.
Worthington Biochemical Corp.
c/o Cambridge Bioscience Ltd., 25 Signet Court, Stourbridge Common Business Centre, Swans Road, Cambridge CB5 8LA, UK.
Worthington Biochemical Corp., Halls Mill Road, Freehold, NJ 07728, USA.

Index

Index

for estimation of total collagen production by tissues 38
identification of collagenous proteins on gels 58–9
cyanogen bromide peptides
cleavage of collagen α chains 62
cleavage of proteins for determination of collagenous peptides 64–5

decorin
core protein size and GAG type 13 (*table*)
member of leucine-rich family 12 (*table*)
Descemet's membrane
purification of type VIII collagen from 137–42
source of type VIII collagen 132, 133 (*table*)
desmosine 242 (*fig.*)

Ehlers–Danlos syndrome 22
EHS sarcoma
extraction of extracellular matrix (Matrigel) 282
preparation of laminin-entactin complex 109–11
elastase 268
elastin 20–1, 241–60
amino acid composition 245 (*table*)
elastase 268
identification of lysine-derived cross links 294–51
desmosine structure 242 (*fig.*)
isolation of insoluble elastin 241–5
solubilization
alkaline degradation for preparation of κ-elastin 247–8
enzymatic digestion 248
oxalate hydrolysis for preparation of α-elastin 246–7
tropoelastin
isolation 251–5
quantitation 255–9
entactin (nidogen) 19; *see also* laminin–entactin complex
size, location and function 16 (*table*)
epidermolysis bullosa 22
extracellular matrix
involvement in disease 22–4
acquired diseases 23–4
inherited diseases 22–3
regulation of cellular biology 2, 3 (*fig.*)
unique features 21–2
use in cell culture
assays for cell attachment to fibronectin and its fragments 184
cell attachment to laminin–entactin complex 118–19

use for measuring matrix degradation 278–84
extracellular matrix proteoglycans 14; *see also* 12 (*table*)

FACIT collagens 7 (*table*), 9
fibrillin 20
association with Marfan syndrome, ectopic lentis and congenital arachnodactyly 23
size, location and function 16 (*table*)
fibrils, collagen from cartilage 88–9
fibrin, use in assay for plasminogen activators 269–74
fibromodulin
core protein size and GAG type 13 (*table*)
member of leucine-rich family 12 (*table*)
fibronectin 18, 22 (*table*), 175–85
cell-adhesive properties
assay for cell attachment 184
characteristics of intact molecules and fragments 176 (*table*)
preparation of substrates 183
preparation of coated coverslips for measuring degradation 279–80
preparation of monomers 197
preparation of thermolysin fragments 181–2
preparation of tryptic fragments 178–81
production by cultured cells 192
purification of plasma fibronectin 177–8
size, location and function 16 (*table*)

glycosaminoglycans
analysis of disaccharide composition 205–8
depolymerization techniques 202–5
disaccharide composition 199–201
extraction and purification 201–2
mapping of oligodisaccharide domains 208–17
chromatographic separations 208–9
gradient PAGE 209–16
separation of disaccharides 217
glypican
core protein size and GAG type 13 (*table*)
member of orphan core protein family 12 (*table*)
Goodpasture syndrome
involvement of the α3(IV) collagen chain 23

heparan sulfate and heparin 200–1
hexabrachion, *see* tenascin

400

Index